MINNESOTA WEATHER ALMANAC

MINNE

SOTA

WEATHER ALMANAC

SECOND EDITION

Completely Updated for the New Normals

MARK W. SEELEY

 MINNESOTA HISTORICAL SOCIETY PRESS

The following essays have appeared previously:

"Mother Nature Smiled: The Grand Excursion of 1854," *Minnesota History* 59.1 (Spring 2004): 36–38.

"May Rituals" and "Meteorological Lamentations," *Minnesota Monthly*.

www.mnhspress.org

The Minnesota Historical Society Press is a member of the Association of American University Presses.

Manufactured in the United States of America

10 9 8 7 6 5 4 3 2 1

♾ The paper used in this publication meets the minimum requirements of the American National Standard for Information Sciences—Permanence for Printed Library Materials, ANSI Z39.48-1984.

International Standard Book Number
ISBN: 978-0-87351-977-9 (paper)

Library of Congress Cataloging-in-Publication Data available upon request

I dedicate this second edition of Minnesota's weather and climate history to the many thousands of volunteer weather observers who have provided the rich history of daily measurements and other documentation used in this book. Without them, this book would not exist. Our Minnesota culture has been blessed to have many citizens who care so deeply about the environment that they have committed much time and effort to quantifying it and describing its impacts. Our knowledge of the state's climate behavior and its impacts are foundationally based on the aggregate of daily observations by these volunteers, who in this busy world take up the challenge and commit themselves to making measurements every day, 365 days a year.

Secondly, I am professionally indebted to Earl Kuehnast, former state climatologist for the Minnesota Department of Natural Resources, and Dr. Donald G. Baker, former professor and colleague at the University of Minnesota. These gentlemen were valuable mentors to me, and though both are no longer with us, their work continues to inspire me to make Minnesota's vast weather and climate history relevant to our lives today. I will always be grateful for their mentoring, and this book was written out of a dedication to continuing their good work for the citizens of Minnesota.

CONTENTS

LIST OF TABLES AND MAPS

Tables

Maps

♦ INTRODUCTION

Stricken with Meteorological Affected Disorder

November 1978. I was about to join the University of Minnesota Department of Soil, Water, and Climate as a climatologist. The drought of 1976 (more devastating in some western Minnesota counties than those of the 1930s) had been so disastrous that the legislature, with lobbying from farmers and university administration, had created this new faculty position. Some probably viewed it as a way to deal with future droughts and other weather disasters. After all, climatologists are supposed to study the lessons in weather history and find ways to better cope with the extremes that sometimes occur.

Great expectations aside, I had few regrets about leaving behind all the security-coded, computer-based gadgetry of the NASA Johnson Space Center in Houston, Texas. I was looking forward to connecting with the people of the Upper Midwest, who by reputation were very keen on weather and climatology. Surely they would renew my long-standing fascination with the atmosphere, which dated from a stint as a volunteer weather observer in Utah. More important, as a university faculty member I would be able to share my observations and knowledge—in other words, inflict myself upon the innocent public.

In 1969 the weather bug bit me. I was a VISTA volunteer in Salt Lake City when I became a fan of local TV meteorologist Mark Eubank. I faithfully watched him give the weather, even more so when I moved back to the city in 1971 while my wife, Cindy, pursued her graduate degree. Though he only had four to five minutes to present the weather on the evening news, Mr. Eubank was phenomenally exuberant. How could someone be so enthusiastic, so over the top about the *weather,*

I wondered. Yet he always found something interesting to tell his avid viewers. I joined his local network of weather observers and filed my reports on the happenings in Liberty Park neighborhood, a real microclimate because of its abundant trees and vegetation. A few times I even got mentioned on the air for reporting strong winds and heavy rainfall. After some months, I discovered that the University of Utah had a Department of Meteorology, specializing in mountain weather. My work schedule allowed me to take one class per semester, and so began my study of the atmosphere.

Following Cindy's graduation we moved to Illinois, where I obtained my own master's in meteorology at Northern Illinois University and went to work for DeKalb AgResearch, Inc., studying the impacts of climate variability on its corn breeding program. Based on a paper I presented at a scientific meeting in Chicago in 1975, faculty members from the University of Nebraska offered me a research assistantship to come to Lincoln and earn a PhD. I soon discovered that the study of climatology was even more enjoyable amongst the irrigated cornfields and roaming coyotes of the Great Plains. Finishing my degree in 1977—and by that time scraping the bottom of our bank account—I took the first job offered, which happened to be in Houston working on a satellite-based remote sensing project for the USDA.

Just a year later, my meteorological germination complete, we loaded our belongings into a rented truck and left Houston's mid-November, 80°F weather, traveling straight north on Interstate 35. We had no sooner crossed the Iowa border into Minnesota when we were hit with blinding snow in Albert Lea. We were barely able to see well enough to exit the interstate safely and find a motel. Luckily the room was large enough to accommodate our many houseplants, which we removed from the car so they wouldn't freeze to death. The Texas fleas gave up and fell off our dog. During the two days we were stranded, we fondly recalled that the miserable 70°F dew points in Texas really hadn't been so bad. I feebly attempted to convince Cindy that this November blizzard must be a great aberration for Minnesota. When at last we resumed our trip up I-35, the truck radiator froze near Clark's Grove, where a farm shop owner was kind enough to repair it. Finally we reached St. Paul and moved into our rented duplex—in the middle of a blinding snowstorm, the second during that Thanksgiving week.

Despite a frigid welcome that would have sent a less resolute individual speeding back south, I have lived here for nearly four decades now. It is indeed paradise for a person whose vocation is the weather. Where else can one live in a community in which everyone is afflicted with Meteorological Affected Disorder (MAD)?

A variety of MAD symptoms are evident throughout the population. If you think the baseball fan is obsessed with statistics, contemplate the treatment of weather in Minnesota. Every statistical formula imaginable is used to describe averages, means, modes, medians, extremes, probabilities, threshold events, episodes, streaks, and indices. Climate statistics have been derived for nearly every popular holiday, festival, celebration, and public event. Some are well known for their climatic characteristics: snowy state high school tournaments, windy St. Patrick's Days, cold Memorial Day weekends, sultry Independence Days, stormy Aquatennials, blustery Veterans' Days, and dull, gray Election Days are but a few. True weather exuberance is exhibited by those who keep climatological statistics on their own birthday celebrations. Such statistics were not flaunted in other states I have called home. We are a citizenry that describes our personalities and the weather using the same adjectives: dull, gloomy, drippy, sunny, bright, refreshing, stuffy, blustery, foggy, thick, dense, unsettled, and uproarious, for example.

Another MAD symptom is evident in addictions to National Oceanic and Atmospheric Administration (NOAA) weather radio, local television weather broadcasts, and cable TV's Weather Channel, not to mention pages upon pages of Internet bookmarks for weather websites. Extreme or severe weather events and episodes sometimes mandate continuous broadcast coverage on radio and television, yielding a community-wide sharing of observations and stories, such as the famous helicopter telecast of the 1986 tornado passing across Spring Lake Park; or the devastating flash flood of July 23, 1987, that brought 10 inches of rain in six hours to the Twin Cities, flooding thousands of basements; or the Red River Valley spring snowmelt floods of 1997. The images depicted in media coverage of the state's 1988 drought are branded in our memories: withered corn stalks, large cracks in the soil, parched streambeds, kids sleeping on their porches.

How many of your friends routinely start their morning with the newspaper, a cup of coffee, and the Weather Channel or NOAA weather radio? During the Clinton administration, vice president Al Gore, an enthusiastic fan of NOAA weather radio, proposed that it should be available to every American 24 hours a day. It is perhaps one of the best ways to keep informed of threats from severe weather or to keep abreast of changeable weather, a common occurrence in the North Star State. Of course, some Minnesotans go overboard by memorizing the daily broadcast schedule of NOAA weather radio or taking one with them wherever they go. In the not-too-distant future, we'll probably see NOAA weather radio built into our watches or cell phones.

Perhaps the most obvious symptom of MAD is the dominance of weather as a topic of conversation. Try counting the number of times the weather is mentioned conversationally in a single day. Many people use it to acknowledge one's presence ("Good morning, kinda blustery out there") or to be cordial ("Brush that snow off and come on in and warm up"). It is a perfect fit for the Minnesotan tendency to converse in a nonconfrontational manner, showing that one is observant but not judgmental ("Forgot your umbrella? Here, we'll share mine. It looks like it's going to clear up soon anyway"). One cautionary note, however: be careful with whom you converse about the weather. I have seen Minnesotans go on for an hour in response to the benign opening remark, "Nice weather we're having." (Guilty as charged!)

Everyone can connect with the weather. It has inspired us, humbled us, scared us, disappointed us, tortured us, and restored us. It bookmarks significant events in our lives as we remember the weather from the day of a wedding, of a child's birth, of a family reunion, or of that Vikings/Packers game at the old Met. MAD may be considered an infection, even a character flaw by some, but to me it is a Minnesotan characteristic that binds us all together. This book—a collection of essays and stories and facts about the state's weather—is dedicated to my many compatriots afflicted with MAD.

★ ACKNOWLEDGMENTS

Although it is customary to thank book publishers and editors in an author's final words of acknowledgment, in this case I want them to be first. Staff at the Minnesota Historical Society Press encouraged me to write this second edition, knowing that our climate in Minnesota is changing and that it would take a long and comprehensive effort to update the original volume. I am grateful for their support. I especially thank Shannon M. Pennefeather for lending her skills to this project. She edited the first edition and has been a patient and thoughtful editor for the second one as well.

The genesis of the original *Minnesota Weather Almanac* came from my collaboration with Minnesota Public Radio (MPR), which started with an invitation to be on the *Morning Edition* news program back in 1992, when it was produced by Bruce MacDonald and hosted by Bob Potter. MPR continues to be a highly valued partner in my professional career. Being a member of the MPR community has allowed me to gain and share a much broader and richer knowledge of weather- and climate-related topics than I would have ever experienced as a university professor. I have learned a great deal and continue to learn from MPR listeners about why weather and climate are important to them, and I have heard many stories of how their lives have been directly or indirectly affected by the state's environment. My appreciation for the level of effort and competence embodied by the MPR newsroom and programming staff has only grown. *Morning Edition* host Bob Potter encouraged me to write the original *Minnesota Weather Almanac,* and it took me six years to do so. In the years since its publication in 2006, we have witnessed so many dramatic and traumatic weather events and climate episodes and broken so many thousands of climate records, I guess a second edition of this book was inevitable.

I am most grateful to all members of the MPR *Morning Edition* team, those from the past and those in the present: Cathy Wurzer (a wonderful host), Jim Bickal, Eugene Cha, Julie Siple, Rick (Scooter) Hebzynski, Patti Rai Rudolph, Sasha Aslanian, Randy Johnson, Marianne Combs,

Clifford Bentley, Phil Picardi, Perry Finelli, John Bischoff, Larissa Anderson, Stephanie Curtis, Greg Magnuson, John Wanamaker, and many others who have filled in from time to time. Also, my thanks go to Sara Meyer, Curtis Gilbert, Steven John, and Gary Eichten for so many years of doing the State Fair Weather Quiz broadcasts, along with the Santa Forecast during the holiday season. The *All Things Considered* program staff—Jayne Solinger, Tom Crann, Jeff Jones, Sam Choo, and others—have shown great trust in allowing me to occasionally provide commentary on topics other than weather. Reporters Stephanie Hemphill and Elizabeth Dunbar have always been thorough and precise in their coverage of weather and climate research and stories I shared and thought should be covered. Paul Huttner and Craig Edwards have been real friends and colleagues in collaborations related to broadcasting content, addressing listener questions, and providing web-based educational materials. Many other friends and colleagues at MPR too numerous to mention have shared their knowledge and interest with me, for which I am most appreciative. I should also acknowledge how much I have benefited from my interactions with the MPR listener audience as well over the past 23 years.

Twin Cities Public Television (TPT) has consistently offered a broadcast format for me to educate citizens about our weather and climate. It has been a privilege to work with Mary Lahammer on documentaries about Minnesota's historical tornadoes, floods, and fires. I am most grateful, too, for the opportunity to share my knowledge and perspectives on the *Almanac* program, and I thank Brendan Henehan, Kari Kennedy, Mary Lahammer, David Gillette, and hosts Eric Eskola and Cathy Wurzer for their collaboration over the years.

In preparing the second edition of this book, I have benefited greatly from the research, creative ideas, and constructive criticism of colleagues, friends, and family. My colleagues at the University of Minnesota provided a good deal of tangible material for this book as well as inspiration and encouragement. Thanks go to Professors Tim Griffis, Peter Snyder, John Baker, Richard Skaggs (retired), and Kathy Klink as well as to chief technology guru Dave Ruschy. All have contributed immensely to the understanding of Minnesota's climate and to my own education in particular. My appreciation also to Allison Serakos for her assistance with data assimilation for the tables, graphics, and maps as well as for some editing. I particularly want to thank her for her patience and diligence in dealing with such a high volume of data.

I have also benefited a great deal from the support of my colleagues in Extension, who have helped keep my perspective on the real world and what affects Minnesota citizens the most. Extension is still the university interface for bringing scientific knowledge and pragmatic application of it to the local community and citizens. I want to especially thank Extension Dean Bev Durgan for her support of my work on this book.

The Minnesota Department of Natural Resources State Climatology Office is, in my humble opinion, the best of its kind in the nation. Started by Earl Kuehnast in the 1970s, followed up by Jim Zandlo's leadership from 1986 to 2011, and now with state climatologist Greg Spoden at the helm with assistance from Pete Boulay, this office has developed an outstanding reputation for public service and for expertise in weather and climate matters. The office archives nearly all of the weather data collected in the state and shares it online. I am indebted to them for making so much information available to use in this book. They are diligent stewards of Minnesota's climate database and history, as well as terrific colleagues.

Friends and colleagues at the National Weather Service have kept me up to date on the latest technologies being used in meteorology. We are blessed indeed to have such a competent group of meteorologists forecasting our weather. Special thanks to Craig Edwards (retired), Dan Luna, current meteorologist-in-charge at the Chanhassen office, Michelle Margraf, Steve Buan, and Todd Krause, among many others. One could not expect to work with more dedicated colleagues.

I express my appreciation to fellow Minnesota citizens who have shared their weather stories and knowledge with me over the years. The weather is a common thread in our lives, a bookmark to our most vivid memories of significant events. Sincere thanks to the thousands of weather observers across the state for their contribution to our understanding of the climate in which we live. It is a real privilege for me to know so many of you by name.

Lastly and most importantly, I thank my family for putting up with my "Meteorological Affected Disorder" (MAD). It is not easy living with someone who is afflicted with MAD, but they have developed a high tolerance for my infatuation with weather and climate and even encouraged me to share my enthusiasm and knowledge with others. I consider myself a lucky man indeed to have the love and support of my wife Cindy, daughter Emma, sons Adam and Alex, daughter-in-law Mandy, granddaughter Clara, and grandson Drew.

MINNESOTA
WEATHER ALMANAC

Crossroads of Climate

Minnesota's climate is heavily influenced by its geographical position in the middle of North America generally, and on the western edge of the Great Lakes particularly. It is well downstream from the north-south oriented mountains of the American West and also from the low-relief plains lying to the south all the way to the Gulf of Mexico. Minnesota's latitudinal position heavily influences its seasonality, producing very short days of eight or fewer hours in winter and very long days of up to 16 hours in summer. The midday overhead sun elevation angle can range from a lowly 20 degrees above the horizon in winter—nearly blinding if one is driving into it—to a very high 68 degrees in summer, when sun block is mandatory to protect one's skin from burning. The state's landforms, soils, and vegetation interact with northern hemisphere air masses and individual weather disturbances, or midlatitude cyclones, to produce a variety of weather as well as very pronounced seasonal changes. Additionally, it can be argued that some of Minnesota's climate has been altered since the nineteenth century by urbanization and land use change, including agricultural drainage, dam construction, and deforestation. However, simple geography still overwhelmingly influences the behavior of climate in the state.

Geography

Minnesota's southern to northern border stretches from roughly 43.5 degrees north latitude to just less than 49.5 degrees north latitude, a distance of approximately 400 miles. The northernmost reaches of the contiguous 48 states lie within the state's boundaries: the Northwest Angle in Lake of the Woods, at 49 degrees, 23 minutes, and 4 seconds north latitude. From east to west, Minnesota's borders reach from 89 degrees, 30 minutes west longitude to 97 degrees west longitude, a distance of about 340 miles. The western shoreline of Lake Superior, largest and coldest of the Great Lakes, stretches along 150 miles from Duluth to the Pigeon River on the Canadian border. The significant climatic influence of this body of water will be described later.

Elevations are fewer than 1,200 feet above mean sea level along each of the three major rivers that run through the state: the Red River of the North, the Minnesota River, and the Mississippi River. There are three areas where elevations exceed 1,600 feet: the Iron Range as it parallels the north shore of Lake Superior, the Coteau des Prairies (Buffalo Ridge) where it extends out of South Dakota into southwestern Minnesota, and a small area near Lake Itasca encompassed by the Alexandria moraine. Extremes of elevation occur along the northern shore of Lake Superior, from a low of 601 feet to a high of 2,301 feet, the latter at Cook County's Eagle Mountain. Discharge of surplus runoff follows a three-directional continental divide: drainage toward Hudson Bay to the north, toward the Atlantic Ocean to the east, and toward the Gulf of Mexico to the south.

Minnesota's landscape encompasses 79,617 square miles, making it the fourteenth-largest state. Approximately one-third of the land area contains forests, estimated at 16.3 million acres. Fifty-six percent of the state's land is used for agriculture, accounting for 28.5 million acres. Nearly one-fifth, approximately 10.6 million acres, is occupied by wetlands. In fact, Minnesota is second only to Alaska in the amount of wetlands within its borders. Known across the nation as the land of 10,000 lakes, Minnesota actually contains at least 11,842 lakes greater than 10 acres in size, some of which influence the local climate.

Q & A

What is a box and whisker diagram?

a. display of data from an instrumented balloon

b. display of the median, distribution, and extremes of climate data

c. graphical depiction of the daily forecast

(answer on page 329)

Minnesota exhibits a continental-type climate characterized by pronounced changes in temperature and moisture conditions caused by strikingly different air masses. Continental polar air from very high latitudes can intrude on the state in any season, though most frequently in the winter months. Such air masses bring very cold, dry air with extremely low dew points: a number of communities have recorded lows colder than -50°F during these polar intrusions. Containing few particulates and aerosols, this usually clean air is manifested in February's climatology, which shows the least atmospheric turbidity (a measure of how easily sunlight is transmitted through the atmosphere) for the year. Conversely, both dry and wet subtropical air masses can move into Minnesota from the south, producing either desert-like dry heat or tropical moist heat that may persist for days, particularly in the summer months, when temperatures have soared above 110°F in many places and the Heat Index (the combined effect of temperature and dew point) has reached above 120°F. Moderate maritime Pacific Ocean air also visits the state with some frequency, bringing comparatively mild temperatures and light precipitation.

DID YOU KNOW?

The Camelot Climate Index is used to determine climate suitability for human habitation. It got its name because Camelot was never too hot, too cold, too wet, or too dry.

The landscape and soils that support the state's tremendous biodiversity were shaped over geologic time scales by the waxing and waning of ice sheets during alternating glacial and interglacial periods. Both water and aeolian (wind) erosive actions have sculpted the landscape. Across the northern tier of the state are the table top–like prairies of the Red River Valley in the northwest, the gradually rolling and lake-dotted terrain of the Alexandria moraine, and the abrupt and magnificent rocky cliffs and peaks of the North Shore (part of the Canadian Shield, geologically some of the oldest exposed rock surfaces on planet Earth). Because the population density of this area of the state is low, the night sky is not contaminated by the many lights of civilization, making it an ideal location to view the northern lights. Many Minnesotans cherish memories of spectacular nighttime skies over the Red River Valley, Voyageurs National Park, or the Boundary Waters Canoe Area. Down the middle of the state are many larger lakes, such as the Lower and Upper Red Lakes, Leech Lake, Lake Winnibigoshish, and Mille Lacs, along with the ever-widening swath of the mighty Mississippi River as it travels from Lake

Itasca to the Twin Cities. In the central counties, agricultural and resort-based economies intermingle: the lakeshores are populated by resorts, cabins, and second homes, while the rolling rural areas are farmed or have livestock grazing in meadows and pastures. In the southern part of the state, the Minnesota River Valley drains a primarily prairie and agricultural landscape from west to east, joining with the Mississippi River at Fort Snelling. Formerly populated by small family farms and more tightly spaced hedgerows and windbreaks, this vast landscape with its deep, rich soil is dominated by large agricultural operations, tile-drained crop fields, or concentrated livestock operations. Today, only drainage ditches and township roads bisect the enormous fields, and the farmsteads are much more widely spaced than they were a hundred years ago. Finally, along the southeastern border with Wisconsin lies the beautiful river bluff country, where the Mississippi River slices through karst, or limestone bedrock. Many of the state's larger apple orchards are located in this significantly milder climate regime, where temperatures only rarely dip below -20°F.

Urbanization

Minnesota's population is just over 5.5 million residents, the majority of whom live in the Twin Cities metropolitan area and other modest-sized cities such as Duluth, Rochester, St. Cloud, Mankato, Moorhead, Winona, and Austin. As a result of this concentration in population, many citizens live in urbanized environments that have modified climate regimes compared to those of nearby rural landscapes. Urbanization's effects on climate are many; some are quite subtle.

. . . .

Twilight
The period of incomplete darkness that occurs after sunset and before sunrise is described as twilight. With the sun below the horizon, light scattered in the upper atmosphere may produce a purple, red, or yellow glow. Three somewhat arbitrary subdivisions of twilight are used to define outdoor visibility:

__Civil twilight__ refers to the interval of incomplete darkness that occurs when the sun's center is approximately 6 degrees below the horizon. The amount of light is sufficient to carry on outdoor work.

Nautical twilight refers to the interval of incomplete darkness that occurs when the sun's center is approximately 12 degrees below the horizon. The amount of light allows for navigation using visible features on water or land.

Astronomical twilight refers to the interval of incomplete darkness that occurs when the sun's center is approximately 18 degrees below the horizon. No discernible horizon glow remains, and sixth-magnitude stars are directly overhead.

Twilight's duration varies considerably with latitude and season. The sun's daily path near the equator differs significantly from that near the poles because of the Earth's rotational speed and the sun's changing angles. At the equator civil twilight may last only 21 minutes and change little from season to season. In Minnesota it may vary from 30 to 40 minutes, while at high latitudes in the summer civil twilight can literally last all night.

Noted climatologist Helmut Landsberg was one of the first scientists to describe how the development of an urban area modifies the local environment. Urban areas host a higher concentration of atmospheric pollutants; present rougher surface terrain, which influences wind velocities; and are composed of different surface materials, which affect heat storage and convection. Some of the climate-related disparities noted by Dr. Landsberg in his studies of urban climates are listed in Table 1.

Local effects related to the Twin Cities urban heat island are not always consistent, but many are worth noting. A longer frost-free growing season by seven days or more, earlier green-up of lawns and gardens in the spring, later freeze-up of lakes in the fall, average wind speeds 10 to 20 percent less, generally higher winter average temperatures (fewer Heating Degree Days), and more snowfall are evident in the Twin Cities in comparison to surrounding rural areas. Paul Todhunter's study of the urban heat island effect in 1989 reported a mean annual temperature difference of three to four degrees between downtown Minneapolis and the southwestern suburban areas and a difference of 16 days in the frost-free growing season. With expanded urbanization around the Twin Cities metropolitan area, recent climate statistics (1981–2010) show that differences in temperature and length of the frost-free growing season have changed somewhat between the downtown areas and the inner- and outer-ring suburban communities. The range in the length of the growing season within the Twin Cities metropolitan area is 23 days.

TABLE 1: Effects of Urbanization on Local Environment	
ELEMENTS	**COMPARISON WITH RURAL ENVIRONMENT**
Cloud cover	5 to 10 percent greater
Fog, winter	100 percent more
Fog, summer	30 percent more
Precipitation	5 to 10 percent more
Snowfall	5 percent less
Rain days with less than .02"	10 percent more
Relative humidity, winter	2 percent less
Relative humidity, summer	8 percent less
Solar radiation	15 to 20 percent less
Ultraviolet radiation, winter	30 percent less
Ultraviolet radiation, summer	5 percent less
Duration of sunshine	5 to 15 percent less
Pollutants, solid particles	10 times more
Pollutants, gases	5 to 25 times more
Mean annual temperature	1 to 2 degrees higher
Annual Heating Degree Days	10 percent fewer
Annual mean wind speed	20 to 30 percent less
Frequency of calms	5 to 20 percent more

Further, outer-ring communities like Stillwater, which has seen significant further growth and development in recent decades, now show a higher mean annual temperature and fewer Heating Degree Days than other sections of the metro (See Table 2).

The annual precipitation around the Twin Cities metropolitan area varies by a little under 5 inches; regardless, the amount of pavement generates a much higher degree of runoff, presenting a challenge to the storm water runoff system. The city's heat storage capacity is much more evident in the winter than in other seasons (note, for example, in Table 2, the difference in February mean temperatures among communities is greater than the difference in the annual mean temperatures). This difference in heat storage capacity is seen in higher overnight minimum temperatures in the more densely populated areas. Under some winter circumstances, the overnight minimum temperatures across the Twin

TABLE 2: Minneapolis–St. Paul (Urban) Climate Statistics Compared to Nearby Communities (1981–2010)						
LOCATION (COUNTY)	ANNUAL MEAN TEMP. (°F)	AVERAGE FEBRUARY TEMP. (°F)	ANNUAL HEATING DEGREE DAYS	MEDIAN LENGTH OF GROWING SEASON (DAYS)	AVERAGE SEASONAL INCHES OF SNOW	ANNUAL MEAN PRECIP. (INCHES)
MSP Airport (Hennepin)	46.2	20.8	7580	167	54.4	30.61
Buffalo (Wright)	43.6	17.1	8343	152	41.7	30.02
Chaska (Carver)	46.9	21.0	7376	160	45.9	31.85
Farmington (Dakota)	45.8	20.7	7631	160	42.0	31.30
Forest Lake (Washington)	44.6	18.4	8042	147	51.1	32.05
Rosemount (Dakota)	44.5	18.1	8061	154	40.4	34.97
St. Paul (Ramsey)	45.8	20.7	7662	159	50.6	33.45
Stillwater (Washington)	48.0	22.5	7088	170	NA	34.15

Cities metropolitan area can vary considerably, with forested parks and hiking areas showing much cooler readings.

Climate Observations and Climate Stations

The earliest climatic data available for Minnesota come from the diaries and observations of explorers and pioneers, whose sporadic notations diverge significantly in types of measurement and level of detail. Military outposts established during the early and mid-nineteenth century also provide some records of climate, kept in a systematic way on a daily basis. The government, interested in learning more about the weather and climate in the western territories, used military personnel at locations including Fort Snelling and Fort Ridgely to collect such data.

By 1849 Joseph Henry, secretary of the newly created Smithsonian Institution, had established a network of some 150 volunteer weather observers across the states and American frontier, including some locations in Minnesota Territory. The Smithsonian supplied observers with instructions and forms and in some cases instruments, and the observers mailed in monthly reports that included records of temperature, winds, cloud conditions, and precipitation. In 1861 the first of two volumes of climatic data and storm observations was published, based on reports submitted between 1854 and 1859. In the mid-1850s Henry worked out an arrangement with telegraph stations in major cities from New York to New Orleans. Daily telegraph reports enabled Henry to update weather conditions on a large color-coded weather map displayed in the Smithsonian. Henry also shared the dispatches with the *Washington Evening Star* newspaper, which began publishing daily weather for nearly 20 cities in 1857. Henry foresaw the need for a much broader system of Smithsonian storm warnings and in 1870 convinced Congress to pass a bill putting storm warnings (and, later, weather forecasts) in the hands of the U.S. Army Signal Service. In the 1870s and 1880s state weather services were formed to manage local observational networks and to distribute data and forecasts via telegraph and railroad networks. Professor William Payne of Carleton College in Northfield (Dakota/Rice counties) was the first director of State Weather Services for Minnesota.

Readings from weather stations, this one atop Minneapolis's post office in 1890, contribute to the overall picture of Minnesota's climate.

Congress, seeing the value of the observer network, established the Weather Bureau within the U.S. Department of Agriculture in 1890, charging it with placing volunteers every 25 miles in order to estimate rainfall more accurately. Today there are more than 11,000 volunteer Cooperative Weather Observers taking readings in all 50 states, including more than 200 in Minnesota. For well over a hundred years these observers have filed monthly reports with the National Weather Service and had their data published by the National Climatic Data Center. They still file monthly, but many report weather observations more frequently, even daily or weekly, by using fax machines, voice messaging, mobile device text messages, e-mail, or Internet websites that host various state climate databases.

Beyond climate observations made by official National Weather Service Cooperative Observers, many more are taken to meet the needs of federal, state, or local government agencies. Some agencies and organizations that operate networks of either volunteer observers or automated weather stations include the Community Collaborative Rain, Hail and Snow Network (CoCoRaHS), University of Minnesota Agricultural Experiment Station, the Minnesota Department of Natural Resources, the Metropolitan Mosquito Control District, Deep Portage Conservation Reserve, Soil and Water Conservation Districts, the Minnesota Department of Transportation, the Federal Aviation Administration, and the Minnesota Pollution Control Agency. All of the climate data are archived by the Minnesota State Climatology Office of the Department of Natural Resources, and they are available in various forms, including online via its website.

There are several classes of climate stations. A first-order station is where hourly records of barometric pressure, temperature, humidity, wind, sunshine, and precipitation are kept. In addition, synoptic observations (taken every six hours) of the amount and type of cloud cover and weather conditions are recorded. First-order stations are typically staffed around the clock, like the National Weather Service Forecast Office in Duluth.

Second-order climate stations record much of the same data that first-order stations do but without 24-hour personnel coverage. They record hourly data such as temperature, humidity, barometric pressure, dew point, wind direction and speed, cloud cover, and type of precipitation (if any), along with daily maximum and minimum temperatures. These stations, located at most community airports, are known as Automated Surface Observing Systems (ASOS) or Automated Weather Observation Stations (AWOS). More than 90 second-order stations operate in Minnesota. Though they provide meteorologists with valuable data in a timely manner, climatologically speaking they record neither accurate precipitation data nor reasonable snowfall or snow density measurements, all of which are extremely difficult to automate. Even today the manual measurement of rainfall, snowfall, or snow density remains the most trusted by the meteorological and research communities.

Third-order climatological stations, sometimes called substations, include most of the National Weather Service cooperative volunteer

Q & A

When the Minnesota State Climatology Office conducted a survey to rank the most significant weather events of the twentieth century, which of the following was not among the top five?

a. 1930s dust bowl era

b. 1940 Armistice Day blizzard

c. July 1987 Twin Cities flash flood

(answer on page 329)

observers, who record daily maximum and minimum temperatures along with precipitation. Some also take special observations (such as snow depth, snow density, evaporation, wind) but usually only once per day. Minnesota boasts more than 200 cooperative weather stations. This large number is associated with the state's diverse landscape and a widespread interest in gathering representative measurements that can be extrapolated over county-sized areas.

At precipitation stations volunteers use some type of gauge to record the total amount of liquid precipitation received each day. Some will also note the type and duration of precipitation. Minnesota has around 1,400 observers of precipitation, all of whom report their data to the State Climatology Office. These data are most highly valued for quantifying soil moisture status, assessing droughts, forecasting flood potential, estimating river flows, and even making mosquito forecasts.

Automated weather stations use electronic instrumentation wired to a small microprocessor or computer that records hourly values of temperature, barometric pressure, humidity, sky cover, precipitation type, ground temperature, wind, and visibility. These data are transmitted to a central location on an hourly or daily basis to be used operationally or to be archived into a climatic database. Most automated weather stations in Minnesota are deployed by the National Weather Service (which uses ASOS), by the Federal Aviation Administration (AWOS and ASOS), or by the Minnesota Department of Transportation's Road Weather Information Systems (RWIS). A small number are organized by the University of Minnesota Agricultural Experiment Station, located where the university operates research and outreach centers.

Climate Characteristics

Minnesota is noted for its warm, humid summers; cold, dry winters; and truly extreme weather events and climatic episodes. Temperatures have been as high as 115°F and as low as -60°F, a 175-degree range that is unrivaled over most of the Earth. Minnesota Territory was known as the American Siberia, based primarily on the climate measurements and writings of soldiers stationed at Fort Snelling. Winters were described as long and very cold, with temperatures as low as -41°F. Lakes and rivers froze solidly, allowing for horse-drawn sledding. Blizzards could rage for days, prohibiting outdoor activity of any kind. Heavy seasonal snowfall usually brought spring snowmelt flooding to the major watersheds. Conversely, summers could be hot and oppressive, with frequent thunderstorms accompanied by hail and tornadoes. In fact, a tornado hit the barracks at old Fort Snelling in April 1820. More unpredictable than spring snowmelt flooding, which could be anticipated based on the size of the winter snowpack, were flash floods from intense summer thunderstorms that produced as much as 3 inches of precipitation per hour. The transitional seasons—spring and fall—developed reputations for wild swings in weather. Dry periods tended to produce high fire danger for both the native prairie grasses and the forests. The spring and fall seasons lasted anywhere from days to months as polar and subtropical air masses wrestled for dominance over the Minnesota landscape. With a change in air mass, accompanied by a shift in wind direction, temperatures could vary as much as 70 degrees over periods as short as a day.

More than 70 Twin Cities daily weather records—maximum/minimum temperatures, precipitation, snowfall, and so on—set in the last decade of the nineteenth century (1891–1900) still stand today.

Any discussion of climate raises certain questions. What are its most important elements? How do they influence the way people live? Which aspects of the Minnesota climate are changing? How have the state's residents adapted to these climate characteristics? The following sections consider the character of Minnesota's major climate features, the first step toward shaping answers to these important questions.

Temperature

The mean annual temperature across Minnesota ranges from 45–48°F in the Twin Cities metro area and southernmost counties to 35–38°F in the north-central and northeastern counties. Based on the mean annual temperature for the years 1981–2010, Winona is the warmest spot in the state, with a value of 48.5°F, and Embarrass the coldest, with a value of 34.5°F. Beyond these bare facts there is a great deal more to understand about this subject.

The American Meteorological Society's *Glossary of Weather and Climate* defines temperature as a measure of the average kinetic energy of individual atoms or molecules that compose a substance. In climatology this substance is typically the atmosphere around us, but sometimes it may be the soil, the ocean, or a lake. Temperature is measured with a liquid-in-glass thermometer or an electronic thermometer, such as a thermistor or thermocouple, using a scale of reference based on the defined fiducial points of water, notably when it freezes (the ice point, 32°F) and when it turns to vapor or steam (the boiling point, 212°F). The temperature data used to define Minnesota's climatology have historically been recorded with liquid-in-glass thermometers, containing mercury to measure maximum temperature and alcohol—more accurate in the low range because of its very low freezing point—to measure minimum temperature. Most of the temperature data were taken by observers using standard National Weather Service instrument shelters prescribed since the 1880s: a louvered, double-roofed, wooden structure mounted on four legs, with a base resting about 4.5 feet above the ground. The hinged door of the shelter, used to access and read the thermometers, always faced north to prevent the sun from shining directly on the instruments. In the post-1990 era of National Weather Service modernization, temperature readings are more often taken from thermistors sheltered from sunlight by a small shield of stacked plastic plates. Temperature readings are shown on an LCD display, usually located in the observer's home, office, garage, or barn.

The observation of daily maximum and minimum temperature most often occurs at 8:00 AM, 5:00 PM, or midnight. At these times, observers note the maximum and minimum temperatures for the previous 24-hour period and reset the instrument to the current temperature value, ready to record another 24-hour cycle. Consequently, an inherent time-of-observation bias is introduced into any series of temperature data, as morning observers tend to double record minimum temperature and afternoon observers tend to double record maximum temperature. True calendar date observations of the National Weather Service Offices reflect a midnight-to-midnight observing cycle. When comparing temperature data from state or regional networks with differing observation times, climatologists typically employ correction factors (statistical algorithms) that adjust to a common time of observation such as 8:00 AM or midnight. For the data described in this book, such a correction scheme has been used.

Though temperature measurements in Minnesota date to the early nineteenth century in some locations, the time series commonly used to define the climatology of a given location or community is the most recent three complete decades of data. This practice is mandated by the World Meteorological Society and used by nearly all government weather services. Thus currently for a climatological description of any place on Earth, the statistics are likely to be derived from the 1981–2010 period. This span of years, called the normals period, is used to characterize the climate of Minnesota.

According to Donald G. Baker, Earl L. Kuehnast, and James A. Zandlo in *Climate of Minnesota 15: Normal Temperatures (1951–1980) and Their Application,* four factors affect temperature significantly when comparing values over large geographic regions: latitude, microenvironment, topography, and urbanization.

Mean annual temperature tends to decrease with higher latitude positions. This trend is visible on most global- or continental-scale climate maps as isotherms (lines of equal temperature value), which are commonly oriented west to east. Departures from this orientation may be due to the presence of large bodies of water such as oceans and their associated behavioral features or to topographic features such as mountains, high platcaus, or deep valleys. The range in latitude across Minnesota from south to north—nearly six degrees—contributes to the range in mean annual temperature. Lower sun angles, shorter day length, and greater snow cover duration foster colder wintertime temperatures in the northern portions of the state.

The air temperature recorded in an observation shelter is influenced by the underlying surface—its microenvironment. Whether the shelter is over a grass surface or pavement, located near snow or ice cover, or surrounded by water, such local-scale features affect temperature measurement. A sheltered temperature measurement in the presence of actively growing vegetation—turf, for example—with ample soil moisture will be affected by a process called evapotranspiration. Both the solar energy and the advected sensible energy, the heat transported by the air, are mostly consumed by evaporation or transpiration, the moisture released as water vapor from growing plants. If the instrument shelter is near a body of water such as a lake, the evaporation process will consume the majority of energy, leaving little to heat the air. Similarly,

MAP 1: Average Annual Temperatures, 1981–2010

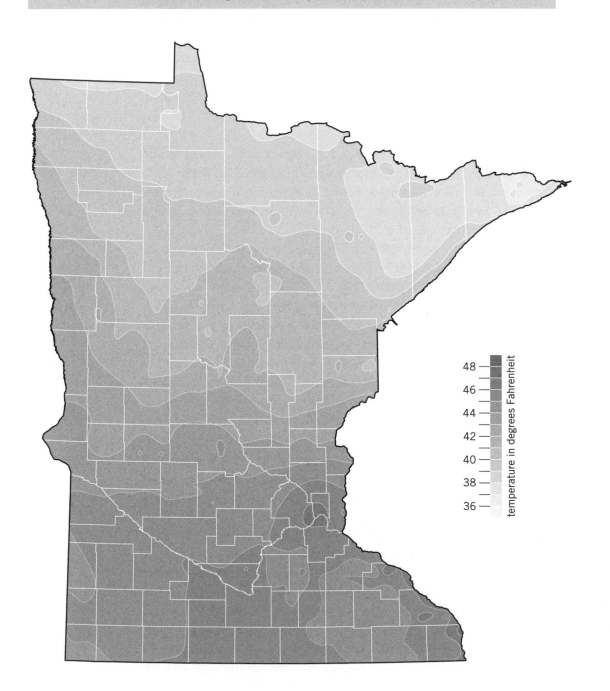

temperature in degrees Fahrenheit

48
46
44
42
40
38
36

instrument shelters in close proximity to buildings and paved surfaces will be affected by them, as they absorb solar energy and reradiate it, heating the surrounding air.

The forested landscapes of northern Minnesota exhibit a twofold effect on the microenvironment. In the early winter the coniferous canopies absorb the sun's radiation and also inhibit the loss of long-wave radiation from the soil. Under these conditions, overnight temperatures can be higher than those from more exposed landscapes where bare soil or dormant vegetation exists. Later in winter, after snow cover has built up, the forested canopies can prevent the melting and loss of snow, which has a higher albedo and reflects more incoming solar radiation, keeping near-surface temperatures colder than those of more exposed landscapes that have lost snow cover to melting and evaporation.

In the Twin Cities metropolitan area, the simplest, quickest, and most accurate source of weather information is NOAA weather radio, broadcast around the clock from the National Weather Service Forecast Office in Chanhassen, KEC 65 at 162.55 Mhz.

Topography and its interaction with major weather systems as they cross the North American continent can have complex effects on the climate. One well-known effect is the forced ascent or descent of air as it passes over variable topographic features. On the upslope, air moves to lower pressure regions, expands, and therefore diminishes in temperature. When all else is held constant and there is no condensation of water vapor, air cools at a rate of 5.5 degrees per 1,000-foot rise in elevation. Descending air compresses, with an associated increase in air pressure, and warms at the identical rate, 5.5 degrees per 1,000-foot drop. Called the adiabatic process, in meteorology it is accounted for in forecast models to predict temperature at any given place and time.

Minnesota's topographic features affect temperature distributions across the state. One of the state's most prominent features is the Buffalo Ridge (also known as the Coteau des Prairies), located in its southwestern corner. Oriented northwest to southeast, the ridge bisects Lincoln, Lyon, Murray, and Nobles counties. The elevation rise on the ridge's western side amounts to about 800 feet, while the descent on its eastern side toward the Minnesota River Valley is about 1,000 feet. In certain seasons, the effect of the Buffalo Ridge is apparent as the temperature distribution is mapped across this part of the state. A cool "tongue" of air emerging out of South Dakota runs down the western side of the

TABLE 3: Average Mean Temperatures by Month (°F) (1981–2010)

STATION	JAN	FEB	MAR	APR	MAY	JUN	JUL	AUG	SEP	OCT	NOV	DEC	ANNUAL
Ada	6.8	11.8	25.7	43.1	55.8	65.3	69.8	68.4	58.1	44.2	27.5	12.4	40.9
Aitkin	9.9	14.7	26.8	41.8	53.4	62.8	67.5	65.2	56.3	44.2	29.3	15.2	40.7
Albert Lea	14.1	18.9	31.2	45.8	58.0	68.0	71.9	69.5	60.9	47.9	33.2	18.5	44.9
Alexandria	10.2	15.6	27.8	43.6	56.2	65.4	70.6	68.4	58.8	45.3	29.4	14.8	42.3
Argyle	5.0	10.2	24.0	41.5	54.0	63.5	67.9	66.5	56.7	43.1	26.2	11.0	39.3
Austin	13.4	17.9	30.6	45.6	57.4	67.4	70.7	68.4	60.0	47.7	32.5	17.9	44.2
Baudette	5.0	11.1	24.2	41.4	53.9	64.0	68.6	66.8	56.2	42.9	26.7	10.8	39.4
Bemidji	5.9	11.2	24.2	39.1	53.1	62.6	67.2	64.7	55.1	41.9	26.1	11.5	38.7
Big Falls	7.9	14.5	27.5	42.8	54.8	62.9	67.4	65.6	56.7	44.3	27.7	12.4	40.5
Brainerd	8.3	13.5	26.6	41.7	54.9	64.1	68.7	66.5	56.6	43.9	28.8	13.9	40.8
Browns Valley	11.7	17.1	28.9	44.0	56.8	66.1	71.3	69.0	59.2	45.9	30.5	15.9	43.2
Caledonia	15.1	20.0	31.6	45.5	56.8	66.5	70.8	68.6	59.9	47.7	33.4	19.3	44.7
Cambridge	12.6	17.3	29.3	45.0	56.9	66.4	70.8	68.4	58.6	46.5	31.4	16.9	43.5
Campbell	7.6	13.7	26.2	43.1	57.2	65.8	70.7	68.6	58.7	44.6	28.3	12.5	41.5
Canby	14.5	19.5	30.6	45.2	58.0	67.8	72.7	70.3	61.0	47.6	32.1	18.6	44.9
Cloquet	10.9	16.0	27.2	41.0	52.7	61.5	67.0	65.5	56.7	43.9	29.1	15.1	40.7
Collegeville	13.7	19.2	31.1	46.3	58.6	67.5	72.4	71.0	61.2	48.2	32.2	17.8	45.0
Cook	5.0	10.9	23.3	38.9	52.7	61.6	66.8	65.1	55.4	42.4	27.3	11.3	38.5
Cotton	5.4	11.2	24.2	38.8	51.0	59.9	64.4	62.6	53.9	41.4	26.3	10.8	37.6
Crookston	6.1	11.3	25.0	42.1	55.2	64.6	69.1	67.5	57.2	43.3	26.7	11.6	40.1
Detroit Lakes	8.6	14.2	27.2	43.5	57.3	66.4	70.8	69.4	59.2	45.3	29.3	13.7	42.2
Duluth	10.2	15.1	25.9	39.6	51.4	60.1	65.8	64.3	55.6	43.2	28.8	14.8	39.7
Ely-Winton	5.3	11.7	24.1	38.8	51.9	61.5	66.1	64.9	55.3	41.6	26.7	10.3	38.3
Fairmont	15.3	19.9	31.6	46.6	59.6	69.4	73.4	71.2	62.3	49.2	33.5	19.2	46.1
Faribault	13.8	18.3	30.3	45.0	56.8	67.1	71.2	68.7	59.9	47.2	32.5	18.7	44.2
Farmington	15.5	20.7	32.7	47.5	58.9	68.2	72.2	69.7	61.5	48.9	33.5	19.2	45.8
Fergus Falls	9.0	14.0	27.4	43.6	56.4	65.6	70.3	68.7	59.0	45.4	29.3	14.3	42.0
Fosston	5.7	11.0	24.4	41.0	53.9	62.6	66.9	65.1	55.6	42.7	26.4	11.7	39.0
Grand Marais	15.6	19.1	27.7	38.8	47.0	53.8	61.5	63.2	55.8	44.2	32.3	20.0	40.0
Grand Meadow	13.3	17.9	29.3	43.6	55.6	66.0	69.4	67.2	58.8	46.3	31.9	17.9	43.2
Grand Rapids	8.4	14.3	27.3	42.0	54.2	63.3	68.1	66.2	56.4	43.6	28.4	13.3	40.6
Gunflint Lake	3.3	9.0	21.1	36.4	50.0	59.7	64.9	63.0	53.6	40.3	25.7	10.0	36.5
Hallock	0.8	6.3	21.3	39.7	53.2	64.2	68.4	66.5	55.0	40.6	23.4	7.0	37.4
Hibbing	6.7	12.0	24.4	38.8	50.6	59.4	64.2	62.4	53.1	40.7	25.9	11.3	37.6
Hutchinson	12.7	17.9	30.6	45.8	58.0	67.7	72.0	69.2	60.5	47.1	31.6	16.8	44.3
Int'l Falls	4.4	10.3	23.6	39.3	51.8	60.8	65.2	63.3	53.6	41.1	25.5	9.7	37.5

STATION	JAN	FEB	MAR	APR	MAY	JUN	JUL	AUG	SEP	OCT	NOV	DEC	ANNUAL
Itasca	6.1	11.4	23.9	39.2	52.3	61.9	66.7	64.9	54.7	41.9	26.2	11.6	38.5
Leech Lake Dam	7.7	13.7	26.6	41.7	54.4	63.7	68.4	66.5	57.1	43.9	28.3	13.4	40.6
Luverne	16.4	21.7	33.3	47.2	59.4	68.6	72.6	70.3	61.8	48.8	32.9	19.3	46.1
Marshall	14.3	19.1	30.4	45.4	58.4	68.1	72.5	70.1	60.9	47.6	32.0	18.3	44.9
Milan	14.1	19.5	31.9	47.4	59.9	68.7	72.7	70.3	61.6	48.4	32.3	17.9	45.5
Montevideo	12.1	17.0	29.0	44.7	57.6	67.2	71.6	69.3	60.0	46.6	30.9	16.4	43.7
Mora	10.2	15.7	28.4	43.3	55.5	64.7	69.2	67.1	57.7	44.9	30.3	15.3	42.0
Morris	10.0	15.0	27.8	43.9	56.9	66.2	70.4	68.1	59.0	45.4	29.6	15.0	42.4
MSP	15.6	20.8	32.8	47.5	59.1	68.8	73.8	71.2	62.0	48.9	33.7	19.7	46.2
New Ulm	14.7	19.8	32.0	47.0	59.4	68.9	73.1	70.4	61.5	48.6	33.3	18.4	45.7
Olivia	12.3	17.2	29.4	44.8	57.7	67.4	71.3	68.6	59.5	46.9	31.2	16.7	43.7
Park Rapids	6.8	12.4	25.7	41.2	53.8	62.6	67.6	65.7	56.0	42.6	26.3	11.6	39.5
Pine River Dam	8.5	13.9	26.6	41.6	54.7	64.2	69.1	67.1	57.4	44.5	28.8	13.7	41.0
Pipestone	13.3	18.5	30.4	45.0	57.3	66.7	71.4	69.2	59.9	46.5	31.0	17.0	44.0
Pokegama Dam	6.6	12.8	26.3	41.5	54.2	63.2	67.9	66.3	56.9	43.3	27.7	12.3	40.0
Red Wing	14.9	19.4	31.4	46.5	58.4	67.6	72.4	70.1	61.6	48.5	33.9	19.5	45.5
Redwood Falls	12.9	18.6	30.2	45.6	58.6	68.2	72.6	70.5	61.3	48.4	32.7	18.0	44.9
Rochester	15.7	20.3	32.4	46.7	58.1	67.6	71.4	69.1	60.9	48.4	33.8	19.7	45.4
Roseau	3.0	8.5	23.0	40.0	52.9	62.8	67.2	65.4	55.6	41.7	24.8	9.2	38.0
St. Cloud	11.6	17.0	29.4	44.5	56.5	65.6	70.3	67.8	58.6	45.7	30.4	15.9	42.9
St. James	14.5	19.3	31.0	45.6	58.6	68.2	72.1	69.5	60.8	47.7	32.6	18.5	45.0
St. Peter	14.6	19.3	31.9	46.2	58.8	68.8	73.2	71.0	61.5	48.6	33.5	18.7	45.6
Sandy Lake Dam	8.4	13.9	26.4	41.7	54.5	63.5	68.2	66.7	57.6	44.5	29.0	13.6	40.8
Stillwater	13.4	17.5	30.5	46.2	57.1	67.0	71.3	69.2	60.5	47.8	32.3	17.6	44.3
Tower	5.0	10.2	23.1	38.2	50.3	59.4	63.9	61.8	53.0	40.7	26.0	11.2	37.0
Two Harbors	15.7	19.7	29.0	40.1	48.6	56.5	64.6	65.3	57.4	45.8	33.0	20.3	41.4
Wadena	8.0	12.8	25.6	41.4	54.1	63.2	67.6	65.7	56.4	43.5	28.0	13.5	40.1
Warroad	4.4	10.0	23.4	39.2	52.2	62.2	67.1	65.1	55.2	42.1	26.1	10.3	38.3
Waseca	13.2	18.5	31.2	46.1	58.7	68.5	72.0	69.8	61.3	48.2	32.7	17.8	45.0
Waskish	4.2	10.1	23.3	39.9	52.3	61.9	66.2	64.4	54.6	41.2	26.3	10.0	38.0
Wheaton	10.0	15.2	27.4	43.6	57.1	66.7	71.6	69.4	59.3	45.7	29.5	14.4	42.6
Willmar	11.8	17.0	29.6	45.3	58.1	67.6	71.9	69.3	60.3	47.2	31.3	16.5	43.9
Windom	17.0	21.9	33.5	47.3	59.4	69.3	73.5	70.8	62.2	49.2	33.6	20.2	46.6
Winnebago	14.7	19.7	31.9	46.4	58.7	68.4	72.3	69.9	61.4	48.6	33.4	18.9	45.5
Winona	18.7	23.9	35.3	49.8	61.1	70.1	74.5	72.4	64.2	51.4	36.7	22.7	48.5
Worthington	14.5	19.3	30.5	44.8	57.4	67.5	71.5	68.8	60.3	47.6	32.0	18.3	44.5
Wright	10.3	15.5	27.6	41.8	53.3	61.8	66.5	65.1	56.6	44.3	29.1	14.6	40.6
Zumbrota	12.9	17.6	30.3	45.0	56.1	66.3	70.2	67.8	59.1	47.2	31.8	17.5	43.6

slope as the lifted air mass cools adiabatically. Conversely, with summer's southwesterly winds a warm "tongue" of air can be detected in the temperature pattern on the leeside of the ridge as descending air is warmed adiabatically. Quite often the state's warmest summertime temperatures are reported from the downwind counties of Brown, Lyon, Martin, Redwood, Watonwan, and Yellow Medicine.

Though not as obvious as the Buffalo Ridge, a second topographic feature of importance to the state's temperature pattern is the Alexandria moraine in the northwest. It begins south of Alexandria and continues north through Itasca State Park. The moraine's western slope starts in the Red River Valley and rises by about 800 feet by the time it reaches Itasca State Park. To the east on the downslope side it drops off less dramatically, by about 400 feet. The adiabatic cooling process results in lower temperatures across Becker, Clearwater, Hubbard, and Mahnomen counties, particularly at Itasca State Park. The downslope warming is not as evident to the east, however, because of a smaller change in elevation and because of irregularities in the moraine itself.

To the west of the Red River Valley, in eastern North Dakota, there is also a rise in elevation. Thus air masses moving across the region from west to east will be warmed by downslope compression as they cross the Red River Valley toward the Alexandria moraine, where they are cooled by upslope motion. This adiabatic process produces a temperature pattern that shows a warm tongue of air oriented north to south along the Red River Valley. Indeed, Ada in Norman County has recorded some of the state's warmest temperatures.

In northeastern Minnesota's Cook, Lake, and St. Louis counties the north shore ridge runs essentially parallel to the Lake Superior shoreline. Elevations from the highlands to the shoreline range from 500 to 1,200 feet. Air descending to the lake would normally exhibit a significant warming trend; however, Lake Superior's very cool summertime water temperatures negate this effect. Thus there is a sharp temperature gradient in this region, showing cooler temperatures near the lake and warmer temperatures in the highlands. Conversely, in the winter months when the lake is still unfrozen, the water temperature, along with additional water vapor released into the lower atmosphere, has a stabilizing influence, producing warmer temperatures along the lakeshore and colder temperatures in the highland areas.

Finally, the temperature pattern exhibited across the state is influenced by river valley topography. Both the Minnesota River Valley (across the south) and the Mississippi River Valley (of the southeast) act to restrict large-scale air motions. Further, their southern-exposed slopes act as a trap for daily solar radiation, amplifying its heating of the surrounding landscape. This enhanced warming is especially notable during seasons with lower sun elevation angles. In addition, overnight low temperatures often remain high as a result of the elevated concentration of water vapor released by the rivers themselves. The higher isotherms depicted on temperature maps often run parallel to these valleys, and the state's highest temperatures in winter will sometimes be reported from Chaska, New Ulm, and St. Peter along the Minnesota River Valley or Winona along the Mississippi River Valley.

Topography can amplify the microclimate effect by enhancing cold air drainage and producing frost pockets, relatively low-lying areas where colder, denser air can accumulate overnight in the absence of strong winds. Such pockets are prone to later occurrences of frost in the spring and earlier occurrences in the fall. Such is the case for areas around Jordan just southwest of the Twin Cities metropolitan area and Zumbrota in southeastern Minnesota. Jordan's climate data show a mean frost-free growing season of only 145 days compared with Chaska's growing season of 160 days and Minneapolis–St. Paul's 167 days, even though all three cities are within 40 miles of each other. Similarly, Zumbrota's climate data show a mean frost-free growing season of 133 days, while nearby Rochester claims a 159-day growing season and Faribault a 144-day growing season, all three cities being within 25 miles of each other. The general mapped pattern showing frost-free growing season length ranges from 160 days or more in the southern counties to 110 days in the far north. Northern frost pockets such as Embarrass and Tower in St. Louis County have average growing seasons of fewer than 75 days.

Though urbanization has been shown to affect local climate even with a population change as small as 3,000 people, the monthly, seasonal, and annual temperature patterns exhibited across Minnesota primarily show only the effect of the Twin Cities metropolitan area. It is quite difficult to pick out effects from other population centers such as Rochester or St. Cloud. The temperature patterns exhibited in the Twin

Cities area match those along the state's southern border, suggesting that urbanization has offset approximately 150 miles of latitude displacement. In other words, although the distance from northern Iowa to the Twin Cities should reduce the mean annual temperature, all other things being equal, the heat island effect of the Twin Cities compensates so that, with respect to temperature, the metro area appears equivalent to northern Iowa. The average frost-free growing season in the Twin Cities (based on MSP airport data) is 167 days, similar to that shown by climate stations along Iowa's border (e.g., Fairmont). The urbanization effect on temperature is evident in the Twin Cities' winter temperatures, especially minimums that are higher than the average values seen in southern Minnesota's climate records.

The state's mean monthly temperatures range in value from the low seventies during July—Minnesota's warmest month—in the southern counties to the low single digits during January—Minnesota's coldest month—in the far north. The temperature range is often maximized during the spring (March-May), when differences in snow cover can be most pronounced. For example, the state record high in March is 88°F at Montevideo (Chippewa County) on March 23, 1910, when the landscape was dry and there was no snow cover, while the state record low for March is -50°F at Pokegama Dam (Itasca County) on March 2, 1897, a day when more than 10 inches of fresh snow covered the ground. Another remarkable example comes from a single moment in time, 4:00 PM on May 19, 2009, when Montevideo and Madison in western Minnesota were reporting 100°F with relative humidity of less than 15 percent (dry landscape), while Grand Marais harbor on Lake Superior reported 37°F with 79 percent relative humidity (wet landscape).

Many southern and western counties have recorded temperatures above 100°F, while some locations in the north and east have no such temperatures in their climate records. Conversely, no southern Minnesota counties have seen temperatures as cold as -50°F, though a number have recorded -40°F. Virtually all of the state's low temperature records during the winter months have been set with the presence of snow cover, which provides such a high albedo that most of the incoming solar radiation is reflected rather than absorbed and converted into heat.

Precipitation

Like temperature, precipitation's geographic pattern and variability is influenced by latitude, urbanization, and topography. In addition, patterns by season or for individual precipitation events may be affected by the position of the polar jet stream. The trajectory of weather systems across the state, the relative abundance of water vapor in the atmosphere, and the atmosphere's depth interact to dictate the potential type and amount of precipitation.

Although total annual precipitation is an important climate feature, the nature of Minnesota's landscape and vegetation is primarily influenced by the distribution of moisture across the year. Annual precipitation generally ranges from over 36 inches in the southeast (Lanesboro and Caledonia) to under 20 inches in the far northwest (Caribou), with 60 to 70 percent of it falling during the growing season, May through September. Based on historical averages, the driest month for most communities is February, while the wettest is June. Most communities' climate histories show that there have been months with virtually no precipitation or just trace amounts (less than 0.01 inches). On a statewide basis the two driest years were 1910 and 1976, which reported a statewide average precipitation of fewer than 16 inches. From a growing season standpoint (May-September), these years were also the driest of record statewide, averaging 10 or fewer inches. The wettest years were 1965, 1968, 1977, and 2010, averaging more than 33 inches each statewide. The wettest growing seasons (May-September), however, were 1905, 1944, 1993, and 2010, all averaging more than 23 inches of rainfall statewide.

What is occult precipitation?

a. precipitation that occurs at night

b. an underground irrigation system

c. moisture from dew, frost, or fog

(answer on page 329)

Severe thunderstorms can bring rainfall in great excess, challenging the discharge capacity of municipal storm water runoff systems and agricultural tile-ditch drainage. Thunderstorms can occur in any month, but the peak season is May through July. The average number of days per year that bring thunderstorms ranges from 30 in northern counties to 45 in southern counties. In the extreme, rainfall rates of as many as 3 inches per hour have been recorded, as well as total daily amounts in excess of 10 inches. Individual thunderstorm cells or lines of thunderstorms

MAP 2: Average Annual Precipitation, 1981–2010

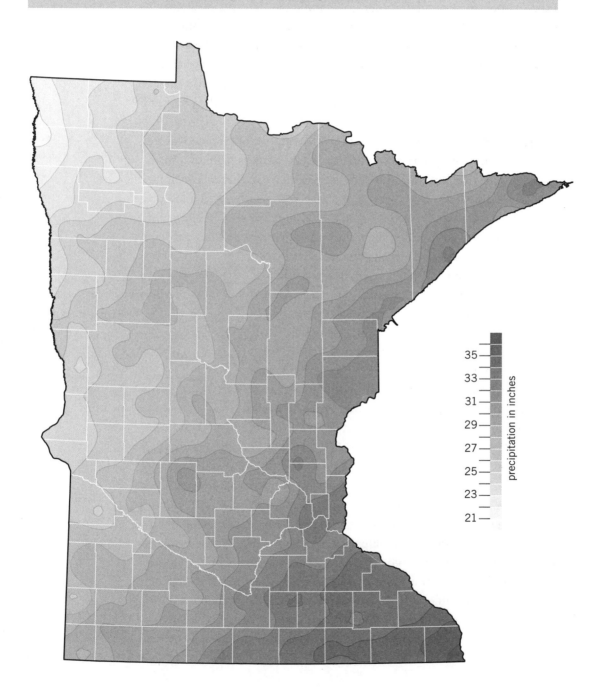

precipitation in inches

35
33
31
29
27
25
23
21

(mesoscale convective systems) most commonly spread across the state in orientations west to east. Other trajectories, though possible, are relatively rare.

Occasionally thunderstorms bring hail and tornadoes. The annual average number of hail days ranges from less than one in the north to three in the southwest. Hailstones with diameters greater than 4 inches have been observed in the state, and each year modest-sized hailstorms cause crop losses for Minnesota farmers, most of whom carry insurance. Though hail has occurred historically across the state in every month of the year, the hail season reaches its peak around June 1, the storms' arrival most frequently falling between 4:00 and 8:00 PM.

Pluviometric coefficient

Taken from the Latin roots pluvio, *"rain," and* metre, *"to measure," the pluviometric coefficient is a means of expressing rainfall as a ratio of the average monthly value to one-twelfth of the normal annual amount. Sometimes called an "isomer," the pluviometric coefficient enables climatologists to express each month's share of a location's annual precipitation. Values below one represent a less-than-equal share or a relatively dry month, while values above one represent a greater-than-equal share. In Minneapolis–St. Paul, for example, February's 0.79-inch average represents a less than 0.3 share, while June's 4.34-inch value represents a nearly 1.5 share. Such variability is typical of midlatitude continental climates, while equatorial and marine climates often have values uniformly close to one.*

Minnesota's yearly average number of tornadoes is in the range of 35 to 40, with a record 113 occurring in 2010, the most in the nation that year. Although much has been made of increasing tornado frequency, this phenomenon can be explained by the deployment of both Doppler radar systems that can detect tornadoes and better weather spotter networks, like Skywarn, that provide more comprehensive ground observations. These complementary systems offer such detailed coverage that few if any tornadoes are unobserved these days, though weaker EF-0s—tornadoes of lowest intensity—still present challenges because they can be so small and short lived. Thus comparison with historical tornado counts is unfair, since many were neither observed nor verified before the advent of these modern spotter networks.

TABLE 4: Average Monthly Precipitation (inches) (1981–2010)

STATION	JAN	FEB	MAR	APR	MAY	JUN	JUL	AUG	SEP	OCT	NOV	DEC	ANNUAL
Ada	0.89	0.65	1.00	1.41	3.24	4.48	3.67	2.74	2.64	2.24	0.96	0.81	24.73
Aitkin	0.99	0.75	1.68	2.61	3.12	4.46	4.15	3.18	3.32	2.86	1.58	1.02	29.72
Albert Lea	0.73	0.69	1.85	3.66	4.44	4.71	4.56	4.45	3.58	2.53	1.75	1.11	34.06
Alexandria	0.72	0.48	1.26	2.11	2.80	4.05	3.42	3.54	2.90	2.31	1.04	0.48	25.11
Argyle	0.66	0.58	0.89	1.01	2.74	3.70	2.75	3.06	2.40	1.98	1.01	0.78	21.56
Austin	0.92	0.84	1.89	3.55	4.34	4.87	4.75	4.39	3.47	2.38	1.91	1.24	34.55
Baudette	0.55	0.37	0.63	1.21	2.89	4.35	3.54	3.24	2.89	2.12	1.03	0.62	23.44
Bemidji	0.73	0.66	1.09	1.78	3.05	4.38	3.95	3.21	3.10	2.54	1.35	0.79	26.63
Big Falls	0.82	0.57	1.16	1.69	2.91	4.28	4.00	3.45	3.04	2.51	1.29	0.83	26.55
Brainerd	0.77	0.67	1.52	2.35	3.37	4.39	3.78	3.07	3.18	2.93	1.54	0.81	28.38
Browns Valley	0.86	0.72	1.58	2.30	2.55	3.78	3.47	2.97	2.67	2.07	1.11	0.71	24.79
Caledonia	1.04	1.03	1.94	2.89	4.02	5.22	4.57	5.02	3.80	2.50	2.34	1.56	36.93
Campbell	0.75	0.60	1.28	2.18	3.02	3.83	3.68	3.03	3.17	2.45	1.03	0.57	25.59
Canby	0.80	0.83	1.70	2.66	2.99	4.00	3.43	2.76	3.00	2.41	1.57	0.95	27.10
Caribou	0.58	0.45	0.79	1.02	2.56	3.34	3.19	2.51	2.33	1.63	0.87	0.57	19.84
Cloquet	0.95	0.83	1.49	2.32	3.23	4.24	4.21	4.00	4.34	3.12	1.96	1.14	31.83
Collegeville	0.77	0.74	1.79	2.70	3.43	4.72	3.62	3.90	3.52	2.77	1.66	0.92	30.54
Cook	0.86	0.63	1.09	1.98	2.71	3.97	3.66	3.68	3.77	2.61	1.49	0.96	27.41
Crookston	0.50	0.52	0.80	1.20	2.91	3.80	3.00	3.29	2.44	2.09	0.91	0.66	22.12
Detroit Lakes	0.79	0.50	1.08	1.85	3.51	4.44	3.82	2.96	3.57	3.05	1.20	0.69	27.46
Duluth	0.96	0.81	1.49	2.43	3.23	4.23	3.85	3.70	4.11	2.85	2.09	1.21	30.96
Ely-Winton	1.05	0.66	1.19	2.23	3.03	3.95	4.26	3.66	4.02	2.83	1.70	1.25	29.83
Fairmont	0.92	0.80	1.97	3.39	3.85	4.52	4.33	4.04	3.47	2.34	1.80	1.28	32.71
Faribault	0.94	0.77	1.98	2.87	3.82	4.39	4.26	4.73	3.44	2.39	1.86	1.18	32.63
Farmington	0.78	0.75	1.87	2.66	3.56	4.35	3.96	4.60	3.31	2.53	1.80	1.13	31.30
Fergus Falls	0.91	0.62	1.26	1.73	3.19	3.95	3.37	3.21	2.79	2.45	1.01	0.60	25.06
Fosston	0.69	0.65	0.89	1.60	2.98	4.62	3.94	3.23	2.87	2.40	0.96	0.64	25.47
Grand Marais	0.81	0.50	0.87	1.67	2.53	3.61	3.12	2.71	3.00	3.10	1.70	0.98	24.60
Grand Meadow	0.84	0.74	1.74	3.62	4.19	4.95	4.79	4.98	3.91	2.51	1.99	1.16	35.42
Grand Rapids	0.93	0.64	1.32	2.07	3.09	4.37	4.29	3.41	3.32	2.81	1.61	1.07	28.93
Gunflint Lake	0.94	0.71	0.97	1.86	2.76	3.83	3.70	3.47	3.33	2.57	1.85	1.13	27.12
Hallock	0.65	0.48	0.88	0.90	3.16	3.73	3.08	2.84	2.08	1.76	1.07	0.75	21.38
Harmony	0.86	0.96	1.84	3.43	3.86	5.30	4.22	4.76	3.64	2.34	2.16	1.26	34.63
Hibbing	0.64	0.44	0.93	1.64	2.55	4.16	4.38	3.14	3.07	2.46	1.17	0.69	25.27
Hutchinson	0.63	0.60	1.54	2.43	3.11	4.64	3.55	4.41	3.15	2.09	1.39	0.84	28.38
Int'l Falls	0.62	0.57	0.95	1.53	2.86	3.92	3.70	2.81	2.99	2.08	1.38	0.81	24.22
Itasca	0.75	0.57	1.28	1.82	3.09	4.80	3.84	3.35	3.30	2.88	1.35	0.86	27.89

STATION	JAN	FEB	MAR	APR	MAY	JUN	JUL	AUG	SEP	OCT	NOV	DEC	ANNUAL
Leech Lake Dam	0.67	0.51	1.12	1.70	2.93	3.57	4.24	3.38	3.14	2.58	1.26	0.84	25.94
Luverne	0.48	0.78	2.05	2.95	3.32	4.57	3.62	3.44	2.87	2.39	1.93	1.10	29.50
Marshall	0.99	0.70	1.73	2.67	3.39	3.82	3.68	3.39	3.09	2.13	1.52	1.17	28.28
Milan	0.76	0.68	1.40	2.40	2.80	3.72	3.92	3.16	3.16	2.43	1.13	0.58	26.14
Montevideo	0.87	0.96	1.81	2.51	3.02	4.24	3.65	3.39	3.28	2.33	1.69	0.86	28.61
Mora	0.72	0.75	1.57	2.43	3.12	4.20	3.99	3.78	3.52	2.74	1.64	0.99	29.45
Morris	0.74	0.71	1.43	2.31	2.83	4.01	3.88	3.33	2.90	2.53	1.08	0.72	26.47
MSP	0.90	0.77	1.89	2.66	3.36	4.25	4.04	4.30	3.08	2.43	1.77	1.16	30.61
New Ulm	0.64	0.64	1.86	2.86	3.44	4.82	3.98	4.10	3.21	2.26	1.62	0.86	30.29
Olivia	0.58	0.55	1.46	2.60	2.96	4.57	3.56	3.49	3.27	2.16	1.39	0.84	27.43
Park Rapids	0.52	0.47	1.11	1.85	3.10	4.20	4.10	3.55	3.08	2.65	0.96	0.57	26.16
Pine River Dam	0.78	0.65	1.64	2.08	3.52	4.21	3.87	3.09	3.14	2.98	1.55	0.89	28.40
Pipestone	0.59	0.57	1.75	2.90	3.29	4.21	3.34	3.14	3.24	2.33	1.45	0.79	27.60
Pokegama Dam	0.76	0.56	1.10	1.80	2.99	4.22	4.28	3.61	3.36	2.68	1.38	0.91	27.65
Red Wing	1.14	0.82	1.94	3.18	3.75	4.05	4.26	4.67	4.00	2.21	2.26	1.13	33.41
Redwood Falls	0.70	0.61	1.65	2.51	3.11	4.09	3.80	3.61	2.48	1.88	1.60	0.60	26.64
Rochester	0.94	0.75	1.88	3.01	3.53	4.00	4.61	4.33	3.12	2.20	2.01	1.02	31.40
Roseau	0.68	0.50	0.58	1.15	2.24	3.71	3.33	3.09	2.57	1.49	0.80	0.62	20.76
St. Cloud	0.65	0.59	1.55	2.57	2.95	4.17	3.31	3.79	3.46	2.49	1.38	0.82	27.73
St. James	0.77	0.59	1.64	3.15	3.55	4.47	4.27	3.86	3.42	2.17	1.62	0.99	30.50
St. Peter	0.68	0.65	2.10	2.90	3.48	4.84	4.29	4.22	2.90	2.49	1.80	0.91	31.26
Sandy Lake Dam	0.75	0.56	1.23	1.95	3.03	4.74	4.43	3.28	3.38	2.60	1.14	0.79	27.88
Spring Grove	0.92	0.82	1.79	3.60	3.84	5.03	4.33	4.69	3.62	2.40	2.11	1.37	34.52
Stillwater	0.88	0.76	1.72	2.93	3.74	4.42	4.30	4.81	3.72	2.81	1.90	1.43	33.42
Tower	0.63	0.57	0.84	1.75	3.08	3.98	4.10	3.83	4.00	2.84	1.43	0.67	27.72
Two Harbors	0.96	0.92	1.43	2.56	3.13	3.98	3.89	3.74	4.06	2.90	2.20	1.34	31.11
Wadena	0.81	0.64	1.50	2.33	3.30	4.46	3.62	3.11	2.86	2.89	1.38	0.74	27.64
Warroad	0.65	0.56	0.79	1.31	2.87	4.41	3.78	3.14	2.77	2.24	1.10	0.87	24.49
Waseca	1.25	1.00	2.49	3.21	3.93	4.69	4.42	4.75	3.67	2.67	2.16	1.48	35.72
Waskish	0.59	0.54	1.07	1.60	2.97	4.29	3.88	3.36	2.94	2.40	1.22	0.72	25.58
Wheaton	0.82	0.61	1.48	2.24	2.68	3.98	3.20	2.82	3.09	2.17	1.13	0.72	24.94
Willmar	0.65	0.64	1.49	2.62	3.10	4.98	3.82	4.09	3.36	2.40	1.47	0.84	29.46
Windom	0.90	0.71	1.97	3.24	3.55	4.56	4.05	3.52	3.29	2.18	1.67	0.98	30.62
Winnebago	0.83	0.77	1.89	3.40	4.02	4.61	4.23	4.29	3.32	2.45	1.77	1.14	32.72
Winnibigoshish Dam	0.72	0.58	1.19	1.81	3.03	3.79	4.25	3.37	3.28	2.81	1.46	0.96	27.25
Winona	0.86	0.89	1.75	3.43	3.56	4.13	4.23	4.46	3.64	2.33	2.13	1.17	32.58
Worthington	0.66	0.62	1.62	3.07	3.47	4.88	3.64	3.77	3.25	1.98	1.41	0.82	29.19
Wright	0.79	0.68	1.36	2.44	3.31	4.27	4.19	3.75	3.73	3.28	1.46	0.88	30.14
Young America	0.79	0.69	1.85	2.84	3.32	5.00	4.37	4.43	3.13	2.23	1.69	1.10	31.44
Zumbrota	0.90	0.84	1.92	3.35	3.75	4.67	4.25	4.88	3.80	2.46	2.02	1.13	33.97

TABLE 5: Monthly Maximum Precipitation Extremes (inches)

STATION	JAN	FEB	MAR	APR	MAY	JUN	JUL	AUG	SEP	OCT	NOV	DEC	ANNUAL
Ada	2.46 2010	1.56 1948	3.12 2009	6.01 1986	9.19 1985	8.22 1925	8.49 1962	10.72 1941	8.06 1973	7.49 1971	4.87 1977	2.36 2006	33.99 1941
Aitkin	3.17 1993	2.08 1971	5.27 2009	6.27 2001	11.09 2012	11.37 1946	15.52 1952	10.22 1978	9.10 1986	6.48 1973	4.54 2000	3.08 2008	39.97 1965
Albert Lea	3.02 1897	5.06 1915	4.59 1951	8.61 2006	9.39 1902	10.09 1993	13.52 1943	12.28 1928	9.70 1926	7.42 1941	6.40 1909	3.47 1911	50.67 1993
Alexandria	4.16 1975	2.17 1971	3.03 1977	6.51 1986	7.23 1959	7.98 1945	9.68 1962	8.23 1957	7.26 2005	8.19 1984	4.06 1977	2.58 1968	40.03 1986
Argyle	3.04 1969	2.02 1948	2.61 1950	5.33 1924	6.81 2004	9.74 1925	7.89 1942	9.45 2002	7.74 1941	4.98 2007	3.87 2008	2.01 2009	31.74 2010
Austin	4.68 1995	3.79 2005	5.75 2006	7.09 1999	10.98 2013	11.07 1954	10.20 1999	10.23 1980	11.32 1965	7.53 2009	5.46 1975	3.43 1975	46.01 1993
Baudette	3.03 1969	1.30 1932	3.05 1966	5.04 1937	6.28 1962	14.48 2002	12.50 1937	7.85 1974	9.07 1991	7.60 1971	3.73 2000	2.00 1916	31.26 1964
Bemidji	2.20 1969	2.30 1930	3.54 1904	5.41 1964	10.09 1962	11.00 1915	13.44 1949	9.49 1942	7.35 1973	9.56 1901	3.85 1896	3.16 1951	41.03 1901
Big Falls	3.36 1969	1.70 1955	3.88 2012	4.30 1986	7.02 1977	9.46 1944	7.72 1963	11.88 1944	7.54 1977	7.44 1971	4.16 1977	3.22 2013	35.87 1944
Brainerd	2.31 1969	1.88 1981	3.54 2009	6.58 1986	9.40 1959	8.13 1957	10.49 1952	9.55 1953	8.04 1965	7.06 1984	4.45 1971	2.34 1968	37.45 1986
Browns Valley	2.72 1975	1.76 1998	4.63 1977	5.89 1986	5.66 2004	10.39 2003	7.05 1993	5.53 1989	8.08 2004	7.36 2009	4.14 2000	1.98 2009	30.88 1986
Caledonia	3.92 1967	3.58 1951	5.91 1959	9.30 2008	11.63 2004	13.06 1952	10.76 1967	18.96 2007	11.80 1965	9.16 1900	6.00 1991	4.53 1984	52.85 2007
Cambridge	2.99 1975	2.59 1989	3.90 1977	9.04 2001	9.19 1938	12.78 1993	11.37 1972	8.58 1989	9.00 2001	7.58 1968	5.37 1991	3.06 1968	42.66 1991
Campbell	3.29 1937	3.52 1922	3.90 1897	7.44 1986	8.64 1962	12.07 2005	9.62 1949	7.14 1966	10.08 2004	6.44 1984	4.03 1922	2.69 1909	36.29 1916
Canby	3.30 1997	2.77 1922	5.18 1977	8.30 2001	8.08 1965	10.22 1993	13.61 1963	9.33 1940	9.95 2010	7.35 2009	5.45 1983	4.04 1968	38.30 1993
Caribou	1.88 1969	2.42 1998	1.95 1983	2.93 1967	7.20 1977	7.45 1944	7.83 1997	6.83 1942	7.66 1991	4.88 1949	2.83 2000	2.21 1942	28.26 1991
Cloquet	4.17 1969	4.45 1922	4.30 1977	7.96 2001	8.02 1991	12.64 2012	8.60 1955	9.13 1972	10.61 1990	7.90 1984	7.69 1983	3.00 2013	42.53 1991
Collegeville	2.99 1975	3.21 1971	5.52 1965	8.16 2001	8.95 1962	11.19 1953	15.16 1913	9.86 2010	9.38 1926	7.84 1971	4.62 2000	3.51 1936	44.95 1977
Cook	3.96 1969	1.98 1971	2.82 1966	3.89 2001	5.14 2012	7.92 2011	10.52 1999	9.84 1988	8.56 1977	5.54 1970	3.82 1977	2.67 1968	34.33 1977
Cotton	3.13 1975	2.52 2001	3.54 1965	6.15 2001	5.89 1987	9.51 1984	14.14 1999	10.52 1988	7.74 1965	6.37 1971	3.89 1991	2.24 1969	37.01 1999
Crookston	1.97 1916	2.30 1897	2.81 1904	5.10 1924	8.13 1896	9.66 1895	8.83 1919	9.41 2002	6.69 1973	5.32 1998	4.42 2000	2.25 1920	32.87 1941
Detroit Lakes	2.58 1895	3.86 1922	3.07 1951	5.15 1986	8.84 1985	10.89 1915	12.15 1952	10.88 1900	9.89 2004	6.64 2008	3.88 2000	2.15 1922	38.56 2010
Duluth	4.70 1969	2.72 1998	5.12 1965	8.18 2001	7.67 1962	10.03 2012	8.74 1999	10.31 1972	9.38 1991	7.53 1949	5.08 2000	3.70 1968	43.44 1991
Ely-Winton	3.80 1916	1.82 1962	3.33 1916	4.92 2008	7.67 1970	9.09 1990	6.92 1993	12.15 1988	11.67 2007	6.45 1973	4.16 1915	3.20 1918	38.46 1970
Fairmont	3.65 1983	6.26 1915	5.75 1946	8.15 2006	10.41 1959	14.52 1993	10.57 1963	13.90 1979	10.16 2005	7.96 1998	5.61 1909	3.93 1982	52.36 1993
Faribault	2.91 1975	2.55 2012	4.54 1979	7.35 2014	9.26 1991	12.96 2014	8.72 1997	11.15 1924	11.52 2010	6.49 2009	5.92 1991	3.41 1982	42.20 1951
Farmington	3.06 1967	2.82 1971	4.83 1949	6.22 1986	9.28 1938	10.00 1967	11.57 1997	11.76 1959	12.68 1942	8.75 1911	5.26 1996	4.92 1982	41.48 1938
Fergus Falls	3.08 1975	3.08 1922	5.27 1997	6.84 1896	7.06 1962	10.21 1923	9.77 2002	8.44 1900	9.66 2004	6.33 1984	4.75 1922	2.38 1968	38.10 1896
Fosston	2.42 1989	3.01 1930	3.06 1966	11.16 1912	7.76 1962	11.56 2002	13.38 1909	11.41 1942	10.04 1973	10.14 1984	4.78 2001	2.25 1929	33.18 1962
Grand Marais	4.90 1935	3.55 1939	4.43 1979	6.85 2001	5.82 1964	9.17 1944	8.30 1993	7.82 1988	9.05 1900	9.21 2007	4.28 1938	3.68 1936	39.76 1970
Grand Meadow	3.32 1967	3.75 1915	4.89 1951	8.16 1999	14.64 2013	14.45 1914	13.48 1999	13.34 1911	11.51 1965	9.04 2009	7.82 1909	3.44 1887	51.53 1911
Grand Rapids	3.16 1975	3.72 1922	4.24 2009	5.59 2008	6.91 2012	10.66 1994	9.43 1978	10.03 1988	7.53 1965	6.82 1973	5.44 1938	2.76 1968	38.00 1977
Gunflint Lake	2.72 1975	2.32 1981	3.52 1979	5.45 1967	6.08 2012	7.77 1990	9.43 1987	10.84 1988	9.81 2007	6.00 1949	4.04 1991	3.05 1984	38.31 1977
Hallock	2.35 1907	2.22 1930	3.51 1942	5.63 1937	8.12 2010	10.02 1925	8.22 2001	8.26 1942	14.40 1900	4.76 1949	3.95 1944	1.59 1903	30.92 1941
Harmony	3.85 1949	3.92 1981	6.27 1945	7.49 2008	10.42 2013	12.07 2000	11.58 1999	12.21 2007	13.43 1965	6.59 2009	6.96 1991	2.93 1968	47.46 1983
Hibbing	2.34 1969	1.62 1971	2.48 2009	4.36 2001	7.63 2014	8.92 1994	13.51 1999	10.32 1988	6.38 1965	5.68 1970	3.58 1977	1.67 1995	38.64 1999
Hutchinson	2.75 1917	2.51 1919	5.26 1965	6.55 2001	8.67 1991	10.51 1912	9.33 1997	8.44 2002	7.27 2005	6.65 2009	4.96 1991	3.70 2010	41.52 2010
Int'l Falls	3.03 1975	2.90 1911	3.80 2009	4.53 1925	6.67 1985	10.24 2014	9.52 1966	11.26 1942	7.36 1961	5.49 1909	3.63 1919	2.00 2004	34.35 1941

STATION	JAN	FEB	MAR	APR	MAY	JUN	JUL	AUG	SEP	OCT	NOV	DEC	ANNUAL
Itasca	2.69 1969	1.88 1979	5.82 1933	5.47 1964	8.70 1985	10.80 1957	13.15 1949	9.90 1988	8.59 2004	6.67 1971	4.24 2000	3.68 1951	35.64 1985
Leech Lake Dam	2.29 1975	2.11 1893	3.24 1901	5.25 1896	9.71 1962	11.64 1915	12.27 1949	10.43 1909	8.66 2004	6.89 1968	5.31 1906	2.32 1906	39.09 1906
Luverne	2.11 1994	2.17 1951	5.50 1977	5.73 1984	9.80 1959	13.84 2014	10.49 1993	8.30 1902	8.80 1986	6.95 1968	5.25 1996	2.91 1982	39.82 1993
Marshall	2.31 1982	2.47 1952	5.36 1977	8.24 2001	8.05 2012	13.83 1957	12.48 1963	8.52 1994	13.05 2010	6.91 2009	4.10 1983	3.09 2010	43.22 2010
Milan	4.80 1897	2.70 1992	4.33 1977	7.38 2001	10.67 1942	11.46 1953	13.35 1995	11.15 1957	7.25 1929	7.03 1971	4.02 1922	3.25 1909	39.58 1995
Montevideo	3.12 1982	3.30 1984	5.89 1977	7.33 1896	8.86 1965	11.01 1957	8.62 1955	9.39 1957	8.94 1942	7.48 2009	14.92 2001	3.34 1977	41.85 1991
Mora	2.91 1975	3.18 1922	4.32 1977	5.86 1954	9.99 2012	12.96 1944	9.02 1972	13.41 1995	7.67 1986	8.49 1971	6.19 1996	2.98 1968	41.63 2010
Morris	2.69 1975	3.20 1922	3.82 1940	8.54 1954	8.89 1942	12.53 1914	9.77 1949	11.68 1899	7.49 1921	9.21 1984	4.05 1930	3.48 1968	34.10 1984
MSP	3.63 1967	2.14 1981	4.75 1965	7.00 2001	9.34 2012	11.36 2014	17.90 1987	9.32 2007	7.53 1942	5.68 1971	5.29 1991	4.27 1982	39.94 1965
New Ulm	2.67 1917	4.12 1919	7.11 1899	7.24 2001	12.39 2012	11.15 1925	12.58 1968	10.56 1928	10.74 1930	8.71 1911	4.88 1975	3.71 1918	43.43 1968
Olivia	1.44 1996	2.27 2012	3.51 2011	6.24 2001	7.23 2004	12.97 1993	8.90 1991	7.82 1979	8.11 2010	6.14 2009	4.23 1983	2.85 2010	36.21 1993
Park Rapids	2.90 1950	2.04 1939	3.60 1933	5.85 1970	7.99 1899	9.53 1915	11.60 1949	11.79 1944	7.43 2004	6.54 1971	4.34 1922	3.16 1933	40.51 1985
Pine River Dam	4.09 1975	2.57 1897	4.22 2009	6.56 1896	10.83 1902	12.64 1914	12.35 1972	10.32 1899	7.95 1965	7.70 1984	5.65 1988	2.94 2008	45.86 1902
Pipestone	2.58 1917	2.65 1969	4.12 1977	7.64 2003	11.06 2012	9.86 1993	7.75 1989	10.60 1902	10.25 1985	6.38 1911	4.39 1983	2.30 1927	39.02 2010
Pokegama Dam	3.18 1975	1.85 1951	4.11 1901	5.66 1896	7.26 1896	11.75 1994	10.53 1999	10.15 1988	8.10 1965	7.39 1973	4.42 1932	2.65 1887	38.48 1977
Red Wing	3.09 1967	3.34 1981	4.06 1998	7.75 1990	9.20 1938	11.53 2002	12.66 1978	9.69 1979	11.51 1985	9.60 1911	7.55 1991	3.90 1982	46.95 2002
Redwood Falls	3.40 1917	3.29 1919	4.31 1977	7.55 2001	10.56 2004	14.24 2014	8.52 1963	8.37 1981	9.06 2010	7.53 1911	5.64 1983	3.10 1927	41.27 1979
Rochester	2.92 1888	2.30 1915	4.02 1888	7.30 2001	12.26 2013	12.51 2000	12.33 1978	14.07 2007	10.50 1986	9.11 1911	5.91 1909	3.68 2010	43.94 1990
Roseau	3.07 1969	2.34 1983	2.47 1942	4.27 1937	8.48 2004	9.25 2002	9.13 1919	10.97 1974	8.31 1941	4.42 1998	3.26 2000	1.76 1967	29.98 1968
St. Cloud	2.75 1897	2.94 1922	4.66 2009	8.42 2001	9.68 1912	10.56 1920	12.81 1897	9.28 1900	10.72 1926	7.95 1899	4.16 1922	2.56 2010	41.01 1897
St. James	2.69 1943	2.04 2012	4.27 1985	7.15 2001	10.11 2012	10.64 1993	8.93 1986	10.72 2007	13.66 2010	7.46 2009	4.40 1975	4.10 1997	42.72 2010
St. Peter	3.40 1917	2.41 2012	5.36 1977	6.51 2014	9.06 1908	12.10 1908	9.52 1943	10.62 1968	9.58 1986	6.73 1911	4.43 1975	3.00 1982	45.19 1993
Sandy Lake Dam	3.56 1975	9.98 1893	3.39 2009	6.65 1896	8.42 2012	12.42 1944	13.41 1897	12.96 1953	10.67 1986	6.80 1995	3.80 1905	2.22 1893	44.36 1953
Spring Grove	3.31 1996	3.05 1998	4.53 1992	7.94 1999	9.06 2004	13.26 2013	11.99 1999	19.87 2007	12.35 1965	7.00 2009	7.70 1991	2.14 2007	47.37 2007
Stillwater	3.05 1982	2.87 1981	3.52 1966	5.84 2008	9.82 1906	10.55 2014	14.04 1987	10.92 1993	9.04 195-	7.31 1911	6.53 1991	3.65 1982	44.90 1991
Tower	3.67 1969	3.04 2001	4.18 1977	7.28 2014	7.65 1944	8.71 1931	9.81 1993	12.26 1988	12.54 2007	8.11 1911	4.33 1965	2.57 1968	41.37 1971
Two Harbors	4.33 1969	3.88 1998	5.18 1966	8.83 2001	7.43 1950	13.86 2012	10.16 1999	10.86 1939	8.63 1895	7.67 2007	8.02 1991	3.13 2013	43.24 1909
Wadena	3.39 1997	3.37 1922	3.84 1966	5.54 1937	8.90 1965	10.46 1968	9.26 1944	9.81 1995	7.63 1965	8.55 1998	4.75 1922	2.72 1951	38.00 1965
Warroad	2.16 1996	5.25 1996	2.89 1935	4.13 1937	8.23 2004	13.57 2002	9.87 1919	7.59 2000	9.88 1941	5.84 1998	3.96 1922	2.12 1916	33.30 1996
Waseca	3.04 1969	3.19 1971	5.61 1985	6.27 1999	7.78 1942	12.93 2014	9.64 1991	11.89 1924	12.66 2010	7.05 2009	7.42 1991	3.69 2010	50.46 1991
Waskish	2.55 1969	1.31 2001	3.87 2012	4.00 2001	6.34 1999	8.28 2014	7.64 1995	7.11 2001	8.77 1925	6.67 1971	3.32 2005	1.55 1969	31.34 2001
Wheaton	3.14 1997	2.05 1962	4.10 1995	6.33 1937	8.35 1962	10.54 1914	8.01 2011	7.15 1940	7.89 2004	7.03 2009	5.06 2000	2.37 2010	33.31 2007
Willmar	3.67 1969	4.89 1922	5.04 1965	7.43 1896	8.68 1965	12.94 1957	9.89 1986	9.76 1926	11.13 1942	7.10 1968	4.93 2001	3.08 1927	40.66 1965
Windom	2.80 1917	2.90 1915	5.55 2008	7.46 2001	10.90 2012	11.06 1914	8.48 1968	9.33 1940	14.14 2010	6.67 2009	4.28 1975	3.46 2009	44.84 2005
Winnebago	2.91 1975	4.56 1915	4.33 1946	7.66 1999	11.10 1908	13.91 1899	9.61 1963	10.18 1940	11.97 2005	7.30 1998	5.18 1909	3.20 2009	46.34 2005
Winnibigoshish Dam	3.44 1969	2.03 1930	3.18 1979	5.16 1940	8.62 1962	12.67 1915	10.70 1949	10.63 1900	9.40 1900	6.80 1971	4.10 2005	2.36 1968	35.60 1999
Winona	5.60 1886	3.50 1971	5.30 1951	10.30 1999	10.57 2004	14.12 1914	10.01 1978	18.84 2007	11.04 1986	7.54 1911	5.80 1991	3.44 1968	47.89 1938
Worthington	2.32 1975	3.74 1919	4.76 1979	8.24 2001	8.77 2012	2.29 1993	11.19 1993	12.29 2007	10.53 2010	6.28 1977	4.94 1975	2.43 1982	43.45 1993
Wright	2.71 1975	2.00 2014	3.42 1979	7.04 2001	8.75 2012	13.03 2012	9.06 1996	8.82 1999	8.94 1986	8.70 2007	3.99 1996	2.70 2013	40.63 1991
Young America	3.14 1975	3.25 1971	6.61 1965	6.00 1986	9.38 1960	10.92 1968	8.53 1968	8.33 1985	9.41 1991	6.09 1968	5.38 1975	3.04 1982	45.17 1968
Zumbrota	3.18 1967	2.77 1981	4.03 1998	6.41 1999	8.78 2004	12.94 1914	10.66 1968	12.78 2007	14.57 2010	10.23 1911	7.69 1991	2.97 1982	45.52 2010

Tornadoes are classified for strength only after assessment of the damage left along their path. This system is called the Fujita Scale, named for Dr. Theodore Fujita, a highly respected research meteorologist at the University of Chicago who first used the scale in the 1970s. Since 2007 the National Weather Service has used an Enhanced Fujita Scale, ranging from EF-0 to EF-5. Most tornadoes—more than 70 percent nationwide—fall within the EF-0–EF-1 classes on this scale, but Minnesota has documented the occurrence of at least eight EF-5–class tornadoes (winds over 200 mph), the last one striking Chandler (Murray County) in June of 1992.

Freeborn, Polk, and Stearns counties have reported the most tornadoes historically. This higher frequency is difficult to explain, but it may be the result of random distribution or it may have something to do with county size and greater visibility to all horizons in these primarily rural counties. On the other hand Cook County—historically reporting the fewest tornadoes—is far to the north and immediately next to the relatively cool Lake Superior, two attributes that would tend to stifle convection and therefore tornado development. Minnesota tornadoes have struck as early in the spring as March and as late in the fall as November, but their peak occurrence is in the months of May and June.

Mean seasonal snowfall ranges from more than 80 inches along the highlands of Lake Superior's north shore to less than 35 inches in portions of southwestern Minnesota. Snow cover of 1 or more inches averages about 110 days across the state, from 85 days in the south to 140 days in the north. Snow has fallen in every month except July: only a trace amount has been recorded in August—in Duluth on August 31, 1949—while June has produced measurable snowfalls in some northeastern communities and even a trace in the Twin Cities metro area. Heavy snowfalls of 4 or more inches typically occur from the middle of November through early April. Blizzards—defined as winds exceeding 35 miles per hour, snow and blowing snow persisting for at least three hours, visibility reduced to one-quarter mile or less, and falling temperatures, usually with dangerous wind-chill conditions—often present serious hazards in Minnesota's open prairie landscape. On average the state experiences about one to

Q & A

What is precipitable water?

a. the liquid collected in a rain gauge

b. a measure of water vapor in the lower atmosphere

c. the ponded water left by a thunderstorm

(answer on page 329)

TABLE 6: The Enhanced Fujita Scale

EF-SCALE	OLD F-SCALE	TYPE OF DAMAGE
EF-0 (65–85 mph)	F0 (65–73 mph)	• Peels surface off some roofs • Breaks branches off trees • Pushes over shallow-rooted trees • Some damage to gutters or siding
EF-1 (86–110 mph)	F1 (73–112 mph)	• Loss of exterior doors • Roofs severely stripped • Mobile homes pushed off foundations or overturned • Windows and other glass broken
EF-2 (111–135 mph)	F2 (113–157 mph)	• Foundation of frame homes shifted • Roofs torn off well-constructed houses • Mobile homes demolished • Cars lifted off ground • Large trees snapped or uprooted • Light object missiles generated
EF-3 (136–165 mph)	F3 (158–206 mph)	• Entire stories of well-constructed houses destroyed • Severe damage to large buildings such as shopping malls • Trains overturned • Trees debarked • Heavy cars lifted off the ground and thrown • Structures with weak foundations blown away some distance
EF-4 (166–200 mph)	F4 (207–260 mph)	• Well-constructed houses and whole frame houses completely leveled • Cars thrown • Small missiles generated
EF-5 (>200 mph)	F5 (261–318 mph)	• Strong frame houses lifted off foundations and carried considerable distances to disintegrate • Automobile-sized missiles fly in excess of 100 meters • High-rise buildings have significant deformation
EF-No rating	F6 (319 mph to speed of sound)	• Should a tornado with the maximum wind speed in excess of EF-5 occur, the extent and types of damage may not be conceived • A number of missiles such as iceboxes, water heaters, storage tanks, automobiles, etc., will create serious secondary damage on structures

two blizzards each winter season, but there have been cases of extreme winters, like 1996–97, when as many as 11 have been recorded statewide.

Snow's density, or water content, varies tremendously depending on cloud and surface temperatures, both of which affect the formation and structure of ice crystals. Ice crystals may form in clouds as plates, as hollow or solid columns, or as dendrites. The crystals in turn aggregate to form snowflakes of various shapes, sizes, and densities. The ratio of snow to water equivalent has varied historically from 4:1—wet, dense snow associated with very moist air near the freezing mark—to 40:1—light, powdery snow that typical falls in drier, cold air, with temperatures sometimes as low as -40°F. On occasion, very dense snowfalls place heavy snow loads on roofs. Weighing as much as 30 to 60 pounds per square foot, these loads can literally stress a building's roof beyond its bearing capacity. In addition, heavy, dense snowfalls often cause ice dams to form along roof edges, allowing meltwater to accumulate and seep through to the underlying structure, sometimes causing great damage. Because of the risks posed by snow loads and ice dams, many Minnesotans routinely use snow rakes or shovels to clear their roofs of excessive snow.

When surface wind speeds exceed 15 miles per hour, snow usually begins to drift over the landscape, causing reduced visibility and hazardous driving conditions. Blowing and drifting snow present considerable difficulty during most winters; severe winters require repeated and expensive snow plowing to keep roads and highways open. In the winter of 1996–97 more than $215 million was spent to maintain access to Minnesota roads. Most snow-plowing contractors work based on a threshold of 3 to 4 inches of snowfall occurring within a 24-hour period. The Minnesota Department of Transportation estimates that more than 4,000 sections of roads and highways routinely require plowing.

The persistence of snow cover on the landscape is an important feature of Minnesota's climate. Snowfall deposits and persistent snow cover are highly governed by topography, vegetation, and latitude, making for great variation across the state. The presence of snow cover, with its high reflection coefficient—or albedo—for incoming solar radiation, greatly alters the state's temperature patterns. Duration of snow cover into the spring months also reduces fire danger. Minnesota's forests and prairies see the greatest frequency of fire during April, a period that usually coincides with disappearing snow cover and still-dormant vegetation. Thus a longer duration of snow cover reduces the risk of fire from dry or dead vegetation carried over from the fall.

The average duration for snow cover of 1 or more inches as reported by former state climatologist Earl Kuehnast varies from 81 days in Albert Lea near the Iowa border to 175 days along the Gunflint Trail in northeastern Minnesota. The state's maximum snow depth is 75 inches, measured on March 28, 1950, near Grand Portage, while the maximum snow cover duration is 190 days at Gunflint Lake during the winter of 1978–79, when other parts of the Lake Superior highlands probably exceeded 200 days.

Wind

The climatology of wind in Minnesota has been studied rigorously only over the past three decades, due to poor historical spatial coverage and inconsistent measurements. Anemometers (measuring speed) and vanes (indicating direction) have historically been located only at National Weather Service Offices or at larger airports. In addition, the instruments' height above ground varied over space and time, ranging from about 21 feet to well over 70 feet, when placed atop buildings and towers. Since wind speed and direction can vary considerably with height, such disparities in instrument exposure greatly confounded the data. However, since the 1990s modernization of the National Weather Service (NWS) and the deployment of Automated Surface Observation Systems (by the NWS, the Federal Aviation Administration, and the Minnesota Department of Transportation), anemometers and wind vanes have regularly been set at a height of 10 meters. Working only with data resources obtained since the mid-1990s allows researchers to eliminate concerns over variations in exposure height.

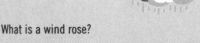

What is a wind rose?

a. a circular, graphical depiction of the frequency of wind speed and direction over a period of time

b. a prairie wild flower

c. an instrument used to measure wind speed

(answer on page 329)

Both wind speed and direction have shaped Minnesota's landscape. Except for the north shore region along Lake Superior, wind direction is primarily bimodal, prevailing from the south-southeast during May through October and from the northwest during November through April. Highest average wind speeds are measured for the months of April and November, while lowest average wind speeds occur in July and August. Maximum daily wind speeds are usually reached in the afternoon hours from noon to 5:00 PM; lowest wind speeds and highest frequencies of calms can be found between midnight and 5:00 AM.

Wind erosion in the late fall or early spring can affect agricultural landscapes when the soil becomes too dry and there is an absence of vegetation or crop residue to protect the surface. Wind speeds in excess of 16

The Rughton Wind Farm generates power from southwestern Minnesota's Buffalo Ridge.

miles per hour can lead to significant soil erosion in open, dry landscapes, piling up soil in drainage ditches, along fence rows, or near shelterbelts and buildings. Fortunately, according to a study by former University of Minnesota climatologist Don Baker, such speeds occur less than 20 percent of the time for most locations in the state. However, wind speeds exceed 25 miles per hour about one to three percent of the time, with higher frequencies during the months of April, May, and November, when agricultural soils are especially exposed due to lack of vegetative cover or to fresh tillage breaking the soil into particles readily transported by the wind. In fact, most locations also show wind speeds that exceed 50 miles per hour at least once annually, sometimes reaching 70 or more miles per hour with strong winter storms or extreme thunderstorms. Thus, severe wind erosion tends to occur with specific weather events, making it highly episodic rather than a continuous problem.

• •

Zonal and meridional

Upper-level wind patterns that indicate the trajectory of air masses and weather systems can be described as zonal or meridional. For the Midwest, zonal *refers to a west-to-east trajectory—along parallels of latitude—and brings moderate or quiet weather and not much change in air mass.* Meridional, *on the other hand, refers to wind flow running southerly or northerly—parallel to meridians of longitude— bringing large changes in air mass and very active weather to the Midwest.*

• •

March and April's typically higher wind speeds sometimes produce a spectacle on Minnesota lakes known as rafted ice. As the lakes thaw, the ice cracks, leaving a pattern of large floes scattered across the surface. When the wind blows strongly from one direction for a period of time, it drives the ice toward one end of the lake, sometimes piling one floe on top of another and even pushing them well onto the shoreline. On occasion the piles of rafted ice grow so large that they block roads running parallel to the lakeshore, particularly stretches of Highways 47 and 169 near Mille Lacs and along Highway 1 near Lower Red Lake.

DID YOU KNOW?

A squall is a sudden strong wind of at least 18 miles per hour that lasts several minutes.

In recent decades the magnitude and constancy of higher wind speeds in western Minnesota have been used to generate electricity. Especially along the Buffalo Ridge of southwestern Minnesota, large wind-powered generators contribute significantly to the regional electrical grid. Dubbed "wind farms," these systems have proven valuable in providing a natural energy source and another avenue of income for the farmer. As more efficient wind turbines continue to be developed, deployment of wind generators across the Minnesota landscape, especially in the western counties, will likely grow in the coming years.

Sky

Minnesota's location upwind from the Great Lakes provides an advantage over neighboring states to the east: more sunshine. Cloud cover fluctuates throughout the year: the least amount can be found in the summer, June through August, while the highest frequency comes in November and December. As the higher degree of cloud cover coincides with late autumn's shorter days, Minnesotans experience a relative spike in the incidence of Seasonal Affective Disorder (SAD). Another light-driven malady, sundowning, more commonly affects the elderly, producing an irritable attitude as the shorter day length yields an earlier sunset centered around the winter solstice.

• •

Sky conditions

In aviation meteorology, sky conditions, based on the percentage of sky obscured by clouds, are described as clear, scattered, broken, and overcast.

Clear   *less than 10 percent of the sky covered by clouds*

Scattered   *10 to 50 percent of the sky covered by clouds*

Broken   *60 to 90 percent of the sky covered by clouds*

Overcast   *greater than 90 percent of the sky covered by clouds*

• •

Two cloud systems prevail depending on the season. With the higher sun elevation angle and summer's longer, warmer days, convective or cumuloform clouds are most commonly visible, some attaining vertical heights of more than 50,000 feet. Rather than obscuring the sky, they more often occupy only a fraction of the horizon and may have a base elevation, or ceiling height, of several thousand feet. Spectacular in shape and size, they sometimes form and dissipate with such speed that their birth, growth, and death can be observed by the naked eye. With winter's shorter days and colder air, layered stratoform clouds dominate and more often than not occupy the entire sky with very low ceilings, sometimes just hundreds of feet in height. These clouds are associated with large midlatitude cyclones that may be slow moving and therefore take several hours or even days to pass.

Towering summertime cumulonimbus clouds, sometimes called "turkey towers," accompany thunderstorms and can be seen sunlit in the upper atmosphere well after sundown on most days.

. . . .

Billow clouds

Billow clouds appear to be a series of breaking ocean waves. They are produced by the interaction of a saturated stable air layer, usually an inversion, and a pattern of vertical wind shear that results in somewhat evenly spaced zones of updraft— where cloud tops crest in a wavelike pattern—and subsidence—where cloud droplets evaporate as they descend back to the stable layer. Their height may vary from tens to hundreds of feet, and they may be hundreds to thousands of feet apart horizontally.

Echelon clouds

Echelon clouds produce the illusion of stair steps or a terrace. Aligned clouds with the same base elevation are viewed at a lower elevation angle with distance toward the horizon. The observer sees a stair-step effect, as if the cloud base grows successively lower as it approaches the horizon. These formations—named with the French term for "ladder"—can occur over Minnesota with fair-weather cumulus or cirrus clouds.

. . . .

Besides cloud forms, fog, haze, and smoke may obscure views of the sky, usually during the shorter days of winter. They are amplified by an atmospheric condition known as an inversion, when cold, dense air settles near the surface under a high-pressure system. When an inversion is present, air temperature increases with height, thereby limiting vertical mixing of the atmosphere by convection currents. When an inversion combines with absence of wind, a usual attribute of a high-pressure system, both horizontal and vertical mixing are limited, making the air stagnant. When the microscopic droplets that compose fog or the aerosols and particulates that make up haze or smoke cannot be thoroughly diluted by the atmosphere, they remain suspended in the surface layer. Such conditions not only obscure the sky and limit horizontal visibility; they also can lead to episodes of poor air quality, when people with respiratory problems must limit their exposure to outside air.

Minnesota's Hydrologic Cycle

What is the fate of the precipitation that falls on the Minnesota landscape? The answer to this question fluctuates with the seasons. One of the best sources on this subject is a study published by Don Baker that describes the four stages of soil water in southwestern Minnesota in the context of the hydrologic cycle.

In summertime, most precipitation is consumed by growing vegetation, evaporates from the surface before fully infiltrating the soil, or, in the case of sloping landscapes or intense rainfalls, may be discharged by surface runoff. In fact, the summer water needs of Minnesota's vegetation, especially agriculture, are so great that the soil moisture reserves—water stored from the previous fall, winter, and spring—are drawn down on the order of 4 to 6 inches across a 5-foot profile depth. In other words, of the water held by tension in the pore space of the soil root zone, the equivalent of approximately 4 to 6 inches is extracted during the summer growth period. Average annual evapotranspiration, a combination of transpiration by crops and evaporation from the soil surface, ranges from 16 to 23 inches for the state. This range matches or exceeds growing season precipitation in most areas, making the soil moisture reserve an essential resource to ensure a good crop. The average annual storage of soil moisture ranges from 4 to 8 inches, alleviating the need to fallow in most traditional farming practices, unlike techniques followed by farmers in more arid western states.

Fall and spring are the best times to recharge soil moisture. In the fall, vegetation is dying off or going dormant or has been harvested; thus, negligible water is consumed by plants. Fall soil condition varies depending on the summer, but because of the abundant seasonal water needs of vegetation the soil surface is often quite dry, a sponge ready to absorb moisture. This condition is amplified by fresh fall tillage, which increases the soil's surface roughness and the area exposed to the atmosphere. In addition, the season of convective or intense rainfall draws to an end and precipitation rates brought by weather frontal passages more closely match the infiltration rates of Minnesota soils. Because of these assorted factors, fall soil moisture recharge efficiency is quite good, sometimes as high as 80 percent.

A secondary recharge period occurs during late winter and early spring. As snow cover melts and discharges across the landscape, the frozen soil begins to thaw from both the top and the bottom, soaking up some of the winter's precipitation. March, April, and early May bring predominantly frontal precipitation, with rainfall rates that closely match soil infiltration rates. However, increasing day length and higher sun angles cause substantial surface evaporation. Storage efficiencies of 15 to 30 percent are frequently seen in the spring season. Depending on when vegetation greens up or crops get planted, the spring recharge season can be short or long, affecting the amount of moisture retained by the soil.

While 60 to 70 percent of annual precipitation falls as rain during May through September, only about 12 to 16 percent falls as snow, sleet, or freezing rain during the winter season. The fate of this frozen precipitation is quite variable, dependent on how early the soil freezes and how long it remains frozen. Much of it remains on the surface of the frozen soil as snowpack, subject to melting and runoff, evaporation, or sublimation, which may go on for a period of time. The storage efficiency of this form of precipitation is very poor, generally less than ten percent. There have been years, however, when the soil was so dry going into the winter that much of the frozen precipitation was absorbed as it melted.

Minnesota's landscape and climate characteristics, viewed in context of the hydrologic cycle, produce both positive and negative consequences for its citizenry. The combined effects yield a productive agriculture, ranked fifth nationally among states in cash receipts from crops and sixth nationally in cash receipts from livestock and livestock products. The rainy season coincides with the crop-growing season, and the soils' water-holding capacity is sufficient to sustain vegetative growth even through summer's irregular dry periods. Tourism flourishes on many Minnesota lakes in both summer and winter with activities such as boating, sailing, fishing, swimming, ice fishing, skating, and snowmobiling. Thanks to sufficiently high flows during the summer months, barge commerce is sustainable on the state's major watershed, the Mississippi River. During infrequent drought situations, the Army Corps of Engineers manages low flows using Big Stone and Lac qui Parle reservoirs on the Minnesota River and the headwaters lakes (Bemidji, Cass, Leech, Pokegama, and Winnibigoshish) for the Mississippi. On occasion saturated frozen soil can indicate potential spring floods if abundant snowfall is followed by a rapid melt or rainy early spring. During other seasons some soils' limited water-holding capacity combined with heavy thunderstorms can generate high volumes of runoff and cause flash flooding, something that occurs more regularly than spring snowmelt floods but usually over smaller geographic areas. Hydrologically speaking, Minnesota is viewed by other states as one with surplus water, draining toward Hudson Bay off the Red River Valley in the northwest, to the Atlantic Ocean via the Lake Superior and the Great Lakes watershed in the northeast, and to the Gulf of Mexico via the Mississippi River watershed in the central and southern counties.

Minnesota is blessed with abundant natural resources and a beautiful and varied landscape. However, to make a living, raise a family, and prosper as a community, its citizens have learned to cope with a highly erratic and challenging climate. Climatological statistics, documented severe storms and climatic episodes, landscape changes, even stories and legends—all gathered here—provide historical evidence that the weather has shaped who we are and how we live. Read on to discover how.

Community Foundations for Climate History

The documentation, study, and application of Minnesota's climatic data and weather history would not be possible without the dedicated and diligent work of volunteers. Thousands have served their communities as daily weather observers, many for much of their adult lives—30, 40, even 50 years. In some localities, single families have provided this service for more than a century; in others, staff members from the public works department, water treatment plant, hospital, or university extension office have been the dutiful recorders.

Our understanding of climate is important because it affects just about every facet of our lives. The realms of agriculture, energy, transportation, water supply, recreation, construction, and even health care have adapted to local climate conditions based on knowledge gained from lifetimes of daily recordkeeping. Though Twin Cities weather records garner much of the public attention, other communities around the state have become singular reference points for understanding the local and regional climate. They deserve recognition for their considerable contribution to this essential body of knowledge.

Ada (Norman County)

This climate station was established in November of 1892 on the farm of Fred Andrist, who kept daily observations for more than ten years. Later, other local residents and farmers maintained the records. This is a valuable station for describing the weather and hydrology of the Red River basin, an area of the country that sees the highest frequency of blizzards. Though snow rarely accumulates to depths of 2 feet or greater there, it moves around the landscape a great deal because of strong winter winds. Winter temperatures have been as cold as -53°F (February 5, 1936), and wind-chill values have been in the negative sixties. Conversely, Ada has seen summer temperatures of 111°F (July 1936). Average annual precipitation at Ada is only 24.73 inches, one of the lowest values in the state. In the drought of 1936 only 12.25 inches fell for the entire year, and during the drought of 1980 only a trace of rain occurred during the month of April; the dry conditions produced a good deal of blowing soil.

Q & A

What is Jevon's Effect?

a. the hypnotic experience of viewing complex weather maps

b. the effect of atmospheric humidity on violin strings

c. the disturbance in airflow around a rain gauge

(answer on page 329)

Albert Lea (Freeborn County)

The earliest weather observations at Albert Lea—from March 1885 through December 1893—were taken by R. B. Abbott of Albert Lea College. A constant daily climatic record has been maintained by the city and by many volunteers; since August 1970 the responsibility has fallen to the staff of the city's wastewater treatment plant. Located in a relatively wet region, Albert Lea has a reputation for heavy rains and thunderstorms. This city is one of the few places in the state to record five separate storms that delivered 6 or more inches of rainfall in a 24-hour period. Indeed, in 1993 it recorded 50.67 inches of precipitation, one of the state's highest annual amounts. About 8:30 PM on June 9, 1935, a tornado moved along the west shore of Albert Lea Lake, destroying many cottages and unroofing a number of businesses. Albert Lea shares with the city of Theilman in Wabasha County one of the greatest snowfalls on Christmas Day: 14 inches in 1945, a storm that shut down all major roads and highways.

Argyle (Marshall County)

This climate station was established in October of 1887, and assorted community residents have contributed daily climate observations for well over a hundred years. Argyle lies in the midst of the Middle-Snake-Tamarac rivers watershed, and therefore its climate data are especially valued for understanding the hydrology of those rivers and the larger Red River Valley watershed. Prior to the severe spring snowmelt floods of 1950, 1979, and 1997, the seasonal snowfall totals at Argyle had exceeded 70 inches, generating large volumes of runoff. During the Christmas blizzard of 2009 (December 24–26), more than 15 inches of snow formed large drifts and closed roads around Argyle for a time, with winds blowing at more than 40 miles per hour. Significant spring snowfalls have occurred, attested to by the record 15.5 inches reported on April 1, 2014. During the extreme year of 1936, Argyle reported a low-temperature record of -43°F on February 15 and a high-temperature record of 107°F on July 12. The former record endured until February 1, 1996, when -48°F was reported.

Baudette (Lake of the Woods County)

Located in the far north along the Rainy River and just southeast of Lake of the Woods, this climate station was established in December of 1908. For a number of decades, observers included employees at the local airport. This station's climate record is valued for its use in understanding the hydrology and ice behavior of the Rainy River and Lake of the Woods. On occasion snowfalls are enhanced when cloud systems pass over Lake of the Woods to the north of the city, as was the case on October 18, 1916, when 16 inches of snow fell. The largest-ever seasonal snowfall total was 124 inches in 1949–50. Maximum snow depth in the station record is 57 inches, which occurred during the second week of March in 1966. Winter temperatures have gone as low as -49°F, and in the winter of 1935–36 Baudette reported 89 days with subzero minimum temperatures. The last time the temperature reached the century mark was August 7, 1983, when the afternoon high was 101°F. The worst flash flooding in this community occurred over June 9–11, 2002, when 9.65 inches of rain fell, washing out some county roads and closing some highways. The monthly total of 14.48 inches of rain for June 2002 is still the all-time monthly record there.

Beardsley (Big Stone County) and Browns Valley (Traverse County)

Located in Minnesota's westernmost thumb, the Beardsley and Browns Valley stations are known as the hottest places in the state. Separated by only eight miles of prairie landscape, these two climate stations account for 58 statewide maximum temperature records, including the all-time high of 115°F at Beardsley on July 29, 1917, recorded by observer J. L. Fitzgerald as 114.5°F. Mr. H. B. Gelting made daily observations at Beardsley from May 1893 until October 1901. A series of observers followed, the last one being William L. Kirkey, who served from 1953 to 1973, when the Beardsley station closed. In order to maintain a continuous climate record, Keith Duffield of nearby Browns Valley then became the daily observer. In 1936 Beardsley reported 21 days of 100°F or higher, while in 1988 Browns Valley recorded 18 such days, topping out at 109°F on July 6. Warm nights are common: Beardsley has recorded overnight lows as high as 84°F, and Browns Valley has seen lows of 83°F. Local lake-effect snowfall occurs on occasion at Browns Valley as cold air passes over Lake Traverse, the nearly 11,000-acre body of water just to the north. There have been some remarkable snowfalls, including 14 inches at Beardsley on December 4, 1955, 15 inches at Browns Valley on March 20, 1982, and 17.5 inches at Browns Valley on January 5, 1997.

Bemidji (Beltrami County)

Located near the Mississippi headwaters and home to Bemidji State University and Minnesota Public Radio stations KCRB and KNBJ, Bemidji can trace daily weather observations back to January 1896. Paul Bunyan is an apt symbol for the area, which is dominated by heavily forested landscape and by Lake Bemidji, one of the headwaters reservoirs. Numerous volunteers and staff members at the Bemidji airport have contributed to the daily climate record, the second longest in Beltrami County, behind Red Lake, which began daily observations in November 1893. Snow and cold are routine here: Bemidji's most recent reading of -50°F was on

January 30, 1950, and during the severe winter of 1935–36 it recorded seven days with lows of -40°F or colder as well as a temperature of -7°F on April 7, its record low for that date. That winter also produced continuous snow cover for 165 days. In the weeks preceding Christmas 1983, Bemidji recorded nine consecutive days when the temperature never rose above zero, and as recently as the winter of 2013–14 this site saw 85 nights of subzero temperature readings.

· ·

Fair weather

Fair weather *commonly refers to visibly pleasant though not necessarily comfortable conditions, usually within the normal range for a particular location and time of year. When the National Weather Service includes this purely subjective term in its forecasts it is supposed to satisfy at least some of the following criteria: no precipitation expected; less than 0.4 sky cover of low clouds; very good visibility; absence of strong winds. Air temperature is usually not a consideration: a very warm or very cold day may still be referred to as a fair-weather day. As technology allows forecasters to be more precise about specific weather elements and events, they are less likely to use the term* fair weather.

In the early days of the U.S. Weather Bureau, when forecasts were provided to local communities either by mail or by telegraph, a system of flag signals was used to post the details for local residents. A designated person, maybe the postmaster, local weather observer, sheriff, banker, or train station manager, would receive the Weather Bureau forecast and display the appropriate flag or flags, plain white for fair weather, blue for precipitation, and various combinations of squares, triangles, and colors for other conditions. Even as late as the 1960s and 1970s, a Minneapolis bank building displayed a colored ball (the fondly remembered Weatherball) to indicate the expected weather.

· ·

Campbell (Wilkin County)

One of western Minnesota's oldest climate stations, Campbell has accumulated more than a hundred years of daily observations kept by various volunteers, starting with Arthur Metcalf in 1894. Because Campbell lies between the Otter Tail River to the north and the Rabbit River to the southwest, its measurements help scientists understand the hydrology of the massive Red River floodplain. Its late winter snow cover and snow-water equivalence observations are important as hydrologists begin to forecast snowmelt discharge that may produce flooding on the Red. For example, the spring of 1951 snow depth at Campbell was nearly 3 feet, with water equivalence of several inches, predictors of the historic flood that inundated the Fargo-Moorhead area that year.

Canby (Yellow Medicine County)

In August 1887 P. C. Scott began the climate record for Canby. More than a dozen others followed, with Darold Snortum serving the longest tenure there, recording daily temperatures and precipitation from 1948 to 2005. With an average July maximum temperature of 87°F, Canby is one of Minnesota's hot spots. On the mornings of June 28 and 29, 1931, for example, the overnight low temperature was 87°F, the two warmest nights in the state's climate record. Located just nine miles from the South Dakota border, Canby can be very dry with exceedingly low relative humidity—factors that lead to some large daily temperature ranges, as on March 7, 1950, when after a morning low of 3°F the afternoon temperature climbed to 60°F. Canby's all-time records include 111°F on July 12, 1936; -33°F on January 22, 1936; 5.35 inches of rainfall on August 19, 1926; and 16 inches of snow on November 20, 1975. The famous October 16, 1880, blizzard that affected southwestern Minnesota and eastern South Dakota with drifts up to 20 feet and snow cover lasting through April is also part of Canby's record.

Cloquet (Carlton County)

Cloquet's daily climatic observations began with the Northwest Paper Company in 1902 and were taken over by the Cloquet Forestry Center in 1911. A number of forestry center employees have served time as the official climate observer there, including Ron Severs and John Blanchard, who shared duties for decades. This site maintains one of the state's best-quality records of snowfall and snow depth and is one of the few locations to report frost in every month during the same year. Both 1912 and 1915 are examples of the latter record, yielding paltry growing seasons of only 48 days and 32 days, respectively. It has been as cold as 24°F in June, 30°F in July, and 26°F in August. Cloquet's all-time records are 105°F for four consecutive days, July 10–13, 1936; -45°F on January 7 and 12, 1912; and rainfall of 8.44 inches on September 6, 1990. In fact,

A Chicago, Milwaukee and St. Paul Railway engine makes its slow way through a snow block-ade near Sleepy Eye during the blizzard-filled winter of 1880–81.

Cloquet has seen two devastating flash floods: nearly 8.5 inches of rain on September 6, 1990, brought a huge flood crest to the St. Louis River and damaged State Highway 210 and the local high school; and rainfall of 9.39 inches over June 18 20, 2012, caused a complete washout of State Highway 210, closed Jay Cooke State Park for months, and sent an all-time flood crest down the St. Louis River.

Collegeville/St. John's University (Stearns County)

Many staff members at St. John's University in Collegeville have been making daily weather observations since October 1892, providing a valu-able climatic record for central Minnesota. This lengthy history includes a single-storm record snowfall of 26.6 inches from the Armistice Day Blizzard, November 10–12, 1940, and a monthly record of 66.4 inches of snowfall set in March 1965—also a state record, and far more snow than most Minnesota communities see in an entire year. That March produced two state daily record snowfalls at Collegeville: 18.8 inches on March 1 and 23.6 inches on March 17, leaving a snow depth of 41 inches. The 109.2 inches of snowfall that occurred over the winter of

1964–65 contributed to flooding in the upper Mississippi River basin when much of it melted over a one-week period, April 4–12. In addition to these remarkable snowfall records, Collegeville also holds two state records for total daily precipitation: 1.75 inches on February 9, 1909 (the liquid equivalent of a 14-inch snowfall) and 5.84 inches from a thunderstorm on May 22, 1962. Collegeville's monthly record for total rainfall, 15.16 inches, set in July 1913, is one of the state's highest as well. With its elevation of 1,242 feet above MSL (mean sea level), Collegeville sits relatively higher than most of the central Minnesota landscape, a factor that may contribute to its larger precipitation totals.

Automated Surface Observing System (ASOS)

The National Weather Service's Automated Surface Observing System uses weather sensors, data-collection hardware, acquisition control modules, communications devices, and displays to monitor local conditions, primarily at airports. Of the more than 1,700 around the country, the 14 ASOS units in Minnesota report, among other things, air temperature, cloud ceiling, visibility, dew point, and wind. Meteorologists use the frequent ASOS reports to refine and update forecasts, and in some locations pilots can dial in for computer-generated voice reports on airport conditions. Still under evaluation, ASOS systems have been criticized for omitting important climate observations such as snowfall, snow depth, snow-water equivalence, cloud ceilings above 12,000 feet, and visibility obstructions like smoke or dust. At some busy airports, ASOS data is augmented by manual reports on other climatic details. It may be a number of years before data users are satisfied that ASOS reports document local conditions comprehensively.

Crookston (Polk County)

The Northwest School, now known as the University of Minnesota Northwest Research and Outreach Center, began making daily climate observations at Crookston in November 1885. Its record describes the climate of the northern Red River Valley, particularly the spring snowmelt flooding that frequently visits the region. The Red Lake River, which passes through Crookston, is one of the larger watershed tributaries to the Red. Though its soil is rich in nutrients and has a high water-holding capacity, the Crookston area has often been subjected to drought. In 1980 it saw 59 consecutive days—from March 28 to May 25—without measurable rain, recording only a trace during the entire month of April. The severe summer drought of 1980, though not as serious as those of the 1930s, devastated local crop production. The winter of 1996–97 brought 100.6 inches of snow to Crookston, setting up one of the worst local spring snowmelt floods in history. Among Minnesota climate stations, Crookston is peculiar in that its all-time temperature extremes occurred in a single year, 1936, when it recorded a low of -51°F on February 15 and highs of 105°F on July 10, 11, and 12.

Detroit Lakes (Becker County)

This climate station was established in December of 1895. George Peoples, whose farm overlooked the lake, was the observer for the first 23 years of record keeping. A number of others contributed to the measurements in the climate record, including employees of radio station KDLM. Climate records show five dates when the temperature fell to -50°F or colder, including one of the coldest days ever measured anywhere in the state: February 9, 1899, when Mr. Peoples reported a low of -53°F and an afternoon high of only -32°F. January and February of 1899 and 1936 brought 19 days with low temperatures of -30°F or colder. The range of ice-out dates for Detroit Lake is one of the most extreme in the state, with the earliest being March 22, 1910, and the latest being May 18, 1908. From September 1942 to August 1943, snowfall was observed at Detroit Lakes in every month except July and August, totaling more than 89 inches, a local record. During the "Storm of the Century" blizzard of January 10–12, 1975, more than 16 inches of snow fell at Detroit Lakes, accompanied by wind gusts up to 54 miles per hour. All highways into and out of Detroit Lakes were closed for a period of time. The year 2010 saw the record for most precipitation, 38.56 inches, while the famous drought of 1976 brought a yearly total of only 11.95 inches.

DID YOU KNOW?

A Stevenson screen—an instrument shelter for housing thermometers—was the first weather instrument shelter adopted for standard daily climate observations in the nineteenth century. The designer, Scottish engineer Thomas Stevenson, was the father of writer Robert Louis Stevenson.

Duluth (St. Louis County)

The U.S. Army Signal Corps established a climate station in the important port of Duluth in 1870 to supply observations and forecasts for the

Great Lakes shipping industry. Since then, additional climate observations and data have been provided by the Duluth Experiment Farm (1902–29) and by Weather Bureau Offices at the harbor, at the post office, at a downtown office building, and at the international airport, where

the National Weather Service is located today. For a number of years starting in 1940, daily observations were made from both the airport and downtown locations, but the airport became the sole reporting office in March 1950. The National Oceanic and Atmospheric Administration (NOAA) developed a composite record for Duluth daily climate back to 1871 (called the ThreadEx record), using all measurements from the different climate stations over time. Thus measures of the climate extremes in Duluth span a period well over 140 years.

The weather in and around Duluth is highly affected by wind, topography, and the water temperature of Lake Superior. Snow is a dominant feature: the annual average number of days with measurable snowfall is 62, the highest frequency statewide. Easterly and northeasterly winds off Lake Superior have enhanced very large lake-effect snowfalls, including 24.2 inches over five days in December 1969, 25.2 inches over six days in January 1982, 27.3 inches over four days in January 2004, 28.9 inches over four days in December 1950, 29.4 inches over nine days in January 1994, and the all-time record, 36.9

Weather instruments are dwarfed by the fire lookout tower at Brainerd, 1935.

inches over four days during the Halloween Blizzard of 1991. As recently as 1998 Duluth saw measurable snow on ten consecutive days—February 25 to March 6—thanks to persistent easterly winds. Seasonal snowfall has exceeded 130 inches four times: the winters of 1949–50 (131.8 inches), 1968–69 (132.7 inches), 1995–96 (135.4 inches), and 2013–14 (131.0 inches). The snowiest month in Duluth history was surprisingly April 2013, when 50.8 inches fell. Finally, Duluth is the only climate station in Minnesota to report snow in August, a trace on August 31, 1949.

Fairmont (Martin County)

The staff of Fairmont's city filtration plant has made weather observations since October 1982. The station, initially called Rolling Green, after the township immediately to the west, was founded in March 1887 when a farmer living just south of Fairmont began recording the weather. The station name was changed to Fairmont in 1906, and city and various volunteers maintained daily climate observations. The city holds 16 state record high temperatures from the past 120 years, including a remarkable 57°F on December 26, 1936—31 degrees above normal—and an even more remarkable 93°F on April 16, 2002—36 degrees above normal. Known for warm summer temperatures as well, Fairmont recorded 50 days with temperatures of 90°F or greater during the summer of 1934. The wettest year in history there occurred in 1993, when 52.36 inches of precipitation was measured.

Faribault (Rice County)

Located midway between Albert Lea and the Twin Cities along Interstate 35, Faribault has been reporting daily weather observations since October 1890. Its climate record—kept primarily by the state school, the city waterworks, and radio station KDHL—shows a relatively high frequency of thunderstorms with associated tornadoes. Observers have reported record daily rainfalls of 5 or more inches four times and monthly totals greater than 10 inches in seven different growing seasons. Faribault's worst flash flood occurred on August 25, 1983, when many areas in and around the city reported 5 to 7 inches of rain over a six-hour period, from midnight to 6:00 AM. Dozens of roads were closed or blocked, and scores of basements flooded. In July 1999, dew points up to 81°F combined with temperatures in the nineties to produce Heat Index values of 114 to 120°F, some of the highest ever measured in the state.

Farmington (Dakota County)

For well over a hundred years, members of the Akin/Stoffel family made daily observations at the Farmington weather station, which resided within a half mile of its original location. Established in March 1888, this long-standing climate station was once used to compare so-called "rural" climate measures against those taken at the MSP International Airport, where the "urban heat island" has influence. These climate stations, only 22 miles apart, sometimes exhibited significant temperature differences, especially in overnight minimum values. However, in recent decades it is clear that Farmington has come to be affected by the Twin Cities urban heat island as residential development has expanded there. The drought year of 1934 was reasonably kind to Farmington area farmers: though precipitation was measured on only 51 days—about half the normal number—seven of these days produced more than 13 inches, accounting for more than half the year's precipitation. Farmington provides one of the state's best long-term snowfall and snow depth climate records, illustrating great variability in Minnesota's snow seasons. For example, the fewest number of days with snow cover of 1 or more inches, just 21 days, occurred in the winter of 1960–61, while the winter of 1950–51 saw such snow cover on 142 days, making for a long winter indeed.

• •

Meteograms

Meteograms depict various weather elements—expected air temperature, dew point, pressure, wind, sky condition, and precipitation—for a given location over time based on National Weather Service forecast models. Available in a variety of forms for major U.S. cities, meteograms may be plotted hour by hour or in 3-, 6-, 12-, or 24-hour time steps.

• •

Fergus Falls (Otter Tail County)

Fergus Falls began daily climate observations in January 1892. From 1898 to 1936, Charles Kissinger recorded weather details at his residence, including the conditions associated with the famous tornado of June 22, 1919, which destroyed his house along with many others in town. A very rare F5 tornado—one of only eight documented in Minnesota—it caused more than $4 million in damages and 57 deaths, making it the second-most lethal tornado in state history. This storm also produced a daily record rainfall of 3.49 inches, yielding flash flooding among the wreckage.

The junction of Interstate 94 with State Highway 210 at Fergus Falls has barrier gates to close off the highways when blizzards visit Otter Tail County, a not infrequent occurrence. In fact, during many recent winters, multiple blizzards have been the rule. One storm on January 10–11, 1975, brought 14 inches of snowfall, followed two months later by one on March 23–24 that produced 15 additional inches. Similarly, in 1989 a January 7–8 blizzard brought 12 inches of snow and wind-chill conditions of -40°F, trailed by another storm on March 14–15 that yielded 11 additional inches. Perhaps the worst winters in recent memory were 1995–96 and 1996–97. Six separate blizzards occurred during the winter of 1995–96: one in December, two in January, two in February, and one in March. Three produced wind-chill readings in the very dangerous -40 to -60°F range. The following winter was even worse, bringing seven blizzards and many heavy snowfalls. Snow depth reached an all-time maximum value of 68 inches at Fergus Falls, eventually leading to significant spring flooding. All of these details—and more—were recorded by observers at radio station KBRF.

Fosston (Polk County)

This climate station was established on the farm of Mr. D. N. Ham in July 1909. He observed the daily weather until 1940. A series of other volunteer observers helped take measurements through much of the 1940s, and then employees of the local power plant took over in 1947. During the first month of weather observations (July 1909), both daily and monthly rainfall records were set: 8.97 inches on the nineteenth and a monthly total of 13.38 inches. Fosston represents one of the coldest prairie landscapes in Minnesota. In January and February of 1996 the low temperature reached -40°F or colder on ten nights, and on January 30, 2004, the thermometer bottomed out at -50°F. During the winter of 1965–66, Fosston reported 86 inches of snow, the most ever. Normal annual precipitation is 25.47 inches, but the drought year of 1910 brought only 11.40 inches—and widespread crop failure for many area farmers. Fosston reported one of the wettest Octobers in state history in 1984 when 10.14 inches of precipitation was recorded.

What two relatively new Minnesota weather stations are competing with Tower and Embarrass as the coldest spots in the state?

a. Ely and Silver Bay

b. Orr and Flag Island

c. Princeton and Rush City

(answer on page 329)

Grand Meadow (Mower County)

Numerous volunteers have made daily observations at Grand Meadow since July 1887, providing Mower County's longest and one of southeastern Minnesota's most detailed climate records. The station was initially set up on a farm along Deer Creek, near a stand of virgin forest. Grand Meadow's climate history contains four state record high temperatures, and observers have reported daytime highs in the sixties during the months of December and February. Located in a particularly wet area, Grand Meadow is one of the few locations in Minnesota to exceed 51 inches in annual precipitation: 51.53 inches in 1911. Further, they reported 14.64 inches of rainfall in May 2013, the highest value there for any month of the year. This community also recorded a very rare ten consecutive days with rainfall three times: from June 28 to July 7, 1908, from May 15 to May 24, 1944, and from April 6 to April 15, 2013. And, as recently as September 15–16, 2004, it saw 7.60 inches of rain from a single thunderstorm.

Grand Rapids (Itasca County)

The Grand Rapids climate station was established in June 1915 at the North Central Experiment Station of the University of Minnesota. In recent decades, observations have been made by employees of the USDA Forestry Station located there. This station is valuable in understanding the forested landscape of northeastern Minnesota. Seasonal snowfall accumulations have surpassed 90 inches during six winters there, topped by 99.1 inches in 1970–71. In both 1954 and 1971, May brought 9 inches of snowfall, rare for even northern Minnesota. Because the boreal forest landscape can absorb more solar radiation in the winter than most of the snow-laden prairie landscape, sometimes daytime high temperatures in the winter season are the highest values in the state. For example, in January of 1942, 1973, 1981, 2002, and 2012, daytime temperatures reached 50°F or above. The boreal forest landscape serves to mitigate extreme high temperature values in the summer because it provides so much shade and consumes so much solar radiation as evapotranspiration. For this reason, Grand Rapids has reported only 14 days when the daytime high temperature has reached 100°F or higher.

Gunflint Lake and Gunflint Trail (Cook County)

In May 1894 James Brault began taking the first daily observations on Gunflint Lake, one of the state's most remote climate stations, located along the Ontario border 30 miles northwest of Grand Marais. Gunflint Lake is typically one of the last in the state to become ice free in the spring, having an average ice-out date of May 6, which is awfully close to the state's fishing opener. Indeed, some years the lake has been closed to avid early-season fishermen, most recently in 2014, when ice-out did not occur until May 20. Gunflint Trail has the longest snow cover season of any station in the state, averaging between 160 and 170 consecutive days with at least 1 inch on the ground. In 2003 it reported snow cover as early as September 30; in 1963 it reported the same through May 21.

Hallock (Kittson County)

Hallock has kept daily climate observations since January 1899, producing one of the Red River Valley's oldest continuous records. Mr. D. A. Robertson was the first observer, taking daily measurements from 1899 to 1948. The town is located 11 miles west of Lake Bronson State Park, home to a unique park feature: a prairie landscape. Lake Bronson was formed by damming the south fork of the Two Rivers watershed following the drought of the 1930s, during which regional wells went dry and water shortages were widespread. In wetter times, the town has seen a number of spring snowmelt floods. In fact, during April 1997 Hallock was essentially an island community, entirely surrounded by water, with no access roads available for days. Its location on the flat prairielands of the Red River Valley brings large swings in daily temperatures, particularly when the weather is dry and the wind changes directions: on October 9, 1933, after a morning low of 22°F, the afternoon high reached 82°F under sunny skies and southerly winds. Hallock's all-time records include 109°F on July 11–12, 1936; -49°F on December 21, 1916; 5.50 inches of rain on September 4, 1900; and 14 inches of snow on April 1, 2014. Early on September 1, 2011, between 3:00 and 4:00 AM, a thunderstorm passed over Hallock bringing heavy rains and wind gusts over 60 miles per hour. Farther south down Minnesota Highway 75 near the town of Donaldson, the wind gusted to 121 miles per hour, setting a new statewide record for maximum straight-line wind speed.

International Falls (Koochiching County)

In northern Koochiching County sporadic climate observations date back to the late nineteenth century, but it was not until July 1906 that the Weather Bureau established a cooperative climate station along the Rainy River at International Falls. During World War II, the station became a full-fledged Weather Bureau Office. In 2006 the National Oceanic and Atmospheric Administration (NOAA) developed a composite record for International Falls daily climate back to 1895 (called the ThreadEx record), using all measurements from the different climate stations over time. Thus measures of the climate extremes at International Falls span a period well over 115 years. Within the National Weather Service network, International Falls developed a reputation as the "nation's icebox," reporting the lowest temperature in the 48 contiguous states more often than any other location. Indeed, International Falls has recorded temperatures of -40°F or colder more than 60 times, including one of the state's coldest days, January 6, 1909, when the morning low was -55°F and the afternoon high only -29°F, a daily average of -42°F. The snowiest season there took place in 2008–09, when 125.6 inches fell. Though this community can brag about its white Christmases, brutal weather can also define its holiday season. Such was the case in 1992, when December delivered nearly 44 inches of snow, including eight consecutive days—from December 24 to 31—of measurable snowfall. Blizzard-like conditions prevailed on December 22, 24, and 25, with wind-chill readings as cold as -35 to -40°F, and minimum temperatures during Christmas week included readings of -17°F, -21°F, -25°F, and -36°F. As recently as January 2, 2014, International Falls reported a temperature of -43°F.

Q & A

Where was the first Minnesota State Weather Service Office (1883–89) established?

a. Carleton College, Northfield

b. Hamline University, St. Paul

c. University of Minnesota, Minneapolis

(answer on page 329)

Itasca State Park (Clearwater County)

Daily climatic observations were first made at Lake Itasca in May 1911 by Mr. J. Stillwell. In recent decades, personnel of the Lake Itasca Forestry and Biological Station of the University of Minnesota have assumed record-keeping responsibilities. This heavily forested site yields some rather interesting climatic characteristics: only 13 days in the station history with temperatures of 100°F or greater, yet 95 days with lows of -40°F or colder, evidence of the still, calm forest air. Its maximum recorded snow depth is 47 inches (March 5, 1966). On July 4, 1995, the park suffered a devastating derecho, or straight-line windstorm, that felled thousands of old-growth trees and blocked many trails and roads. In addition to studying climate and maintaining the county's only long-term daily temperature and precipitation records, Itasca boasts the University of Minnesota Biological Field Station, which conducts research on terrestrial and aquatic ecosystems, wildlife, and forest management and offers resident instruction programs using onsite classrooms and laboratories.

Leech Lake Dam (Cass County)

Daily climate observations at Leech Lake Dam, begun in June 1886, were taken over in April 1887 by the U.S. Army Corps of Engineers, which has maintained the climate record for nearly 130 years, mostly at the dam tender's residence. At one time, Leech Lake held the state record low temperature reading, -59°F on February 9, 1899. Its climate history still contains seven state low temperature records, including 5°F on April 25, 1909, one of the state's coldest readings for that time of year. Additionally, Leech Lake's average ice-out date is April 28, one of the latest in the state. Though the lake is a playground for winter recreation, seasonal snowfall has varied considerably, from only 12.0 inches in 1980–81 to 104 inches in 1965–66. The Leech Lake climate station, part of the Historical Climate Network of the National Climatic Data Center, clearly shows evidence of climate change in its records. For example, temperature readings of -30°F or colder used to be very common, occurring about six to seven times each year. In recent decades this frequency has dropped to about twice per year.

Milan (Chippewa County)

Members of the Opjorden family have maintained Milan's daily climate record on their farm since August 1893. Spanning more than a century, the resulting high-quality climatic record is the work of three generations whose members have diligently recorded daily high and low temperatures and precipitation amounts 365 days each year. Milan's history shows 12 state record high temperatures, including the state's third-highest ever: 113°F on July 21, 1934. In fact, Milan represents one of Minnesota's warmest and driest climates. During the drought decade of the 1930s, it saw 73 consecutive days without precipitation, from November 26, 1930, to February 6, 1931. In 1976, another dry year, Milan reported only 7.91 inches of precipitation, one of the state's lowest annual totals. Also in 1976, daytime high temperatures soared to 90°F or greater on 58 days. Milan holds the state record for the wettest Fourth of July holiday, with 9.78 inches of rainfall in 1995.

Montevideo (Chippewa County)

Daily climate observations at Montevideo began in July 1890 on a farm overlooking the Minnesota and Chippewa rivers, but KDMA radio has taken on this responsibility since 1985. Before the Lac qui Parle reservoir was constructed in the 1930s, low flows through the Minnesota River Valley upstream from Montevideo were strongly associated with dry years in its climate record, for example 1891, 1894, and 1910, when annual precipitation was less than 17 inches. The dry prairie landscape makes Montevideo one of Minnesota's warmest places: it holds 16 state record high temperatures, including the highest ever measured in March, a remarkable 88°F on March 23, 1910—45 degrees above normal for that date. During the summer drought of 1988, Montevideo reported 16 days with temperatures of 100°F or greater.

Mora (Kanabec County)

This climate station was established in June 1904 at the residence of Mr. Hans Peterson, who was the observer until 1950. A number of volunteers have since contributed to making daily observations for this community. Mora reported one of the coldest Christmas weeks in state history in 1983, with low temperature values from the eighteenth to the twenty-fourth that ranged from -25°F to -52°F. The total snowfall has exceeded 90 inches in two seasons, 1930–31 and 2013–14. In February 1930, Mr. Peterson recorded a daily snowfall of 20 inches on the twenty-fifth. Mora is one of the few climate stations in the state where the all-time daily rainfall record was set in the month of October. On October 5, 2005, rainfall of 5.78 inches was recorded, with a storm total of 6.06 inches. Most of the streets in the city flooded, as did many residential basements. In 1968, 1977, and 2010, total annual precipitation exceeded 40 inches. Conversely, in the worst drought year (1910), Mora reported only 13.86 inches of precipitation.

Morris (Stevens County)

Most early observations at Morris's climate station, established in January 1885, were collected by staff of the University of Minnesota West Central School and Experiment Station, now the West Central Research and Outreach Center. Morris has a well-deserved reputation as one of the state's windiest places, recording wind gusts as high as 85 miles per hour as recently as July 23, 2005. This climate characteristic has been put to good use: the University of Minnesota–Morris is home to the Renewable Energy Research and Demonstration Center. Its large, 1.65 megawatt wind turbine provides more than half the campus with wind-generated power of about 5.6 million kilowatt-hours.

On the negative side, higher wind speeds combined with lack of precipitation can dry the landscape. During the drought year of 1976, Morris recorded only 9.39 inches of precipitation—fewer than 5.5 inches during the crop season—and evaporation of 58.22 inches—the largest seasonal evaporation value ever measured in the state. Morris holds two state daily climate records: 5.20 inches of rain on June 26, 1914, and 6.90 inches of rain on April 26, 1954—the greatest 24-hour rainfall measured in that month statewide.

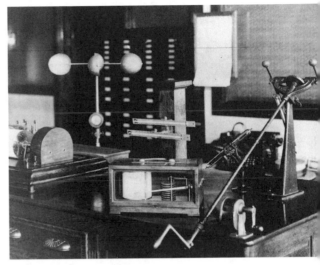

These weather instruments were used at Holman Field in St. Paul during the 1930s.

New Ulm (Brown County)

Daily observations at New Ulm were made from March 1864 through December 1877 and resumed in September 1887. Many volunteers, as well as the *New Ulm Journal* and radio station KNUJ, have contributed to this magnificent climate record, one of the state's longest running. New Ulm's location at a critical point along the Minnesota River Valley makes its record of great value to scientists as they consider the hydrology of this important watershed, which drains western and southern Minnesota's agricultural landscape and discharges into the Mississippi River at Fort Snelling. New Ulm's climate history shows 13 state high temperature records, including a reading of 87°F on October 25, 1927. Residents endured tropical-like climates on July 19, 1940, July 21, 1983, and August 19, 1983, when the temperature never fell below 80°F. On July 15, 1916, a severe thunderstorm brought 7.37 inches of rain to New Ulm between 4:45 and 11:30 PM, a state record for the date. The runoff washed out bridges, roads, railroad tracks, and farm fields, causing widespread and extensive damage.

Olivia/Bird Island (Renville County)

Located along State Highway 212 in the heart of sugar beet country, the Olivia/Bird Island climate station has recorded weather measurements since February 1885. Numerous volunteers, including farmers, extension educators, ministers, and others, have contributed their daily observations to this record, one of the best and most widely used in central Minnesota. The station reported 15 consecutive days with rain from May 23 to June 6, 1965, causing one of Renville County's latest planting seasons. But at the beginning of the dust bowl era, 1930 to 1931, Bird Island saw the opposite extreme, 55 consecutive days without measurable precipitation. During the drought of 2012, Olivia reported 30 consecutive days without rain from September 18 to October 17, and during the heat wave and drought of 1988, it reported 46 days with temperatures of 90°F or greater.

· ·

Socked in

If visibility is so low due to cloud ceiling or fog that it is dangerous to fly, an airport is said to be socked in. *The term refers to the wind sock, a cloth tube mounted on a mast to indicate wind direction along the runway. When the sock was obscured from view, the airport was declared socked in. Improved technologies for instrument flying, airport illumination, and fog dissipation have greatly reduced the frequency of major airports being socked in.*

· ·

Park Rapids (Hubbard County)

Park Rapids climate station was established in January 1885 in a clearing surrounded by forests, near rapids flowing from Fish Hook Lake, 1.5 miles to the south. Today Park Rapids is a gateway to Itasca State Park and a haven for tourists with interests in fishing, canoeing, camping, and biking. It is also a prime area for growing potatoes, thanks to irrigation systems that utilize shallow aquifers under the sandy soil. Snow is a dominant landscape feature: during the months of January and February 1969, this station reported 14 days with a snow depth of 41 inches, while in the winter of 1964–65 it recorded more than 103 inches of snowfall. Park Rapids is also noteworthy for cold temperatures, with 75 historical readings of -40°F or colder and three readings of -50°F or colder, most recently in 1996. During the winter of 2013–14, Park Rapids reported 84 nights with subzero temperature readings, second only to the 85 nights with such readings during the winter of 1916–17.

Pine River Dam (Crow Wing County)

The U.S. Army Corps of Engineers began taking daily observations at Cross Lake's Pine River Dam in March 1887. The dam is a key point along the Mississippi River, where collection of hydrologic—that is, snow and rain—data is important. Like other corps climate stations, daily records were maintained for decades at the dam tender's residence. Pine River Dam holds eight state low temperature records, including May's coldest readings ever, just 4°F on May 1 and 2 in 1909. That year 11 inches of fresh snow fell during the last days of April, and Cross Lake was still frozen over in early May, no doubt testing the patience of local fishermen. The dead of winter can be quite harsh at Pine River: temperatures have dropped below -50°F five times, and overnight minimum temperatures were below zero for 29 consecutive days in the winter of 2013–14 (from January 15 to February 12, 2014). Pine River Dam reported 146 days of continuous snow cover in the winter of 1964–65 (from November 25, 1964, to April 19, 1965), a season that produced widespread spring flooding along the Mississippi River.

The standard number of years used to calculate climate normals like average temperature and precipitation is 30.

Pipestone (Pipestone County)

Pipestone sits on the Buffalo Ridge; at 1,735 feet above sea level, it is one of the highest points in the southwestern part of the state. Mr. D. S. Harris started making daily climate notes in June 1898, and since then more than two dozen volunteer observers have kept a continuous record. The employees at Pipestone National Monument take daily observations, adding to a climate record important for its high landscape data and for being from one of the area's few stations. Records at Pipestone include a high temperature of 108°F on August 3, 1930, May 30, 1934, July 5, 1936, and July 15, 1936; a low temperature of -44°F on December 24, 1983; 6.18 inches of rain on June 17, 1957; and 40 inches of snow cover on March 3, 1982. Another record involves a temperature rise of 63 degrees, from an overnight low of -32°F to an afternoon high of 41°F on January 6, 1924.

Pokegama Dam (Itasca County)

The weather station at Pokegama Dam, along the Mississippi River, was established in April 1887 by the U.S. Army Corps of Engineers, and for many years records were kept by the dam tender. Pokegama's reputation as one of Minnesota's coldest spots is bolstered by more than 30 state low temperature records, including the following: coldest temperature recorded in March, -50°F on March 2, 1897; coldest-ever Memorial Day temperature, 19°F on May 31, 1889; coldest-ever November temperature, -45°F on November 30, 1896; and coldest-ever December temperature, -57°F on December 31, 1898. Temperatures of -50°F and colder have been recorded 30 times, the coldest being -59°F on February 16, 1903. In 1905 Pokegama Dam reported 31 consecutive days with minimum temperatures below zero (from January 20 to February 19). In the winter of 1936–37, it reported 170 days of continuous snow cover.

Rochester (Olmsted County)

Rochester provides the only source of long-term climate records for Olmsted County and is one of just a few such sites in southeastern Minnesota. The observation work has been handled by numerous entities over the years, beginning with the old Smithsonian network in 1869 and followed by the Army Signal Corps in 1885 and a cooperative observer site in 1893, with periodical supplemental data gathered by the *Rochester Post-Bulletin* and the Libby Canning Company. During World War II the National Weather Service established its own office at the regional airport south of downtown, and it provided observations and forecasts for southeastern counties until 1995, when the La Crosse, Wisconsin, office took over that jurisdiction under the NWS modernization plan. Since then, daily observations have been made at the airport by an Automated Surface Observing System (ASOS).

At an elevation of nearly 1,300 feet above MSL (mean sea level), the Rochester airport weather station sits relatively high in the landscape and generally reports greater wind speeds than surrounding areas, including the downtown. The National Oceanic and Atmospheric Administration (NOAA) developed a composite record for Rochester daily climate back to 1886 (called the ThreadEx record), using all measurements from the different climate stations over time. Thus measures of the climate extremes in Rochester span a period well over 125 years.

Rochester's signature weather event was the August 21, 1883, F5 tornado that destroyed much of the town, killed 37 people, and derailed a train. The rescue and recovery effort led to formation of the Mayo Clinic. In 1978 the city suffered from two destructive flash floods: July 5–6 produced 6.74 inches of rainfall, and September 11–12 produced 6.60 inches of rainfall. The Zumbro River spilled its banks and flooded areas in downtown Rochester, causing a great deal of damage. As a result, a flood mitigation project was undertaken to alter the course of the Zumbro through the city and reinforce its banks. This effort has spared the city from significant flash flood damages on several occasions since.

Roseau (Roseau County)

Located west of Lake of the Woods' southern tip, Roseau lays claim to the county's oldest climate record. Daily observations were begun by Mr. O. B. Ekman in February 1894. Home to Polaris Industries, Roseau is also known as the birthplace of snowmobiling, and its long, snowy winters consistently provide an ideal climate for testing and riding. The winter of 1995–96 brought 167 continuous days of snow cover. The Roseau River runs through town, eventually emptying into the Red. Roseau has seen nearly all of the river's top historical crests as spring snowmelt floods that have created serious problems for the town and its residents. However, in June 2002 a severe thunderstorm dumped 6 to 7 inches of rain over a two-day period, resulting in a flash flood crest more than 7 feet above flood stage, setting an all-time high flood mark and damaging nearly every home in the city. Some extremes at Roseau include 107°F on July 12, 1936; -52°F on February 11, 1914, and again on February 2, 1996; and 15.0 inches of snow on November 17, 1996.

Lucille Sjostrom checking temperature and dew point sensors, 1946

St. Cloud (Sherburne County)

St. Cloud's weather has been recorded by assorted groups, beginning with the Smithsonian Institution in 1860, followed by St. Cloud College, which reported data to the U.S. Army Signal Corps from 1881 to 1903, the St. Cloud Reformatory, which took daily observations from

1904 through 1943, and the Weather Bureau, which maintained an office at the airport until 1995. Today climate observations are provided by the airport's Automated Surface Observing System (ASOS), and data reports are augmented by St. Cloud Correctional Facility and St. Cloud State University. The National Oceanic and Atmospheric Administration (NOAA) developed a composite record for St. Cloud daily climate back to 1893 (called the ThreadEx record), using all measurements from the different climate stations over time. Thus measures of the climate extremes in St. Cloud span a period well over 120 years.

St. Cloud's climate station has reported temperatures of -40°F or colder seven times, most recently on February 2, 1996. During the great arctic outbreak of 1899, St. Cloud spent eight consecutive days, January 28 to February 4, below zero. In the winters of 1936–37, 1950–51, and 1964–65, it reported more than 80 inches of snowfall, while in the winter of 1901–02 it recorded only 6 inches. During the drought of 1988, June brought a single day of rain, a measly 0.05 inches, which set a record low value for the month and led to severely parched vegetation and hundreds of dead trees. The worst drought at St. Cloud occurred in 1910, when only 14.64 inches of precipitation was recorded.

Where is the National Weather Service Forecast Office located in the seven-county Twin Cities metropolitan area?

a. Chanhassen

b. Downtown St. Paul

c. Minneapolis–St. Paul International Airport

(answer on page 329)

St. Peter (Nicollet County)

St. Peter's daily climate observations date to May 1893 and are taken at the state hospital, about a quarter mile from the Minnesota River. These readings have been important to scientific understanding of the Minnesota River Valley's hydrology. One of the worst episodes of erosion along the river bluffs occurred on August 7, 1968, when a thunderstorm brought 8.62 inches of rain. The river valley can heat up during the winter months as the steep, sloping bluffs efficiently store low-angle solar radiation. Not surprisingly, this climate station boasts five state high temperature records, including the warmest-ever New Year's Eve, 58°F in 1921. And in the summer of 1901, St. Peter recorded 17 consecutive days with highs of 90°F or greater. As for recent and memorable severe weather, on March 29, 1998, an F3 tornado made a direct hit on St. Peter, producing storm damages estimated at $120 million. Ninety percent of the town's homes were damaged and 25 percent destroyed. Every major building on the Gustavus Adolphus College campus was affected, and hundreds of old trees were broken or uprooted, completely altering the complexion of the landscape. The St. Peter climate station observer recorded a high temperature of only 67°F that day, an unexpectedly cool environment for the formation of such a powerful tornado.

Sandy Lake Dam (Aitkin County)

This climate station was established in August 1892 and has been continuously maintained by the U.S. Army Corps of Engineers. Importantly, it supports greater understanding of the hydrology of the upper Mississippi River watershed. The nearest town is McGregor. The Sandy Lake Dam climate station was used to analyze the Great Hinckley Fire of September 1, 1894. Prior to this disastrous wildfire, observations at Sandy Lake Dam showed a summer (June-August) rainfall total of only 0.76 inches, nearly 12 inches below normal and the driest in the station record. In comparison, in 1953 the summer months brought 26.76 inches of rainfall on the way to a record wet year, with 44.36 inches of precipitation. Abundant snowfall seasons have been numerous at this station, exceeding 75 inches 15 times and 100 inches three times, with a record value of 133 inches in 1995–96. The maximum monthly snowfall, 56.5 inches, fell in November 1995. Sandy Lake Dam recorded a temperature of 102°F on July 28, 1988. Conversely, during the winter season it has reported low temperatures of -40°F or colder more than 65 times.

Tower (St. Louis County)

The Minnesota State Forest Service began taking weather observations at its Tower ranger station in January 1895. Next to Embarrass, this small community, located at the south end of Lake Vermilion, most consistently reports the coldest minimum temperatures in Minnesota during the summer months. Its average frost-free growing season is only 55 days, with median frost dates of June 25 and August 19. Frost has occurred with some frequency in every summer month. Winter's chill is not unheard of, either: on February 2, 1996, radio and television reporters created a media frenzy in Tower as observer Kathleen Hoppa at her location three miles south of town noted a new state record low of -60°F, breaking the record of -59°F set at Leech Lake on February 9, 1899, and at Pokegama Dam on February 16, 1903. Tower's other all-time records include a maximum snow depth of 71 inches on March 13, 1989; 8.70 inches of rain on September 7, 2007; and a maximum temperature of 101°F on July 14, 1901.

Two Harbors (Lake County)

This climate station was established at the residence of Mr. H. L. Holden in February 1894 near the docks of the present-day harbor. Many other volunteers have served as daily weather observers over the past 120 years. Because of nearby Lake Superior and its moderating effects on temperature, the Two Harbors station has not recorded a temperature colder than -34°F in the winter, nor higher than 99°F in the summer. The station has also reported a number of "lake effect" enhanced snowfalls of great quantity, including 22 inches over March 4–5, 1985, and 36 inches over November 1–2, 1991 (associated with the Halloween Blizzard). November 1991 brought a total of 51.5 inches of snow. Six seasons have produced snowfall totals of 100 inches or greater, topped by 127.4 inches in 1996–97. The wettest month in history at Two Harbors was June 2012, with 13.86 inches of rainfall; a slow-moving thunderstorm brought a station-record 10.45 inches of rain on the twentieth, causing widespread shoreline erosion as well as street and basement flooding. Perhaps the all-time worst ice storm occurred there over March 23–24, 2009, when 1 to 2 inches of ice was deposited on power lines and trees. Thousands of residents were without power, and thousands of trees were broken and damaged by the weight of the ice. Highway 61 along the shore of Lake Superior was closed while crews removed broken trees from the pavement.

Wadena (Wadena County)

The city of Wadena, originally a trading post along the Red River Trail, is the county's only climate station with any longevity. Its first observer, Mr. B. F. Buck, Jr., began making daily reports of the weather for the Army Signal Corps in March 1885, and numerous others have since contributed to the century-old climate record. During an average frost-free growing season of about 138 days, Wadena County produces corn, hay, and edible beans, some of which are grown on irrigated sandy soils, which have a limited water-holding capacity. Wadena's historical climate extremes include a maximum temperature of 112°F on July 10,

1936; a minimum temperature of -43°F on February 2, 1996; rainfall of 5.97 inches on August 7, 1995; and snowfall of 19.5 inches on March 15, 1957. The driest year on record, 1976, saw just 13.42 inches of precipitation, while the wettest, 1965, tallied 38. By far the worst weather disaster in the historical record occurred on June 17, 2010, about 5:00 PM, when an EF-4 tornado (winds of 170 mph) tore through the town. The path of destruction was up to one mile wide and 10 miles long. Many homes and warehouses were destroyed or damaged, as was the high school, which had to be rebuilt. Twenty people were injured, but fortunately no deaths were reported.

Warroad (Roseau County)

Members of the MacDonald, Sawyer, Anderson, and Landby families have contributed decades of daily climate observations to the Warroad record, which dates back to March 1901. Other observers included, for a time, employees of the U.S. Customs Office. Located on Lake of the Woods' southwestern shore, Warroad has a deserved reputation for cold and snow. As recently as 2005 it reported an inch of snowfall on October 1, and in 1970 it had measurable snowfall, 0.1 inches, as late as May 15. Warroad tends to receive its heaviest snows when northeasterly winds blow over an unfrozen Lake of the Woods. Little surprise, then, that its record includes snowfalls of 12 or more inches in mid-November. During the winters of 1966 and 1969, Warroad reported more than 40 inches of snow cover. The snowiest winter was 1965–66, when 81.5 inches of snow was reported. Warroad has recorded low temperatures at or below -45°F during six different winters, most recently in 1996. Indeed, the coldest date in the station record is January 29, 1904, when the morning low was -45°F and the afternoon high reached a frigid -30°F. But this northern town has summer records, too: in June 2002, five consecutive days of thunderstorms produced unprecedented rainfall of 11 or more inches, flooding many properties in Warroad and helping set a station record monthly total of 13.57 inches of rain.

Waseca (Waseca County)

Waseca's daily climatic observations, which date to August 1914, were handled by experiment station superintendent Robert E. Hodgson for more than 40 years, from 1919 to 1960. Since that time, staff members of the University of Minnesota's Southern Research and Outreach Center have maintained the record. The city took a direct hit from a tornado on the evening of April 30, 1967, as the storm cut a four-block-wide path of destruction, tearing apart numerous homes.

Waseca is located in prime corn and soybean country, characterized by rich, deep, and often tile-drained soils. Thus, its hundred-year record is valuable to researchers who assess climate influences on agricultural practices. The station boasts one of the state's best long-term, detailed records, including measurements of soil temperature (at 2-, 4-, 8-, 20-, and 40-inch depths), soil tile drainage, solar radiation, evaporation, and wind. In recent years, its tile drainage discharge records have been used to assess nutrients driven from the soil by heavy rainfall, which is common to the area. Waseca has recorded 12 daily rainfalls of 4 or more inches since 1914. August, often the wettest month, saw 11.29 inches of rain in 1924, 10.11 inches of rain in 1935 (a dust bowl year), and 10.76 inches in 2007. One of the wettest growing seasons was 2010, when 34.61 inches of rain fell from May to September—a record for the station. Even more remarkable, that measurable rainfall occurred on 78 days during the 147-day growing season, or better than one day in two. Finally, Waseca has seen a 30 percent increase in average annual precipitation since 1950, one of the largest such changes in the state.

Willmar (Kandiyohi County)

The Willmar Regional Treatment Center, formerly the Willmar State Hospital, provided climate data for central Minnesota for well over a century, its record begun by Axel F. Elfstrum in February 1893. Today a number of other volunteer weather observers provide data from the city. The Willmar area is known for turkey and hog production as well as corn and soybeans, and the climate record has supported agricultural as well as hydrological research. Located downwind from the Buffalo Ridge, the southwestern counties' relatively higher topography, Willmar has recorded some very large thunderstorm rainfalls—including 7.20 inches on June 9, 1895, 6.82 inches on June 17, 1957, and 5.54 inches on June 17, 1992—and a record snowfall—21 inches on November 27, 2001.

Another weather extreme, an exceptionally rare November tornado, occurred southwest of Willmar, near Prinsburg, on November 1, 2000; the parent thunderstorm of this EF-1 dropped a record-setting 1.35 inches of rain. Extremely long-lived—but minimally damaging—tornadoes also occurred near Willmar on July 1, 1928, and June 9, 1947, with 60-mile and 50-mile paths, respectively. A famously photographed tornado passed between Willmar and Kandiyohi on July 11, 2008, destroying a number of farm buildings and injuring two people.

Winnebago (Faribault County)

This climate station was established at the residence of Mr. W. Z. Haight in October 1898. The Haight family made observations until 1918, after which a number of volunteers began contributing daily measurements to the climate record. Located in the heart of the state's agricultural landscape, Winnebago has provided climate records that have been valuable to crop producers in the selection and hus-bandry of crops. At least 15 winter days have produced a minimum temperature of -30°F or colder, while during the summer Winnebago has reported 45 days with temperatures of 100°F or greater, most recently with a new statewide daily record of 103°F on May 15, 2013. Another statewide record owned by the Winnebago station is 8.64 inches of rainfall on September 25, 2005, during its wettest year, which recorded 46.34 inches of precipitation. During the dust bowl era Winnebago received relatively more rainfall than other areas of the state, and local farmers were able to salvage at least some crops. The greatest seasonal snowfall total there was 88.3 inches in 1983–84.

In the 1890s the National Weather Service Offices in downtown Minneapolis and downtown St. Paul ran a rather heated forecasting competition.

Winnibigoshish Dam (Itasca County)

Since April 1887 the Winnibigoshish Dam climate station has been con-tinuously staffed by the U.S. Army Corps of Engineers; for a number of years the readings were taken at the dam tender's residence. Unlike some other Mississippi headwaters stations, Lake Winnie does not claim any state low temperature records, though on February 8, 1933, it was as cold as -49°F. A freak storm dropped 12.5 inches of snow at the dam over the first three days of May 1954, one of the state's largest May snowfalls, probably enhanced by water vapor released from the lake. A favorite fishing spot for walleye, Lake Winnie is often anglers' destination of choice for the fishing opener. It is peculiar that Lake Winnie's earliest and latest ice-out dates were in back-to-back years: March 30, 2012, and May 17, 2013. The median ice-out date there is April 27.

Winona (Winona County)

Winona's long climatic record, one of the few made along the upper Mississippi River Valley, dates to notes taken in August 1885 by Mr. J. M. Halsinger. Staff of the local newspaper, now the *Winona Daily News,* made observations in town for many years, while members of the U.S. Army Corps of Engineers take their readings right along the river's edge. This stretch of the Mississippi River, from Hastings, Minnesota, to La Crosse, Wisconsin, offers one of the state's most hospitable climates for plants. Winona's average frost-free growing season is 174 days, more than two weeks longer than that of Rochester, which lies only 40 miles to the west. In fact, some growing seasons in the Mississippi River Valley have stretched to more than 200 days. In addition, winter temperatures are less extreme. Perennial plants tend to survive for longer periods, their vibrancy made visible in the area's vineyards, apple orchards, and other fruit and nut trees. Winona had a historic flash flood over August 19–20, 2007, when 10–15 inches of rainfall flooded streets and caused some landslides on the neighboring hill slopes.

• •

Parallax error

From the Greek term parallaxis, *meaning to alter or change a little, parallax error refers to the apparent displacement of an object when viewed from different angles. One of a number of potential observational errors, it can occur when reading a thermometer or a mercury barometer if the line of sight is not carefully kept perpendicular to the reading scale. In the case of a liquid-in-glass thermometer, the parallax error might range from a few tenths of a degree to nearly two degrees.*

Personal equation

Personal equation describes a systematic error characteristic of an individual observer. A statistical analysis of his or her data may reveal a particular bias or level of uncertainty. For example, observations of cloud cover in tenths: one observer may see 0.6 cloud cover differently than another; or parallax error in reading a glass thermometer: subtle changes in line of sight—left versus right eye, looking upward versus downward at the instrument—can affect the recorded data.

• •

Worthington (Nobles County)

Worthington's century-old climate record, begun in 1893 with observations by Mr. E. L. Porter, is maintained by city employees today. Located in the heart of corn and soybean country on the exposed loess soils south of the Buffalo Ridge, this community is prone to high winds and, in the winter, blowing and drifting snow. Worthington holds the statewide record for the greatest daily precipitation in November: 4.80 inches on November 26, 1896, a Thanksgiving Day storm that began with lightning and thunder while temperatures were still in the thirties and ended with a 1-inch coating of ice after the mercury dropped. One of the worst ice storms in history occurred on April 9, 2013, when nearly an inch of ice coated trees and power lines and made for hazardous driving conditions. Residents had to endure a power outage for many days during a cold and snowy April that year. A peculiar feature of the Worthington climate record is that of the 10 historical daily rainfalls of 4 inches or greater, 6 have occurred in the month of September, most recently 5.61 inches on September 23, 2010. Worthington often reports very hot conditions: temperatures have exceeded 100°F more than 70 times, 16 of those in 1936 alone.

Zumbrota (Goodhue County)

Established in December 1894 as Goodhue County's only climate station not influenced by the Mississippi River Valley, Zumbrota has recorded some of southern Minnesota's coldest temperatures. Many local residents have served as climate observers there. The rolling landscape has a reputation for being a frost pocket in the late spring and early fall. Spring frosts have occurred on Memorial Day weekends and as late as June 10, as in 1972, while fall frost has come as early as September 1, as in 1974. Zumbrota's all-time coldest temperature—indeed, one of the coldest recorded in southern Minnesota—occurred on January 30, 1950, a reading of -45°F. In January and February 1936, Zumbrota saw 36 consecutive days with a minimum temperature below zero. On December 19, 1983, a minimum temperature of -40°F was reported, and in January 2009 the low temperature hit -36°F on two consecutive days. Over September 23–24, 2010, a massive thunderstorm dropped 7.66 inches of rain, sending an all-time flood crest down the Zumbro River and washing out many local roads. The 14.57 inches of rainfall that September was the most of any month in history at Zumbrota.

Earliest Observations

Alexander Henry, known by some as Henry the Younger to distinguish him from his well-known uncle, was a prominent figure in Canada's North West Fur Company during the late eighteenth and early nineteenth centuries. He spent considerable time exploring, hunting, and trapping in the Red River Valley and helped establish old Fort Pembina just across the river from present-day St. Vincent, Minnesota. A literate man, Henry was one of the first pioneers in the region to keep a journal that contained detailed weather observations. Documents suggest that he established his trading post on the same ground used by Charles Chaboillez in 1797. Chaboillez was a partner in the North West Fur Company who later supervised a company post on Lake Superior. He, too, kept a journal of his activities near the junction of the Pembina and Red rivers, but it lacked the weather and climate detail of Henry's.

Henry traded alcohol, tools, and weapons with the American Indians who lived near Fort Pembina: the Dakota tribes of the plains on the western side of the river (later North Dakota) and the Ojibwe on the eastern side of the river. He was meticulous in keeping daily weather observations from September 1807 through June 1808.

The landscape of northwestern Minnesota was vastly different during Henry's time. Buffalo and beaver were abundant, as were red deer, moose, bear, and wolves. The wide river valley was populated with tall prairie grasses and old trees (oak, birch, maple, pine, poplar, and willow, among others) and dotted with wetland areas (sedge and marsh). A mosaic of trails and paths devoid of vegetation weaved across the prairie where the abundant buffalo herds trampled and compacted the heavy, wet soils. In addition to hunting buffalo and red deer for food, Henry and his men had good luck fishing the Red River, which was full of sturgeon.

Henry prized his daily journal of weather and nature observations, three volumes of which were published in 1897. Though some historians consider the Henry documents to be dry and humorless, for climatologists they are a treasure trove for use in evaluating the weather behaviors of a bygone era. His records are also noteworthy because they include climate descriptions of the area before the end of the Little Ice Age (1850), a relatively cooler period for the northern hemisphere.

The Minnesota State Climatology Office has archived records of

Henry's daily weather observations from September 1, 1807, to June 1, 1808. These include morning, midday, and evening observations of air temperature, wind, and sky condition, along with remarks describing the day's character. The tables on page 75 show comparisons between Henry's observations and the current published climatic normals at Pembina, North Dakota, and at Hallock, Minnesota.

In general the tables show that the climate of the Red River Valley during 1807–08 was similar in temperature to today's climate, with the exceptions of October and April, which were warmer. In fact, April was a remarkably warm month, producing a rapid and early break-up of ice on the Red River, with associated flooding. Henry noted the return of migratory birds as early as April 4 and 5. He also brags of enormously successful sturgeon fishing: on April 24 he "set a sturgeon net and caught 3 instantly." A more detailed look shows that the lowest temperature measured by Henry during the winter was -30°F on February 18, quite close to the modern era's usual lowest winter temperature, which ranges from -30 to -35°F at Hallock and Pembina. The coldest spell in February 1807 also coincided with some of the strongest winds Henry noted, likely producing wind-chill conditions in the -50°F or colder range. However, the winter of 1807–08 saw only 43 days with temperatures below zero. Modern records from Pembina and Hallock both show an average of 57 days each winter with such temperatures. Thus, Henry and his men did not have to endure the persistent cold that Mother Nature usually brings to the area.

The precipitation climate of 1807–08 was quite different, however, with more frequent occurrence of snow and rain during November through March. Snowfall totals were at least 10 and 20 inches, respectively, for the months of November and December, considerably higher than the modern averages taken from Pembina—7.0 inches for November and 8.0 inches for December. In this context the Red River probably flowed at a higher level during Henry's time there, lacking as it did the agricultural drainage and reservoirs or holding ponds that characterize the valley today.

Some of Henry's remarks and climate summaries from 1807–08 are worth noting:

Following early September warmth, with temperatures as high as 89°F, Monday the seventh brought a frost to the valley that ended the growing season for cucumbers and melons. These temperatures helped trigger a change in leaf color on most of the trees. Thunderstorms, with large hail and strong winds, occurred at mid-month, followed by 2 inches of snowfall near where Two Rivers joins the Red River of the North, just west of present-day Hallock. A number of hard freezes, with temperatures dipping into the twenties, were noted in the second half of the month.

Corn and potatoes were harvested the first few days of October. A luminescent star with a long tail was noted in the western sky; this feature (perhaps a comet) lasted until mid-November. Numerous prairie fires occurred during the month, some producing enough smoke to drastically reduce visibility. Fires were a common occurrence on the prairie landscape throughout the nineteenth century: the dry native grasses provided excellent fuels easily ignited by cloud-to-ground lightning strikes.

November brought frequent snows and cold. Snowfall occurred on 15 days, probably totaling more than 10 inches for the month. The Red River was frozen over by the twelfth and could accommodate foot and sled traffic by mid-month as the temperature fell to -1°F on November 14.

Very cold temperatures and abundant snow continued in December. The month's snowfall total exceeded 20 inches. Temperatures fell below zero on seven days, reaching a low of -17°F on the seventeenth. In fact, on four days the temperature never rose above zero. Combined with very strong winds, these low temperatures produced some dangerous wind-chill values, but of course this was long before the concept of "wind chill" had ever been proposed. Nevertheless, one could infer that layered clothing under a buffalo robe was the rule.

January's cold temperatures produced halos around the sun and moon on several occasions. Fifteen days registered below-zero temperatures. Unlike the late-twentieth-century climate of the Red River Valley, which showed a high frequency of January thaws, there was no thaw period—or even a single day above freezing—in January 1808. There were 11 days with snowfall, and a good deal of blowing and drifting.

February 2 (Groundhog Day) brought the first thaw, as the temperature reached an afternoon high of 36°F. But this relative warmth was only a tease, as the rest of the month brought 15 more days of below-zero temperatures, some with presumably dangerous wind-chill conditions

TABLE 7: Mean Monthly Temperatures for Red River Valley Communities of Hallock and Pembina (1981–2010) Compared to the Monthly Values Observed by Alexander Henry (1807–08) (all values in degrees Fahrenheit)

LOCATION	SEP	OCT	NOV	DEC	JAN	FEB	MAR	APR	MAY
Hallock, MN	55.0	40.6	23.4	7.0	0.8	6.3	21.3	39.7	53.2
Pembina, ND	55.2	41.4	24.6	9.3	3.5	8.7	22.3	40.1	53.1
Henry's Camp	55.0	47.2	22.6	10.9	1.4	5.0	22.5	49.1	54.9

TABLE 8: Mean Frequency of Daily Precipitation by Month for Red River Valley Communities of Hallock and Pembina (1981–2010) Compared to the Frequency of Daily Precipitation Observed by Alexander Henry (1807–08) (days)

LOCATION	SEP	OCT	NOV	DEC	JAN	FEB	MAR	APR	MAY
Hallock, MN	6.9	7.0	5.5	6.2	6.9	5.0	5.7	5.8	8.7
Pembina, ND	7.4	7.1	5.7	7.4	6.5	5.7	5.9	4.9	9.6
Henry's Camp	7	3	15	14	10	9	9	3	4

due to high winds. February ended on a warming trend, reaching a high of 42°F on the twenty-eighth and melting quite a bit of snow cover.

Early March saw the warm spell continue, with three consecutive days in the forties. Henry observed migrating swans the first week of the month, followed by buzzards. By March 13, water began to flow over the top of the ice in the Red River. Six inches of fresh snow was noted at mid-month, then alternating freeze-thaw periods. Four nights recorded below-zero readings, the coldest, -7°F, coming on the morning of the twelfth. The month's high temperature, 52°F, was recorded on the twenty-sixth.

April 2 brought a snowstorm, but it was followed by a pronounced thawing period. Several days saw temperatures in the forties and fifties, and on the eighth the ice broke up on the Red River and started to flow north. Henry observed abundant migrating waterfowl, often resting along the river's banks. He caught the season's first fish on April 7. With the prolonged warm spell and the absence of significant precipitation, the ice break-up occurred in relatively short order. The daytime temperature hit 65°F on the twelfth, and the river level began to drop two days later, falling 2 feet in a 24-hour period. The last week of April, Henry and his men enjoyed great success as they fished for sturgeon. April 27–29 saw summer-like temperatures—highs in the mid- to upper eighties (89°F on the twenty-ninth)—and a substantial rainstorm ended the month.

May brought wide swings in temperature and even a snowstorm on the third, the last of the season. Alternating high-low temperatures and very strong winds indicated numerous cold fronts passing through. Temperatures ranged from 28°F on the fourth to 86°F on the eighteenth. Sturgeon fishing remained very good: Henry noted taking up to one hundred fish per day out of the Red River.

Henry's climate records for 1807–08 end on June 1, a sultry day with a high of 92°F and a low of only 70°F. From his journal entries and daily weather observations—one of the few such documents of the region prior to the establishment of Fort Snelling in 1819—we have a picture of the harshness, beauty, and serenity of the Red River Valley's landscape and climate.

Minnesota's Winter

The two clear signatures of winter in Minnesota—snow and cold—have inspired many vivid memories, chilling stories, and even a national reputation to uphold.

Though snow may arrive as early as September in the northern counties, winter weather generally settles in during November and early December. By then even the most stubborn of leaves has turned color and fallen to the ground, ice has begun covering the lakes, and the soil has started to freeze. For those suffering from Seasonal Affective Disorder (SAD), the onset of symptoms usually appears in November, with its higher incidence of cloud cover, lowering sun angle, and shorter day length. SAD is treated with both pharmaceuticals and light therapy, though many Minnesotans simply choose to vacation in sunnier and warmer locales. However, a significant segment of the state's population absolutely loves, even thrives on the winter climate. Ice fishing, snowshoeing, cross-country skiing, snowmobiling, ice skating, sledding and tubing, ice hockey, and many other winter activities occupy people of all ages across the state. One tradition born of this appreciation for snow and cold is the St. Paul Winter Carnival, which runs from late January to early February and celebrates the glories of winter recreation, art, and culture.

St. Paulites have embraced winter with their annual carnival,
which featured an ice palace in 1886.

The climatic traits of snowfall and snow cover vary considerably around the state. Snowfalls have occurred in every month of the year except July. Duluth is the only station to record observable snow (a trace) in August, on August 31, 1949. Significant snowfalls can occur in September and June in the northern counties. In September, locations like Alexandria, Duluth, and International Falls actually show an average snowfall of 0.1 inches, though the frequency of measurable September snow in these areas is only about once every five to ten years. Nearly all of the state's climate stations show an average historical value for October snowfall, the largest being 2.6 inches at Tower in St. Louis County. October snowfall frequency varies from once every three years in the north to about once every six or seven years in the south. The snow accumulation season—when snow remains on the ground for extended periods—does not begin until the soil starts to freeze, typically the second week of November in the far northern counties and as late as the first week of December in the far southern counties and the Twin Cities metropolitan area. The timing of soil freeze-up is closely associated with the development of ice cover on shallow lakes. The formation of thin ice cover does not imply that lakes are safe for foot traffic, however. The Minnesota Department of Natural Resources recommends 4-inch-thick ice for foot traffic and 8 to 12 inches for automobile traffic, levels usually not achieved until December.

The snow accumulation season fluctuates greatly from year to year for any particular location in the state. When former state climatologist Earl Kuehnast examined the Farmington (Dakota County) climate record for the period from 1896 to 1981, he found that continuous snow cover varied from as short as 21 days, in 1960–61, to as long as 142 days, in 1950–51. This remarkably high variation makes planning outdoor recreation around the persistence of snow cover a challenge to say the least, except perhaps in some far northern counties and towns, such as

International Falls and Duluth, where snow cover duration rarely dips below 100 days per year. The spatial or geographic variability of snow cover duration is also quite large. Along the southern edges of the state in communities like Albert Lea, Austin, and Spring Grove, the average duration of snow cover is 75 to 90 days. In colder regions to the far north such as Caribou, Crane Lake, International Falls, and Thorhult, the average is 140 to 150 days. Along the Gunflint Trail in the Lake Superior highlands of Cook County, where elevations rise 1,800 to 2,200 feet, the average snow cover duration is 150 or more days, usually lasting until after the first week of April.

· ·

Wind ripple

A wind ripple, also known as a snow ripple, is a wavelike formation often visible in western Minnesota's snow-covered rural landscapes. A series of small waves, each about an inch high, run at right angles to the prevailing wind—frequently southwest to northeast due to dominant northwesterly winds. Wind ripples— usually following light, fluffy snowfalls—can occur several times over the course of a season if the land is undisturbed.

· ·

Total seasonal snowfall is greatest along the North Shore, especially in the higher elevations above Lake Superior, where averages can exceed 80 inches. Southwestern Minnesota's drier and warmer counties show some of the lowest average seasonal snowfalls, for example about 35 inches in Redwood Falls (Redwood County). Here and elsewhere in southern Minnesota, winter storms may produce moderate to heavy rain instead of snow, thereby reducing the overall number of snowfalls per season. Long-term monthly averages for over 85 percent of Minnesota's climate stations show January and December as the snowiest months (based on 1981–2010 averages). But just under seven percent of the state's climate stations show that average snowfall is greatest in March, while only a handful show February as the snowiest month (see Table 11). The month of March is the most highly erratic when it comes to snow: some years record only rain while others see up to 30 or more inches of snow.

TABLE 9: One-Day Maximum Snowfall Extremes (inches)

STATION	JAN	FEB	MAR	APR	MAY	JUN	SEP	OCT	NOV	DEC
Ada	13.0 26-2004	11.0 11-2013	12.0 3-1966	9.0 19-1970	2.5 2-1954		2.0 28-1899	6.0 30-1972	10.0 28-1960	12.0 30-1972
Aitkin	15.0 22-1982	15.0 27-1971	14.0 15-1957	9.0 12-2013	3.0 3-1954		0.9 24-1985	8.0 31-1951	9.0 28-1983	7.0 31-1981
Albert Lea	11.0 4-1971	15.0 9-1909	20.0 18-1933	16.0 4-1945	7.0 2-2013			4.0 2-1999	12.0 10-1896	14.0 25-1945
Alexandria	12.2 10-1975	7.0 4-1955	10.0 17-1965	8.8 19-1966	2.0 2-1954		2.0 26-1942	3.7 30-1951	10.0 28-1983	6.1 22-1968
Argyle	11.1 7-1989	8.3 25-2007	8.2 25-2009	15.5 1-2014	3.0 3-1950		1.7 26-1972	9.7 25-2001	11.3 19-1998	7.3 26-2009
Austin	13.0 22-1982	9.8 18-1961	16.0 19-2005	12.0 7-2003	10.0 2-2013		4.0 26-1942	3.0 27-1939	11.5 11-2006	11.0 28-1987
Baudette	12.0 11-1923	7.0 24-1977	9.0 12-2013	8.1 19-1970	4.0 2-1954		3.0 26-1941	16.0 18-1916	12.0 21-1977	12.0 15-1933
Bemidji	14.0 18-1996	14.0 13-1897	14.0 25-1914	10.0 25-1950	3.5 2-1950		2.0 27-1899	8.0 18-1917	10.4 25-1936	14.0 14-1927
Big Falls	9.4 3-2011	9.1 2-1987	15.0 4-1966	12.5 7-2008	2.5 22-2001			9.5 28-1932	10.0 16-1988	10.5 27-1988
Brainerd	15.0 10-1983	15.0 28-1951	24.0 4-1985	15.0 11-2008	3.0 2-1950		1.0 26-1942	3.0 31-1972	10.0 27-1988	9.0 15-2008
Browns Valley	17.5 5-1997	12.0 21-2011	15.0 20-1982	12.0 26-2008				2.0 24-1995	10.5 25-1993	9.0 25-2009
Caledonia	15.0 4-1971	20.0 2-1915	22.7 7-1959	10.0 9-1973	6.0 2-1911			7.8 28-1917	8.5 21-1909	20.0 11-1899
Cambridge	11.0 22-1982	11.0 25-2007	14.0 3-1989	8.0 25-1950	2.0 2-1954			5.0 31-1991	17.0 1-1991	12.0 11-2010
Campbell	15.0 29-1916	14.0 7-1946	13.5 31-1944	9.0 22-2001	2.5 5-1944		1.0 25-1912	5.0 30-1929	11.0 26-2001	14.0 4-1927
Canby	12.0 21-1917	12.0 21-2011	20.0 16-1917	14.0 8-1989	1.0 2-1954			4.0 24-1995	16.0 20-1975	14.0 1-1981
Caribou	10.0 8-1998	6.0 24-1977	12.0 27-1975	10.0 8-1989	3.0 2-1966		6.0 29-1947	10.0 23-2001	12.0 18-1996	10.0 27-1988
Cloquet	15.1 12-1972	12.5 10-1965	21.4 4-1985	14.0 25-1950	4.2 3-1950		0.2 30-1985	10.6 30-1951	14.0 1-1991	16.5 6-1950
Collegeville	17.5 15-1900	14.0 9-1909	23.6 17-1965	12.2 3-1937	3.0 9-1902			6.0 21-2002	15.0 22-1915	14.0 6-1909
Cook	12.0 11-1975	15.0 27-1971	12.0 24-1975	12.0 4-1968	6.0 19-1971		0.2 30-1974	7.0 17-1990	29.0 1-1991	10.0 11-1983
Cotton	15.0 5-1997	11.0 25-2001	13.2 4-1985	9.0 14-1983	2.0 1-1970			3.0 30-1984	16.4 1-1991	6.6 28-1982
Crookston	12.0 18-1996	11.0 24-1977	10.0 4-1997	13.0 1-2014	3.0 3-1954			5.0 24-2001	15.0 18-1998	12.0 25-2009
Detroit Lakes	14.0 7-1989	25.0 23-1922	15.0 28-1944	13.0 17-1945	3.4 14-1907		6.0 26-1942	10.0 20-1906	11.0 9-1977	12.0 27-1988
Duluth	18.2 26-2004	13.0 28-1948	18.0 13-1917	12.1 3-2007	5.5 10-1902		2.4 18-1991	10.0 23-1933	24.1 1-1991	23.2 6-1950
Ely-Winton	10.0 23-1982	7.0 22-1979	9.0 11-1976	14.1 6-2008	4.0 1-1984			4.7 30-1971	11.0 24-1983	8.9 2-2007
Fairmont	12.0 23-1982	18.5 1-1915	14.6 22-1952	12.0 21-1893	2.0 9-1924		2.0 25-1942	7.0 29-1905	18.0 16-1909	12.0 28-1982
Faribault	15.0 23-1982	16.0 8-1936	13.0 16-1983	11.0 14-1983	4.0 2-2013		2.2 28-1945	4.0 29-1905	12.0 28-1983	11.0 11-1970
Farmington	17.0 20-1982	13.0 9-1939	16.0 29-1924	16.0 14-1983	3.0 3-1954		1.5 21-1924	12.0 4-1926	17.5 1-1991	16.0 29-1982
Fergus Falls	13.0 18-1996	12.5 15-1945	15.0 16-1943	12.5 16-1945	5.1 2-1935		2.4 25-1912	8.0 24-1883	15.0 25-1993	12.0 31-2010
Fosston	10.0 18-1996	7.3 20-1955	12.0 4-1997	9.0 19-1970	3.0 3-1967			8.0 10-1970	11.0 10-1977	8.0 31-1973
Grand Marais	15.0 13-2008	22.0 14-1936	18.0 14-1917	10.0 5-1933	4.0 4-1954		1.5 26-1942	4.0 24-1933	8.6 26-1965	15.0 31-1937
Grand Meadow	12.0 4-1971	10.2 25-2007	18.0 19-1933	17.0 4-1945	9.0 3-2013		5.0 26-1942	5.0 23-1979	9.0 30-1934	13.0 12-2010
Grand Rapids	13.0 11-1975	10.5 27-1971	11.5 1-1965	14.5 6-2008	8.6 19-1971	0.7 2-1969		7.0 30-1951	13.6 26-1965	9.5 2-2013
Gunflint Lake	12.0 23-1982	12.5 25-2001	9.0 28-1979	15.0 19-2013	3.0 5-1989		0.2 30-2003	8.0 18-1990	17.0 1-1991	9.6 8-1966
Hallock	9.0 2-1907	14.0 14-1915	14.0 15-1902	14.0 1-2014	1.5 3-1954		1.5 28-1939	10.0 18-1916	8.0 10-1919	8.0 8-1966
Harmony	16.0 4-1971	12.0 23-1959	14.0 8-1961	11.0 9-1973	6.0 3-2013		4.0 26-1942	5.0 21-1982	9.5 11-1985	12.0 9-2009
Hibbing	14.6 4-1997	11.0 27-1971	15.4 4-1985	8.1 19-1970	7.0 19-1971		0.3 30-1985	3.6 17-1990	18.6 1-1991	12.8 13-1995
Hutchinson	12.0 20-1982	15.0 25-2007	15.0 4-1985	10.0 4-2014	0.4 2-1971			2.0 15-1966	13.0 30-1991	12.0 15-1996
Int'l Falls	14.1 10-1975	12.1 27-1996	13.1 10-2009	13.9 25-1950	8.0 5-1925	0.3 2-1969	2.0 26-1912	6.0 18-1917	12.0 10-1911	13.7 14-2008

STATION	JAN	FEB	MAR	APR	MAY	JUN	SEP	OCT	NOV	DEC
Itasca	15.0 21-1982	15.0 11-2013	18.0 3-1966	12.5 19-1970	6.0 2-1954			8.0 30-1951	12.0 26-1965	10.0 13-1968
Leech Lake Dam	14.0 12-1988	18.4 2-1903	27.0 20-1903	21.5 30-1903	8.0 15-1915		1.5 26-1942	11.0 30-1951	17.0 26-1930	11.0 31-2010
Luverne	8.0 10-1983	11.0 19-1984	15.0 3-1970	7.0 30-1984	1.5 14-1907			7.0 20-1982	11.0 28-1994	12.0 22-1968
Marshall	8.3 2-1999	16.0 20-1952	15.0 31-1949	12.0 15-1947	1.5 5-1944			4.5 2-1999	14.5 27-2001	12.0 28-1982
Milan	11.0 18-1996	14.0 28-1948	15.0 21-2008	15.0 11-2008	3.0 7-1907		2.0 26-1942	7.5 19-1899	13.0 10-2014	12.0 12-1926
Montevideo	12.0 17-1996	12.0 21-2011	14.0 3-1989	8.0 11-2013	2.0 5-1944		5.8 3-1909	6.0 31-1991	15.0 20-1948	12.0 9-2012
Mora	9.0 6-1967	20.0 25-1930	20.0 15-2002	14.0 19-2013	2.5 8-1924		0.5 26-1942	8.0 30-1951	20.0 25-1930	13.0 22-1968
Morris	9.5 7-1967	11.0 21-2011	19.0 4-1985	11.3 26-2008	2.0 2-1954		0.5 26-1965	5.2 24-1995	11.5 26-1971	9.6 1-1985
MSP	17.2 22-1982	11.8 20-2011	14.7 31-1985	13.6 14-1983	3.0 20-1892		1.7 26-1942	8.2 31-1991	18.5 1-1991	16.3 11-2010
New Ulm	11.0 19-1988	12.0 27-1971	12.0 24-1937	12.0 16-1919	5.0 15-1896	1.1 24-1914	5.5 26-1942	5.0 29-1913	12.2 28-1983	12.0 15-1927
Olivia	10.0 20-1982	11.0 3-2003	17.0 4-1985	9.0 11-2013				2.5 24-1976	12.2 28-1983	12.5 10-2012
Park Rapids	12.0 4-1997	9.0 5-1908	14.0 4-1985	15.0 19-1893	5.0 3-1954		3.5 26-1942	10.5 26-1913	10.0 21-1953	12.0 23-1933
Pine River Dam	17.0 11-1975	14.0 28-1948	15.0 4-1985	13.0 20-1893	6.0 19-1971			9.0 30-1951	10.0 11-1940	11.0 1-1985
Pipestone	12.0 24-1982	17.0 2-2011	11.0 4-1984	25.0 7-1911	0.9 11-1966		0.1 22-1995	6.0 17-1937	15.0 28-2001	12.0 28-1982
Pokegama Dam	12.0 11-1975	11.0 27-1971	14.0 4-1985	13.8 25-1950	6.5 15-1907		2.1 29-1899	10.5 30-1951	13.0 22-1898	9.5 30-1936
Red Wing	13.0 20-1988	12.0 12-1965	15.0 9-1999	9.0 1-1985	1.5 3-1954			2.2 25-1981	11.0 24-1983	10.0 9-2009
Redwood Falls	7.2 19-2000	11.0 26-1971	14.0 4-1984	8.5 4-2014	0.3 2-2013		2.2 25-1942	3.6 31-1991	11.0 7-1943	10.0 14-1996
Rochester	15.4 22-1982	13.5 27-1893	19.8 18-2005	14.0 20-1893	14.0 2-2013		0.8 30-1961	5.0 22-1979	12.0 30-1934	15.0 11-2010
Roseau	12.0 7-1989	8.0 13-1938	10.0 6-1916	9.0 9-1990	10.0 6-1938		1.2 25-1912	8.0 24-2001	15.0 17-1996	10.0 13-1995
St. Cloud	12.0 3-1897	12.0 28-1951	20.0 31-1896	24.0 19-1893	3.2 19-1971		1.8 26-1942	5.8 20-1936	13.2 10-2014	11.0 9-2012
St. James	18.0 19-1988	14.0 25-2007	13.0 22-1952	11.0 29-1956	1.5 2-2013			4.0 31-1995	13.0 1-1991	11.5 28-1982
St. Peter	15.0 21-1917	10.4 21-2011	13.9 22-1952	10.5 14-1928	1.5 8-1938		1.0 26-1942	4.0 29-1905	15.0 12-1940	14.0 28-1982
Sandy Lake Dam	14.0 12-1972	15.0 26-2001	11.5 31-1896	12.5 25-1950	4.0 3-1954		1.0 29-1908	15.0 25-1942	12.0 26-1995	10.0 2-2007
Spring Grove	14.0 27-1996	13.3 23-1959	14.5 8-1961	12.0 8-2003	3.0 3-2013		4.0 26-1942	4.5 23-1938	11.0 18-1957	15.0 1-1985
Stillwater	17.0 23-1982	12.0 15-1918	13.5 23-1952	11.0 28-1907	1.5 2-2013			3.8 30-1905	12.8 24-1983	13.0 12-2010
Tower	14.0 4-1897	12.0 25-2001	15.0 4-1966	27.0 17-1961	10.0 3-1950	4.5 2-1945	2.0 27-1899	12.0 18-1990	18.0 30-1926	12.0 15-1902
Two Harbors	14.5 15-1970	18.0 5-1908	20.0 4-1985	17.0 5-1933	8.0 8-1924			7.0 30-1951	24.0 2-1991	18.4 31-1996
Wadena	18.5 5-1997	11.5 11-2013	19.5 15-1957	12.0 1-2009	5.0 2-1935		3.1 26-1942	9.9 30-1951	12.5 25-1993	11.0 2-2007
Warroad	10.0 10-1916	9.0 1-1911	13.0 27-2003	15.0 3-1989	5.0 6-1938		6.0 25-1912	7.0 13-2006	12.0 20-1977	10.0 15-1933
Waseca	12.0 23-1982	13.0 12-1965	15.0 24-1966	12.0 30-1984	2.0 3-1954			4.0 25-1981	14.0 1-1991	12.0 28-1982
Waskish	8.0 20-1982	8.0 25-2007	13.0 12-1976	10.0 26-1950	9.0 2-2013		0.2 21-1974	7.0 28-2003	10.0 21-1977	12.0 6-2013
Wheaton	24.0 4-1997	11.4 11-2013	15.0 3-1951	9.9 11-2008	4.0 2-1954			6.0 30-1951	10.0 27-2001	8.0 31-2010
Willmar	15.5 3-1943	12.0 10-1965	18.0 16-1917	11.0 26-1893	2.0 9-1924		6.0 26-1942	4.5 19-1976	21.0 27-2001	13.3 10-2010
Windom	13.0 20-1988	15.0 23-2001	16.0 13-1940	14.0 29-1956	12.0 8-1938		1.0 26-1942	6.0 24-1976	15.0 8-1943	15.5 24-2009
Winnebago	12.0 23-1982	18.0 2-1915	12.0 6-1900	12.0 11-1922	6.0 8-1938		2.5 26-1942	5.0 26-1918	13.0 1-1991	11.0 28-1982
Winnibigoshish Dam	14.0 4-1897	12.0 5-1908	15.0 15-1957	12.0 19-1970	7.0 2-1954		0.3 28-1908	8.6 23-1933	12.0 12-1948	10.0 13-1968
Winona	15.0 21-1917	12.0 12-1965	16.0 30-1934	10.0 26-1950	5.0 3-2013			4.8 31-1991	9.5 23-1991	16.0 1-1985
Worthington	12.0 20-1988	10.5 19-1984	14.0 2-2007	10.0 7-2000	3.0 2-2013			4.0 24-1976	12.0 28-1983	13.0 28-1982
Wright	14.0 12-1972	13.8 21-2014	12.0 4-1985	11.7 19-2013	3.2 19-1971		0.1 21-1974	6.1 17-1990	19.2 1-1991	10.5 5-2013
Young America	14.0 23-1982	7.0 16-1967	15.0 4-1985	8.0 1-1985				5.5 13-1959	16.0 1-1991	13.0 28-1982
Zumbrota	14.0 23-1982	9.0 21-2011	15.0 30-1934	11.0 5-1947	8.4 2-2013		2.0 26-1942	4.0 22-1938	8.5 27-1952	16.0 12-2010

TABLE 10: Monthly Maximum Snowfall Extremes (inches)

STATION	JAN	FEB	MAR	APR	MAY	JUN	JUL	AUG	SEP	OCT	NOV	DEC	ANNUAL (JUL-JUN)
Ada	29.0 1989	18.3 1936	26.0 1966	15.2 2008	2.5 1954				2.0 1899	8.2 2012	21.0 1985	25.3 1922	103.9 1996-97
Aitkin	32.1 1982	26.0 2001	40.1 1965	35.8 1950	3.0 1950					10.0 1951	20.5 1993	34.0 2008	93.1 1949-50
Albert Lea	34.0 1929	32.5 1936	31.0 1951	18.0 1928	9.0 2013				2.0 1942	7.0 1925	19.0 1909	34.6 2000	66.2 1974-75
Alexandria	41.5 1975	17.3 1979	36.4 1951	14.6 1950	3.0 1954				3.2 1942	7.1 1951	27.7 1947	23.1 1968	77.5 1974-75
Argyle	33.3 1989	15.5 1987	20.2 2013	15.5 2014	3.0 1950				1.7 1972	14.0 2001	19.2 1977	23.8 2008	74.8 1978-79
Austin	28.0 1999	24.4 1945	27.3 1951	14.0 1945	10.2 2013				4.0 1942	5.0 2009	21.0 1985	36.5 2000	76.2 1961-62
Baudette	32.0 1950	19.3 1930	27.8 1966	25.0 1950	7.0 1954				3.0 1941	19.0 1916	30.2 1926	31.5 2009	124.0 1949-50
Bemidji	27.8 1969	22.6 1939	21.6 2013	38.5 2008	5.0 1950				2.0 1899	9.3 1932	22.7 1936	26.4 1951	82.7 1950-51
Big Falls	36.5 1969	30.0 1955	27.8 1965	29.3 2008	3.0 1970					14.6 2006	32.8 1955	33.6 1992	98.3 1955-56
Brainerd	33.0 1950	25.4 2001	35.5 1965	22.8 2008	5.0 1954				2.0 1942	7.0 1951	28.5 1983	29.1 1968	80.6 1996-97
Browns Valley	30.7 1975	24.2 2001	24.5 1997	26.0 2008	0.2 1945					4.2 2009	26.0 1985	21.7 2009	90.0 2000-01
Caledonia	28.0 1996	33.0 1915	44.8 1959	17.0 1973	6.0 1911					11.1 1917	25.0 1991	30.2 2010	91.6 1958-59
Cambridge	34.0 1975	26.7 1936	35.2 1965	28.0 1893	2.0 1954					5.5 1951	40.0 1991	36.0 1969	81.6 1950-51
Campbell	47.1 1937	27.3 1922	40.0 1951	11.0 1950	3.0 1954				1.0 1912	9.0 2001	22.0 1940	18.7 1914	69.5 1951-52
Canby	29.0 1979	29.0 2011	36.0 1951	31.5 2008	1.5 1954					7.5 1919	25.5 1975	33.5 2009	70.4 1996-97
Caribou	34.3 1969	16.0 1979	32.5 1944	20.0 1989	4.0 1967				1.0 1951	18.2 1947	24.5 1977	21.0 1995	89.0 1988-89
Cloquet	41.5 1969	29.2 2001	39.9 1965	41.0 2013	6.5 1954				0.2 1985	10.9 1951	33.7 1991	37.2 1950	113.4 1950-51
Collegeville	32.9 1951	26.5 1971	66.4 1965	29.00 2011	4.0 1935					8.4 2002	49.1 1940	34.0 1968	109.2 1964-65
Cook	50.5 1969	35.1 2014	28.0 1975	18.0 1966	8.0 1971				0.2 1974	8.0 2006	36.0 1991	27.0 2008	99.0 1974-75
Cotton	33.5 1975	20.9 1995	21.0 1975	12.0 1972	2.0 1970					6.0 1969	32.4 1991	21.0 1964	77.9 1988-89
Crookston	31.0 1969	20.5 2006	19.0 2013	17.9 1950	4.0 1954					8.0 2001	32.5 1947	31.1 2008	100.6 1996-97
Detroit Lakes	28.8 1975	43.0 1922	41.0 1944	32.5 2008	4.5 1938				6.0 1942	15.0 1942	29.0 1977	29.2 2008	89.1 1942-43
Duluth	46.8 1969	33.9 1939	48.2 1917	50.8 2013	8.1 1954				2.4 1991	14.0 1933	50.1 1991	44.3 1950	135.4 1995-96
Ely-Winton	38.0 1916	23.0 2001	29.0 1916	33.0 2008	4.0 1984					17.5 1915	30.3 1965	29.6 2008	88.1 2010-11
Fairmont	28.1 1982	35.0 1962	31.0 1899	24.7 1928	2.0 1924				2.0 1942	7.0 1905	30.5 1909	38.3 2009	81.8 1961-62
Faribault	26.5 1982	30.0 1936	29.5 1951	22.0 1983	4.0 2013				2.2 1945	4.1 1905	27.0 1991	36.5 2010	80.2 1935-36
Farmington	38.0 1982	29.5 1962	43.0 1951	24.5 1983	3.0 1954				3.0 1924	8.2 1959	32.6 1991	31.5 2008	90.0 1983-84
Fergus Falls	38.5 1975	30.8 1922	29.1 1975	15.6 1893	7.9 1935				2.4 1912	9.7 1919	26.0 1993	27.5 1968	105.3 1996-97
Fosston	26.7 1969	15.5 1979	26.5 1966	20.0 2008	9.0 1938				6.5 1912	12.0 1970	30.0 1977	22.3 2008	86.0 1965-66
Grand Marais	42.0 1935	49.2 1939	28.0 1917	24.0 2013	10.0 1954				1.7 1942	7.0 1933	28.5 1933	35.9 1937	100.3 1995-96
Grand Meadow	29.8 1999	29.0 1959	42.5 1951	22.0 1973	14.0 2013				5.0 1942	6.5 1917	19.0 1947	40.6 2010	80.0 1950-51
Grand Rapids	40.0 1975	26.0 1955	31.5 1951	32.7 2008	9.0 1954	0.7 1969				7.8 1972	28.7 1965	30.5 2013	100.4 1970-71
Gunflint Lake	29.0 1989	24.8 2001	27.0 1979	30.8 2013	5.0 1989				0.2 2003	15.0 1990	45.0 1991	26.9 1977	90.0 1991-92
Hallock	25.5 1956	19.5 1955	21.0 1956	15.0 1960	1.5 1954				1.5 1939	14.0 1916	31.0 1955	22.1 1949	101.3 1955-56
Harmony	26.0 1982	30.0 1962	40.5 1951	20.0 1973	8.5 2013				4.0 1942	5.0 1982	18.0 1985	36.2 2010	92.0 1950-51
Hibbing	43.7 1975	30.5 1971	26.5 1965	22.5 1970	7.0 1971				0.3 1985	6.0 1972	33.1 1991	30.4 1995	110.6 1970-71
Hutchinson	31.7 1917	22.6 2007	46.0 1965	18.0 2014	0.4 1971					4.3 1916	29.8 1991	27.0 1969	76.6 1964-65
Int'l Falls	43.0 1975	32.3 1992	31.5 1951	36.3 1925	13.4 1954	0.3 1969			2.0 1912	10.5 1917	29.7 1965	43.9 1992	125.6 2008-09

STATION	JAN	FEB	MAR	APR	MAY	JUN	JUL	AUG	SEP	OCT	NOV	DEC	ANNUAL (JUL-JUN)
Itasca	35.0 1982	28.0 1955	41.0 1966	43.5 2008	8.0 1954					14.5 1970	28.5 1977	26.0 1968	110.0 1965-66
Leech Lake Dam	41.0 1969	31.4 1903	37.5 1965	37.5 2008	11.0 1915				1.7 1942	16.8 1917	24.5 1965	34.0 1968	104.0 1965-66
Luverne	31.5 1979	37.5 1962	36.0 1970	11.0 1962	1.5 1907					8.0 1905	21.3 1991	31.0 1968	86.5 1969-70
Marshall	34.3 1982	32.3 1952	35.0 1951	24.6 2013	2.5 1944					6.0 1970	25.0 1983	32.2 1968	82.2 1968-69
Milan	46.0 1897	27.0 1922	33.5 1951	29.0 2013	8.0 1907				2.0 1942	8.2 2009	21.3 1896	25.3 2010	92.0 1996-97
Montevideo	45.0 1916	28.0 1967	44.0 1951	23.5 2013	4.3 1907				1.0 1942	6.0 1995	27.5 1911	32.5 2010	92.2 1983-84
Mora	32.1 1975	27.3 1953	42.9 1965	27.4 1950	4.0 1924				0.5 1942	8.0 1951	22.5 1930	35.1 1968	94.2 2013-14
Morris	33.7 1975	20.0 2001	46.5 1951	21.0 2008	4.0 1954				0.5 1965	9.5 1951	20.9 1977	34.6 1968	82.2 1950-51
MSP	46.4 1982	26.5 1962	40.0 1951	21.8 1983	3.0 1946				1.7 1942	8.2 1991	46.9 1991	33.6 2010	98.4 1983-84
New Ulm	26.7 1975	31.0 1936	40.0 1951	28.0 1928	5.0 1896				5.5 1942	5.9 1916	28.1 1991	37.2 1950	101.7 1950-51
Olivia	20.1 1982	22.2 2001	24.5 1985	17.0 2013						3.5 1976	20.3 1983	33.0 2010	71.5 1983-84
Park Rapids	32.4 1969	24.1 1979	41.4 1966	27.3 1893	10.0 1954				3.5 1942	-0.8 1906	24.0 1993	29.6 1933	105.4 1965-66
Pine River Dam	30.8 1975	27.1 1897	32.9 1965	37.0 1950	7.5 1907					9.0 1951	22.2 1998	26.2 1968	93.2 1970-71
Pipestone	28.7 2011	35.0 2011	26.7 1951	25.0 1911	0.9 1966				0.1 1995	9.7 1932	17.5 2001	25.0 1968	87.7 2010-11
Pokegama Dam	35.5 1975	26.5 1955	31.8 1951	33.0 1950	9.0 1954				2.8 1912	15.0 1917	26.5 1947	29.7 1936	104.1 1949-50
Red Wing	31.1 1982	52.5 1886	30.7 1951	14.0 1983	1.5 1954					7.0 1916	28.9 1991	31.4 1969	82.3 1983-84
Redwood Falls	34.0 1917	28.0 1936	28.6 1951	23.0 2013					3.2 1942	4.9 1976	22.0 1991	28.4 2010	66.8 1983-84
Rochester	30.2 1996	20.1 1959	35.1 1951	16.4 1983	14.5 2013				0.8 1961	7.9 2009	22.5 1985	41.3 2010	85.1 1996-97
Roseau	34.0 1989	16.0 1930	19.8 1944	18.0 1937	10.0 1938				1.4 1912	11.0 1916	23.0 1996	22.8 1995	77.1 1995-96
St. Cloud	32.6 1965	21.6 1971	51.7 1965	24.4 1951	3.3 1938				1.8 1942	6.4 2002	26.9 1940	25.5 1927	87.9 1964-65
St. James	24.5 1982	30.0 1962	38.0 1951	16.3 2013	1.7 2013				1.0 1908	4.0 1976	26.2 1983	38.3 2000	82.7 2000-01
St. Peter	26.8 1988	25.0 1962	37.4 1951	18.5 1928	2.7 1933				1.0 1942	4.5 1905	21.5 1940	36.5 2010	89.2 1951-52
Sandy Lake Dam	43.6 1975	31.6 1897	50.1 1965	28.7 1950	6.0 1954				1.0 1908	16.5 1951	56.5 1995	26.0 1969	107.2 1964-65
Spring Grove	28.0 1996	27.4 1959	36.7 1959	18.7 1973	4.2 1945					2.0 1992	21.3 1991	30.0 1935	86.1 1996-97
Stillwater	42.5 1982	24.4 1962	44.1 1951	14.5 2013	1.6 2013					3.8 1905	34.4 1991	29.6 1950	85.7 1950-51
Tower	46.0 1975	29.3 2014	28.0 1976	41.8 2008	17.0 1954				2.0 1899	13.0 1995	39.7 1965	29.8 1968	121.8 1949-50
Two Harbors	44.4 1967	30.3 2001	35.0 1965	29.3 1950	8.0 1924					10.0 1933	51.5 1991	48.9 1996	127.4 1996-97
Wadena	36.0 1997	25.2 1922	36.3 1966	24.0 1950	5.7 1954				4.6 1942	11.9 1951	25.5 1993	27.8 2008	92.7 1965-66
Warroad	34.0 1969	17.0 1911	22.0 1966	19.0 1989	6.5 1902					15.0 1917	29.0 1977	26.5 1933	81.5 1965-66
Waseca	33.8 1999	21.8 1962	41.0 1951	20.0 1983	9.9 2013					4.0 1999	36.0 1991	39.1 2010	105.5 1983-84
Waskish	36.0 1969	14.0 2014	23.0 1976	18.5 1950	0.5 2001				0.2 1974	9.0 2009	24.8 1977	25.3 2008	73.5 1995-96
Wheaton	38.3 1997	22.7 2013	42.3 1951	27.3 2008	4.0 1954					7.5 1951	22.5 1977	31.1 2010	78.5 2010-11
Willmar	51.5 1982	29.5 1967	52.0 1965	20.0 1893	3.0 1911				6.0 1942	4.5 1976	32.2 2001	34.5 1927	103.7 1950-51
Windom	20.5 1975	25.7 2001	34.0 1940	18.6 2013	12.0 1933				1.0 1942	8.5 1976	27.0 1991	22.3 2000	77.9 2000-01
Winnebago	29.0 1975	29.4 1936	39.0 1951	17.3 2013	6.0 1933				2.5 1942	7.0 1918	24.5 1983	33.5 2009	88.3 1983-84
Winnibigoshish Dam	34.5 1969	28.5 1955	28.3 1951	25.1 1950	12.5 1954				0.3 1908	12.6 1933	22.5 1965	27.5 1968	100.9 1953-54
Winona	36.5 1971	36.0 1936	36.5 1951	15.6 1973	3.0 1911					6.2 1991	17.3 1991	33.0 1985	90.2 1961-62
Worthington	22.0 1979	40.0 1962	27.1 1951	16.6 1988	4.0 2013				0.2 1985	7.0 1991	19.5 1983	33.0 2009	78.7 1983-84
Wright	32.2 1975	27.7 2001	39.0 1965	35.4 2013	3.2 1971				0.1 1974	10.9 1995	36.6 1991	29.5 2013	95.3 1970-71
Young America	40.0 1982	22.6 1967	48.1 1965	10.0 1962						5.5 1959	40.5 1991	29.0 1969	88.3 1964-65
Zumbrota	26.8 1929	30.7 1936	31.2 1951	18.0 1928	9.9 2013				2.0 1942	6.0 1925	19.5 1991	40.0 2010	70.1 1935-36

TABLE 11: Average Monthly Snowfall (inches) (1981–2010)

STATION	JUL	AUG	SEP	OCT	NOV	DEC	JAN	FEB	MAR	APR	MAY	JUN	ANNUAL
Aitkin	0.0	0.0	0.0	0.6	6.7	8.2	8.8	8.0	9.7	2.4	0.1	0.0	44.5
Albert Lea	0.0	0.0	0.0	0.4	4.1	10.0	7.3	6.6	6.3	3.1	0.0	0.0	37.8
Alexandria	0.0	0.0	0.1	0.8	7.4	6.8	9.5	6.4	8.5	2.8	0.0	0.0	42.3
Argyle	0.0	0.0	0.0	1.3	6.1	9.2	8.8	6.1	5.5	1.7	0.1	0.0	38.8
Brainerd	0.0	0.0	0.0	0.4	8.4	9.7	9.6	6.9	8.4	3.4	0.0	0.0	46.8
Browns Valley	0.0	0.0	0.0	0.5	7.1	7.5	8.1	8.3	7.7	3.7	0.0	0.0	42.9
Caledonia	0.0	0.0	0.0	0.3	3.9	12.4	10.5	9.8	7.8	2.7	0.0	0.0	47.4
Canby	0.0	0.0	0.0	0.6	7.7	8.5	7.9	7.0	9.1	4.4	0.0	0.0	45.2
Caribou	0.0	0.0	0.0	2.4	8.2	8.3	10.9	4.9	3.8	2.8	0.1	0.0	41.4
Cloquet	0.0	0.0	0.0	1.4	11.2	14.2	13.9	11.2	10.6	3.6	0.1	0.0	66.2
Collegeville	0.0	0.0	0.0	0.8	9.1	10.4	11.1	8.5	9.9	3.9	0.0	0.0	53.7
Crookston	0.0	0.0	0.0	0.7	5.0	8.3	7.5	6.4	5.0	1.9	0.1	0.0	34.9
Detroit Lakes	0.0	0.0	0.0	1.1	6.1	10.2	9.2	7.0	7.4	3.8	0.0	0.0	44.8
Duluth	0.0	0.0	0.1	2.3	13.7	17.7	19.4	12.4	13.2	6.9	0.4	0.0	86.1
Ely-Winton	0.0	0.0	0.0	2.1	9.8	13.2	13.8	9.3	8.4	7.7	0.3	0.0	64.6
Fairmont	0.0	0.0	0.0	0.3	5.3	13.5	6.3	6.8	9.6	3.4	0.0	0.0	45.2
Faribault	0.0	0.0	0.0	0.0	6.1	10.2	9.2	7.5	7.3	2.3	0.0	0.0	42.6
Farmington	0.0	0.0	0.0	0.2	6.9	8.0	9.4	7.0	7.5	3.0	0.0	0.0	42.0
Fergus Falls	0.0	0.0	0.0	0.8	7.6	7.4	11.9	7.2	8.1	2.0	0.0	0.0	45.0
Fosston	0.0	0.0	0.0	0.4	6.7	7.1	9.1	5.8	7.2	2.5	0.2	0.0	39.0
Grand Marais	0.0	0.0	0.0	0.1	3.7	10.6	13.6	6.1	5.9	2.1	0.0	0.0	42.1
Grand Meadow	0.0	0.0	0.0	0.4	3.9	14.7	10.2	10.7	7.1	1.6	0.0	0.0	48.6
Grand Rapids	0.0	0.0	0.0	1.5	9.9	12.5	12.5	7.2	7.9	3.8	0.2	0.0	55.5
Gunflint Lake	0.0	0.0	0.0	0.5	7.5	10.6	8.3	10.7	8.7	2.7	0.0	0.0	49.0
Harmony	0.0	0.0	0.0	0.7	4.3	11.3	10.3	9.4	7.0	1.9	0.0	0.0	44.9
Hibbing	0.0	0.0	0.0	1.2	13.2	12.3	15.0	7.1	7.8	3.7	0.0	0.0	60.3
Hutchinson	0.0	0.0	0.0	0.1	7.7	9.1	7.5	7.1	9.2	2.8	0.0	0.0	43.5
Int'l Falls	0.0	0.0	0.1	2.2	13.7	15.0	15.0	10.8	7.6	6.4	0.2	0.0	71.0
Itasca	0.0	0.0	0.0	1.1	8.3	9.5	10.4	6.6	7.9	4.0	0.1	0.0	47.9
Leech Lake Dam	0.0	0.0	0.0	0.7	7.0	10.8	10.0	6.5	6.5	4.1	0.0	0.0	45.6

STATION	JUL	AUG	SEP	OCT	NOV	DEC	JAN	FEB	MAR	APR	MAY	JUN	ANNUAL
Luverne	0.0	0.0	0.0	1.0	8.7	8.0	10.2	6.2	7.2	3.6	0.0	0.0	44.9
Marshall	0.0	0.0	0.0	0.5	7.7	11.3	11.7	6.4	7.4	1.8	0.0	0.0	46.8
Milan	0.0	0.0	0.0	0.9	6.4	7.2	9.3	9.3	9.2	3.3	0.0	0.0	45.6
Montevideo	0.0	0.0	0.0	0.5	6.2	9.9	9.6	7.8	8.1	1.2	0.0	0.0	43.3
Mora	0.0	0.0	0.0	0.5	7.5	10.6	9.2	8.4	7.3	1.9	0.0	0.0	45.4
Morris	0.0	0.0	0.0	1.0	6.5	8.9	9.6	8.4	8.5	3.3	0.0	0.0	46.2
MSP	0.0	0.0	0.0	0.6	9.3	11.9	12.2	7.7	10.3	2.4	0.0	0.0	54.4
New Ulm	0.0	0.0	0.0	0.4	6.4	9.1	7.3	6.7	7.5	1.7	0.0	0.0	39.1
Olivia	0.0	0.0	0.0	0.1	4.8	8.4	6.6	7.0	6.9	1.7	0.0	0.0	35.5
Park Rapids	0.0	0.0	0.0	1.1	8.5	7.4	11.0	5.4	8.1	2.6	0.1	0.0	44.2
Pine River Dam	0.0	0.0	0.0	0.2	2.6	8.0	7.0	5.7	6.5	2.8	0.0	0.0	32.8
Pokegama Dam	0.0	0.0	0.0	1.3	6.2	8.7	8.7	6.5	6.3	3.3	0.0	0.0	41.0
Red Wing	0.0	0.0	0.0	0.1	2.3	7.3	7.9	7.1	6.3	1.7	0.0	0.0	32.7
Redwood Falls	0.0	0.0	0.0	0.1	4.5	8.1	6.1	7.7	8.8	0.5	0.0	0.0	35.8
Rochester	0.0	0.0	0.0	0.8	6.1	12.5	12.0	8.5	8.7	3.3	0.0	0.0	51.9
St. Cloud	0.0	0.0	0.0	0.7	8.9	10.5	7.9	7.1	8.1	2.9	0.0	0.0	46.1
St. James	0.0	0.0	0.0	0.5	6.0	11.4	10.4	7.2	9.2	2.8	0.0	0.0	47.5
Tower	0.0	0.0	0.0	2.6	12.3	12.3	12.7	8.5	7.2	5.4	0.3	0.0	61.3
Two Harbors	0.0	0.0	0.0	0.1	6.1	14.8	13.7	9.5	8.9	3.1	0.0	0.0	56.2
Wadena	0.0	0.0	0.0	1.1	8.0	7.9	10.3	7.2	9.0	3.5	0.0	0.0	47.0
Waseca	0.0	0.0	0.0	0.6	7.8	13.0	9.5	9.0	10.2	2.7	0.0	0.0	52.8
Wheaton	0.0	0.0	0.0	0.7	4.6	8.2	9.0	8.1	8.3	4.4	0.0	0.0	43.3
Willmar	0.0	0.0	0.0	0.4	8.8	9.7	8.6	7.7	8.1	3.0	0.0	0.0	46.3
Windom	0.0	0.0	0.0	0.7	6.0	8.9	8.4	7.0	8.6	3.4	0.0	0.0	43.0
Winnehago	0.0	0.0	0.0	0.2	6.7	12.2	8.9	6.3	7.4	3.5	0.0	0.0	45.2
Winnibigoshish Dam	0.0	0.0	0.0	0.5	6.2	7.4	7.6	4.1	4.8	0.7	0.0	0.0	31.3
Winona	0.0	0.0	0.0	0.1	2.6	11.0	8.7	7.6	5.7	1.0	0.0	0.0	36.7
Worthington	0.0	0.0	0.0	1.0	5.7	7.4	7.0	6.1	6.7	3.9	0.0	0.0	37.8
Wright	0.0	0.0	0.0	1.7	11.1	10.3	11.4	9.4	8.9	4.2	0.2	0.0	57.2
Young America	0.0	0.0	0.0	0.3	10.8	9.0	9.7	6.2	7.9	2.7	0.0	0.0	46.6
Zumbrota	0.0	0.0	0.0	0.7	5.4	11.2	10.3	7.2	8.1	3.4	0.0	0.0	46.3

Snowfall density varies considerably from month to month during the winter season. Typically, the warmer temperatures of November and March produce snowfalls of greater density, with snow-to-water ratios of 5:1 to 10:1, or 5 to 10 inches of snow melting down to 1 inch of water. Conversely, January's cold temperatures can yield snowfalls of very slight density, with snow-to-water ratios of 20:1 to 30:1. This lighter snow moves easily around the landscape, piling up in ditches and drifting around buildings and across roads. Snowfalls in Minnesota have been measured in air temperatures as warm as 39°F and as cold as -40°F. The maximum rate of snowfall under the most intense winter storms—such as the Armistice Day Blizzard of 1940 and the Halloween Blizzard of 1991—has been about 3 inches per hour, but this rate rarely lasts for more than a few hours. Maximum one-day snowfalls (see Table 9) of greater than 20 inches are therefore highly unusual in the climate record: only a small fraction of the climate observers in the state have reported 24-hour snowfalls of 20 or more inches. The all-time state record for 24 hours is 36 inches on January 7, 1994, at Wolf Ridge Environmental Learning Center near Finland in Lake County; the snowfall rate there is affected by both elevation and proximity to Lake Superior.

· ·

Frazil ice

Needle-like ice crystals that form in the supercooled water of river or stream currents that move too fast to create surface ice are described as frazil ice, from the French fraisil, *"cinders." These crystals form under pool surfaces or along channel edges and often build up into masses extending to the stream bottom, sometimes growing so large that they create dams and cause flooding.*

· ·

Winter temperatures depend on a host of factors: the presence or absence of snow cover; cloud cover; wind; air mass type, whether polar or subtropical in origin; and the atmosphere's water vapor content. A number of locations have seen lows of -50°F or colder in the months of December, January, and February. A low of -50°F was recorded at Pokegama Dam in Itasca County on March 2, 1897, the latest seasonal date for such a reading. Even as far south as Zumbrota in Goodhue County the temperature has dropped as low as -45°F. Minnesota's position in the middle of the North American continent allows for significant migration of polar air into the state. These "arctic outbreaks" can even bring cold air masses across the pole from northern Siberia. Their signature is clear skies, calm winds, and intense high pressure, sometimes measuring more than 31 inches on the barometer. Such episodes occurred in 1899, 1912, 1936, 1962, 1970, 1983, 1996, and 2015. An arctic outbreak in January 1912 produced 186 consecutive hours below zero in the Twin Cities area, second only to a spell of 226 hours in January 1864.

With few exceptions, Minnesota's heating season begins in October and ends in May. To evaluate energy consumption for home and commercial heating purposes, climatologists and energy companies rely on a temperature-based measurement called Heating Degree Days (HDD). This unit is calculated by taking the difference between a base value of 65°F, which is considered a tolerable indoor temperature, and the daily average temperature (maximum plus minimum, divided by two). For example, in the Twin Cities on December 31 the average maximum temperature is 24°F and the average minimum temperature is 9°F, making for a daily average of 17°F (rounded up). The difference between the base of 65°F and this value yields an average HDD of 48. This difference steadily increases until mid- to late January, when it starts to diminish. Table 13 shows the average accumulated HDD by month for several locations. Geography plays a big role: all locations with greater than 10,000 annual HDD can be found in the far north and inland from Lake Superior, while all those with less than 8,000 HDD are in the south. Lake Superior's moderating influence can be seen in the lower HDD values for lakeside communities like Duluth, Grand Marais, and Two Harbors. As evidence of the Twin Cities' urban heat island effect, the Minneapolis–St. Paul airport (MSP) climate station shows an annual HDD value (7,580) roughly equal to that of Luverne in southwestern Minnesota (7,566).

The variability of winter temperatures is large; as a consequence, there is also large range in HDD across the years. The Twin Cities record since 1871 shows a lowest-ever HDD value of 5,856 in 2011–12 and a highest-ever value of 9,814 in 1874–75. Given the mean annual value

Q&A

What is a snow pillow?

a. a frozen cocktail served in Anchorage, Alaska

b. a snowdrift in the shape of a pillow

c. a pressure transducer buried in the snow to measure its weight or water equivalence

(answer on page 329)

MAP 3: Average Total Seasonal Snowfall (July 1–June 30) across Minnesota, 1981–2010

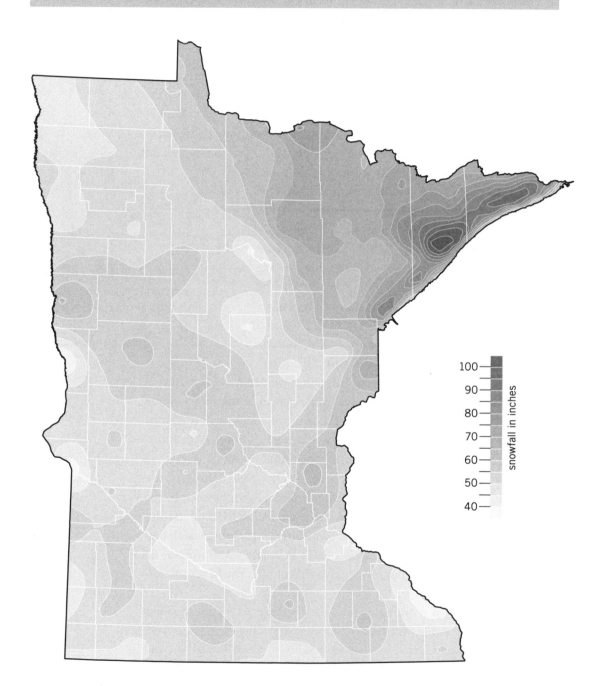

MAP 4: Climate Stations with Four or More Statewide Daily Snowfall Records

TABLE 12: One-Day Minimum Temperature Extremes (°F)

STATION	JAN	FEB	MAR	APR	MAY	JUN	JUL	AUG	SEP	OCT	NOV	DEC
Ada	-48 31-2004	-53 15-1936	-39 10-1948	-12 2-1899	12 12-1918	26 1-1917	37 3-1967	31 31-1895	17 30-1895	-6 26-1919	-35 30-1896	-42 13-1901
Aitkin	-47 15-1972	-43 2-1996	-37 1-1962	-1 4-1995	18 3-1967	27 1-1964	37 4-1972	33 31-1970	19 26-1965	9 27-1976	-22 30-1964	-42 19-1983
Albert Lea	-41 7-1887	-33 2-1996	-29 1-1962	3 6-1979	21 9-1966	34 1-1979	42 30-1903	35 31-1886	22 26-1942	-6 25-1887	-18 27-1977	-29 19-1983
Alexandria	-38 29-1951	-34 1-1951	-34 1-1962	-2 3-1954	18 3-1967	33 1-1946	42 4-1967	38 20-1950	24 22-1974	9 31-1951	-18 30-1964	-32 19-1955
Argyle	-43 20-1996	-48 1-1996	-38 10-1948	-18 6-1979	18 12-1946	26 1-1964	36 1-1939	30 31-1935	19 22-1974	2 26-1936	-34 28-1985	-41 28-1933
Austin	-42 15-1963	-34 2-1996	-34 1-1962	6 4-1995	22 3-1967	35 21-1992	41 1-1948	34 25-1958	23 23-1976	10 29-1952	-25 26-1977	-33 27-1950
Baudette	-49 6-1909	-47 8-1933	-45 1-1962	-19 2-1975	12 3-1967	20 27-1970	34 17-1924	28 31-1935	17 30-1930	-8 26-1936	-29 29-1985	-45 19-1955
Bemidji	-50 30-1950	-48 9-1899	-44 1-1962	-43 3-1951	11 1-1966	24 1-1964	37 1-1939	30 28-1934	18 28-1942	-1 26-1936	-37 30-1896	-45 31-1946
Big Falls	-52 23-1936	-48 1-1996	-42 1-1962	-19 3-1954	12 3-1967	21 1-1964	32 2-1972	25 28-1934	15 26-1965	-12 26-1936	-33 29-1985	-51 28-1933
Brainerd	-48 15-1972	-54 2-1996	-35 11-1948	-12 1-1975	16 3-1967	30 3-1928	36 23-1978	29 15-1976	18 22-1974	4 27-1976	-24 30-1964	-43 19-1983
Browns Valley	-40 9-1977	-41 9-1994	-19 12-2009	-3 3-1975	18 3-2005	35 5-1985	44 1-1984	34 20-2004	19 22-1974	9 17-1976	-19 26-1996	-32 19-1983
Caledonia	-37 15-1963	-35 9-1899	-32 1-1962	2 6-1982	19 9-1966	31 1-1897	42 6-1965	35 14-1964	20 30-1899	13 26-1962	-16 26-1977	-29 13-1903
Cambridge	-41 23-1935	-41 2-1996	-33 1-1962	0 1-1975	18 3-1967	32 2-1946	40 16-1896	33 27-1896	21 25-1947	8 27-1976	-22 30-1964	-41 19-1983
Campbell	-41 16-1977	-41 9-1994	-35 15-1897	-5 1-1975	12 3-1926	25 2-1910	37 16-1911	30 13-1964	17 29-1939	-10 26-1919	-25 30-1964	-38 25-1914
Canby	-33 2-1936	-32 16-1936	-28 1-1962	0 7-1936	10 5-1951	33 2-1946	38 4-1967	35 20-1950	22 28-1942	12 31-1951	-16 30-1964	-27 27-1950
Cloquet	-45 7-1912	-41 15-1939	-35 1-1962	-7 3-1954	8 4-1911	24 4-1945	30 25-1915	26 27-1915	19 30-1947	0 23-1936	-24 20-1921	-37 28-1917
Collegeville	-37 12-1912	-37 2-1996	-30 1-1962	2 4-1975	19 3-1967	33 2-1910	41 1-1969	39 28-1965	23 26-1965	13 21-1913	-19 30-1964	-35 19-1983
Cook	-50 14-1965	-45 18-1966	-46 1-1962	-12 6-1982	5 1-1966	17 1-1964	30 4-1972	25 28-1965	13 26-1965	-3 27-1976	-34 30-1964	-43 19-1983
Cotton	-50 14-1965	-46 2-1996	-40 2-1989	-14 2-1975	8 1-1966	24 20-1969	30 6-1969	23 31-1970	16 26-1965	-2 27-1976	-30 29-1985	-44 19-1983
Crookston	-45 31-1996	-51 15-1936	-39 10-1948	-10 6-1979	6 3-1967	27 1-1964	38 12-1926	31 13-1964	11 26-1893	-2 26-1919	-30 27-1985	-38 26-1990
Detroit Lakes	-51 12-1912	-53 9-1899	-40 10-1948	-12 2-1899	15 5-1968	28 3-1969	33 1-1969	31 31-1910	15 29-1899	-10 26-1919	-32 30-1905	-48 14-1901
Duluth	-41 2-1885	-39 2-1996	-29 2-1989	-5 4-1975	16 3-1907	27 10-1972	35 1-1988	32 27-1986	22 28-1942	8 27-1976	-29 29-1875	-35 28-1917
Ely-Winton	-45 17-2005	-45 12-1967	-42 1-1962	-11 6-1982	15 6-1989	28 1-1964	38 16-2007	34 28-1982	24 27-1991	7 30-1988	-24 29-1985	-40 19-1983
Fairmont	-35 12-1912	-35 2-1917	-30 11-1948	2 1-1924	22 11-1946	29 9-1915	40 9-1895	34 24-1934	18 29-1899	0 29-1925	-19 27-1891	-26 29-1917
Faribault	-40 9-1977	-36 3-1996	-30 11-1948	0 6-1982	16 12-2006	33 4-1945	36 6-2001	33 24-2005	23 29-1945	4 30-1925	-15 26-1977	-36 19-1983
Farmington	-40 23-1935	-40 9-1899	-30 11-1948	-1 1-1924	20 3-1967	26 1-1897	40 6-1984	36 31-1892	20 30-1899	2 30-1925	-20 24-1898	-35 19-1983
Fergus Falls	-38 24-1904	-42 16-1936	-31 14-1897	-2 6-1976	19 2-1967	29 6-1897	37 1-1969	34 30-1931	20 28-1930	2 26-1919	-23 30-1896	-35 14-1901
Fosston	-50 20-1996	-53 1-1996	-35 7-1906	-9 6-1979	6 3-1967	27 1-1917	34 5-1983	28 27-1982	19 22-1974	-4 26-1919	-24 28-1985	-42 20-1916
Grand Marais	-34 23-1935	-34 9-1933	-24 2-1943	-8 1-1975	17 9-1947	25 12-1947	28 4-1950	33 1-1947	23 28-1942	6 26-1936	-14 30-1976	-27 25-1933
Grand Meadow	-38 12-1912	-36 8-1899	-31 1-1962	1 4-1995	20 4-1907	32 8-1901	37 9-1895	34 30-1915	22 30-1939	-5 30-1925	-22 30-1893	-30 13-1901
Grand Rapids	-51 26-1927	-45 21-1939	-39 1-1962	-10 2-1920	11 14-1918	24 3-1928	33 22-1925	27 28-1934	15 26-1939	-10 23-1917	-25 28-1940	-45 31-1946
Gunflint Lake	-44 15-2009	-43 12-1967	-35 2-1989	-17 6-1982	14 10-1966	27 1-1964	35 1-1993	32 29-1976	22 25-1976	5 26-1976	-26 29-1985	-40 19-1983
Hallock	-48 13-1916	-51 11-1914	-42 10-1948	-13 2-1920	11 12-1946	24 3-1936	34 6-1908	28 31-1935	13 29-1899	-7 28-1919	-32 15-1911	-43 31-1967
Hibbing	-50 20-1996	-44 2-1996	-37 2-1989	-4 4-1975	14 3-1967	25 1-1964	32 5-2001	29 28-1986	20 26-1965	0 27-1976	-27 29-1985	-38 19-1983
Hutchinson	-39 18-1994	-36 2-1996	-32 1-1962	3 3-1975	11 3-1967	36 27-2007	44 4-1967	38 14-1964	22 22-1974	12 27-1976	-18 26-1977	-34 19-1983
Int'l Falls	-55 6-1909	-48 8-1909	-38 1-1962	-14 3-1954	11 3-1967	23 1-1964	32 11-1911	27 31-1906	19 15-2011	2 30-1988	-32 28-1985	-41 19-1955

STATION	JAN	FEB	MAR	APR	MAY	JUN	JUL	AUG	SEP	OCT	NOV	DEC
Itasca	-51 22-1922	-52 2-1996	-44 10-1948	-17 3-1954	11 1-1966	24 1-1964	32 12-1926	26 26-1915	16 26-1965	-14 26-1919	-30 30-1964	-47 29-1917
Leech Lake Dam	-52 30-1899	-59 9-1899	-48 15-1897	-13 3-1896	12 3-1907	24 1-1897	35 4-1932	27 21-2004	16 21-1934	1 23-1936	-37 30-1896	-51 31-1898
Luverne	-37 19-1970	-36 28-1962	-23 5-1960	6 6-1982	9 3-1967	31 1-1905	39 4-1967	34 13-1964	20 19-1991	9 19-1972	-16 26-1993	-30 13-1903
Marshall	-30 18-1970	-30 2-1996	-24 1-1962	3 1-1975	21 3-1967	32 8-1995	42 4-1967	38 28-1965	24 22-1974	9 26-1981	-15 26-1977	-27 18-1983
Milan	-38 19-1970	-42 16-1936	-32 14-1897	-3 3-1975	16 13-1997	29 6-1897	36 9-1895	30 30-1931	14 29-1899	-1 28-1919	-21 24-1893	-35 19-1983
Montevideo	-38 12-1912	-39 16-1936	-25 1-1962	2 3-1975	16 27-1909	31 29-1900	36 9-1895	34 30-1931	14 29-1899	4 28-1919	-23 28-1891	-33 29-1917
Mora	-48 12-1912	-46 4-1982	-35 1-1962	-4 1-1975	15 4-1907	23 8-1988	33 6-1984	27 23-1987	13 29-1984	4 27-1976	-26 30-1905	-52 18-1983
Morris	-40 21-1888	-41 16-1936	-30 11-1948	-2 7-1936	18 3-1907	27 1-1929	37 12-1927	31 30-1931	20 29-1899	1 29-1895	-27 28-1891	-34 29-1887
MSP	-34 19-1970	-33 9-1899	-32 1-1962	2 13-1962	18 3-1967	34 3-1945	43 4-1972	39 19-1967	26 22-1974	10 30-1925	-17 30-1964	-29 19-1983
New Ulm	-37 18-1984	-37 19-1929	-18 2-1962	-3 1-1924	19 3-1967	31 3-1945	39 13-1926	34 29-1946	21 29-1945	1 30-1925	-17 26-1977	-36 19-1983
Olivia	-37 9-1977	-51 9-1899	-38 4-1917	5 3-1896	24 3-2013	36 5-1894	38 9-1895	36 21-2004	23 30-1895	5 29-1895	-17 26-1966	-30 19-1983
Park Rapids	-47 12-1912	-51 4-1907	-41 11-1948	-8 2-1899	14 1-1966	27 6-1897	35 4-1972	31 27-1982	17 22-1974	-3 26-1919	-32 29-1896	-46 19-1983
Pine River Dam	-53 12-1912	-51 4-1907	-26 1-1962	-8 3-1954	4 1-1909	24 30-1925	32 23-1925	29 21-1925	17 21-1934	0 30-1925	-26 30-1905	-47 25-1933
Pipestone	-40 12-1912	-38 2-1917	-50 2-1897	-2 3-1975	15 3-1967	30 7-1901	32 30-1971	25 31-1987	15 26-1984	-3 29-1925	-21 29-1985	-34 13-1917
Pokegama Dam	-57 24-1904	-59 16-1903	-23 4-2014	-17 2-1899	14 4-1907	20 5-1891	33 31-1903	24 24-1896	12 25-1894	-1 23-1917	-45 30-1896	-57 31-1898
Red Wing	-39 12-1912	-36 1-1996	-25 11-1948	7 5-1995	25 1-1909	38 12-1903	46 4-1972	41 22-2004	23 22-1913	16 21-1913	-10 8-1991	-25 26-2000
Redwood Falls	-34 19-1970	-33 16-1936	-31 1-1962	4 7-1936	21 3-1909	34 4-1945	43 30-1971	36 20-1950	25 26-1939	12 28-1967	-14 28-1952	-29 18-1983
Rochester	-40 30-1951	-35 2-1996	-44 1-1916	5 6-1982	21 3-1967	31 9-1937	42 22-1947	32 30-1915	22 28-1942	10 23-1936	-20 26-1977	-33 19-1983
Roseau	-49 23-1936	-52 11-1914	-32 1-1962	-18 1-1975	11 10-1981	21 1-1917	31 19-1912	23 26-1915	12 22-1913	-16 26-1936	-36 30-1896	-45 30-1910
St. Cloud	-50 21-1888	-42 9-1896	-26 1-1962	13 2-1954	18 3-1907	32 1-1993	40 7-1972	23 26-1915	18 22-1974	2 25-1896	-23 30-1905	-42 25-1884
St. James	-30 21-1970	-30 2-1996	-30 11-1948	6 5-1995	19 1-1961	36 11-1995	40 1-1995	33 21-2004	25 22-1974	11 27-2007	-13 26-1977	-27 19-1983
St. Peter	-40 12-1912	-36 5-1907	-40 1-1962	-5 1-1924	14 5-1989	32 3-1945	34 12-1910	36 21-2004	21 30-1939	4 30-1925	-22 30-1896	-32 19-1983
Sandy Lake Dam	-52 31-1899	-49 21-1939	-37 1-1962	-9 2-1899	13 27-1895	27 23-1917	35 29-1925	34 30-1931	17 21-1934	3 27-1976	-32 29-1896	-45 31-1946
Stillwater	-40 9-1977	-36 2-1996	-42 2-1989	4 2-1975	20 3-1967	34 3-1985	45 29-1971	27 28-1934	26 22-1974	15 19-1972	-16 30-1964	-39 19-1983
Tower	-57 20-1996	-60 2-1996	-28 1-1962	-22 6-1982	10 1-1966	18 1-1897	24 7-1997	21 28-1986	14 25-1976	-7 27-1976	-39 29-1896	-52 19-1983
Two Harbors	-35 7-1912	-36 9-1899	-33 10-1948	-15 1-1975	9 2-1909	29 9-1913	27 30-1964	37 28-1965	22 19-1976	0 28-1976	-18 30-1896	-29 30-1898
Wadena	-42 23-1935	-43 5-1907	-33 1-1962	-7 3-1975	17 3-1907	30 7-1935	37 27-1904	32 28-1934	17 30-1974	2 31-1951	-35 30-1905	-42 20-1983
Warroad	-48 24-1904	-48 19-1966	-40 10-1948	-18 16-2014	14 1-1958	20 6-1952	32 31-1903	30 22-1964	16 21-2012	2 29-1988	-30 29-1958	-40 31-1967
Waseca	-37 5-1924	-35 2-1996	-27 11-1948	-3 1-1924	19 3-1967	31 6-1916	39 3-1917	34 24-1934	20 30-1939	-1 29-1925	-21 26-1977	-35 24-1983
Waskish	-49 20-1996	-48 1-1996	-36 1-2014	-11 6-1979	15 2-1967	25 1-1964	35 2-1972	32 21-2004	20 26-1965	5 12-2006	-27 28-1985	-36 26-1996
Wheaton	-33 19-1970	-33 2-1996	-22 1-1962	2 1-1975	21 1-1961	34 5-1951	39 3-1972	37 28-1965	22 26-1965	10 19-1972	-17 24-1951	-30 19-1983
Willmar	-41 9-1977	-37 9-1899	-31 11-1948	-5 1-1926	19 3-1967	31 6-1901	41 14-1924	33 30-1931	19 22-1974	-1 30-1925	-20 30-1964	-34 19-1983
Windom	-36 7-1988	-32 2-1917	-25 1-1962	6 3-1975	20 3-1967	30 9-1915	41 3-1917	30 20-1908	18 22-1913	6 21-1913	-17 28-1952	-30 26-1914
Winnebago	-35 12-1912	-32 2-1996	-25 1-1962	0 1-1924	20 3-1967	32 3-1945	44 13-1926	36 11-2004	22 28-1942	0 29-1925	-13 26-1977	-33 23-1985
Winnibigoshish Dam	-46 12-1912	-49 8-1933	-40 10-1948	-11 3-1954	16 1-1966	27 1-1964	38 4-1967	31 26-1915	19 26-1965	4 26-1936	-25 30-1964	-45 31-1898
Winona	-35 12-1912	-33 1-1918	-28 1-1962	-3 1-1924	21 1-1956	35 9-1937	43 1-1924	33 30-1915	25 28-1942	7 30-1925	-11 26-1977	-31 25-2000
Worthington	-31 2-2010	-30 2-1996	-16 3-2014	5 3-1975	23 3-2005	37 3-1985	36 4-2008	36 21-2004	24 22-1974	10 19-1972	-12 28-2014	-28 19-1983
Wright	-52 15-1972	-45 1-1996	-43 1-1962	-12 1-1970	15 3-1967	26 1-1964	33 5-1967	28 31-1970	16 26-1965	1 27-1976	-28 30-1964	-47 19-1983
Zumbrota	-45 30-1951	-37 3-1996	-36 1-1962	-1 6-1982	19 4-1907	28 4-1945	38 22-1947	33 30-1931	20 30-1939	3 30-1925	-20 30-1947	-40 19-1983

TABLE 13: Average Annual Heating Degree Days in Minnesota (1981–2010)

STATION	JUL	AUG	SEP	OCT	NOV	DEC	JAN	FEB	MAR	APR	MAY	JUN	ANNUAL
Ada	23	40	241	648	1,124	1,630	1,804	1,491	1,218	657	307	82	9,265
Aitkin	37	72	275	644	1,071	1,544	1,708	1,408	1,186	697	368	119	9,128
Albert Lea	10	24	175	537	954	1,441	1,576	1,289	1,050	581	248	49	7,934
Alexandria	16	35	221	613	1,068	1,558	1,697	1,385	1,152	644	294	79	8,761
Argyle	36	62	274	678	1,164	1,674	1,860	1,533	1,271	708	355	114	9,729
Austin	15	33	197	542	977	1,461	1,599	1,317	1,065	584	262	52	8,104
Baudette	29	54	282	684	1,148	1,680	1,861	1,508	1,266	707	354	98	9,671
Bemidji	47	91	312	717	1,167	1,658	1,833	1,505	1,266	777	378	127	9,878
Big Falls	39	70	270	643	1,117	1,630	1,770	1,414	1,163	669	332	119	9,236
Brainerd	29	56	273	653	1,088	1,584	1,756	1,441	1,192	698	328	99	9,197
Browns Valley	16	37	215	597	1,037	1,522	1,651	1,341	1,119	633	279	72	8,518
Caledonia	14	31	193	540	948	1,415	1,548	1,259	1,035	590	277	64	7,914
Cambridge	15	35	224	576	1,008	1,489	1,624	1,336	1,107	603	275	66	8,357
Campbell	18	38	224	734	1,100	1,627	1,781	1,436	1,203	660	270	75	9,067
Canby	10	25	179	545	989	1,440	1,567	1,274	1,066	600	252	54	8,000
Cloquet	44	67	270	656	1,077	1,547	1,677	1,373	1,172	720	386	145	9,134
Collegeville	7	20	166	523	981	1,461	1,589	1,282	1,051	563	232	52	7,928
Cook	51	71	305	699	1,131	1,664	1,858	1,515	1,294	784	390	144	9,906
Cotton	85	123	342	730	1,163	1,680	1,847	1,506	1,266	786	441	184	10,154
Crookston	28	51	262	672	1,150	1,655	1,824	1,503	1,240	689	326	95	9,497
Detroit Lakes	16	33	217	613	1,071	1,590	1,748	1,422	1,172	649	266	69	8,867
Duluth	63	86	298	678	1,088	1,556	1,699	1,399	1,210	762	426	179	9,444
Ely-Winton	63	72	305	724	1,149	1,697	1,850	1,492	1,268	788	411	145	9,963
Fairmont	5	15	147	498	945	1,421	1,539	1,263	1,034	556	207	34	7,663
Faribault	13	34	201	558	977	1,435	1,587	1,308	1,074	605	281	63	8,136
Farmington	7	20	161	504	945	1,418	1,534	1,240	1,003	530	223	45	7,631
Fergus Falls	22	36	221	610	1,071	1,572	1,734	1,429	1,166	644	289	80	8,875
Fosston	51	80	303	693	1,158	1,652	1,838	1,513	1,259	722	359	128	9,757
Grand Marais	142	95	280	643	983	1,395	1,531	1,284	1,158	788	560	339	9,196
Grand Meadow	23	46	220	583	993	1,460	1,603	1,319	1,105	643	311	71	8,377
Grand Rapids	35	62	277	665	1,100	1,601	1,756	1,419	1,169	691	348	114	9,238
Gunflint Lake	79	116	353	766	1,178	1,705	1,913	1,569	1,362	858	468	187	10,554
Hallock	30	71	317	757	1,247	1,798	1,988	1,643	1,353	761	382	111	10,458
Hibbing	86	126	367	752	1,173	1,663	1,809	1,485	1,259	786	449	192	10,147
Hutchinson	10	26	184	558	1,002	1,493	1,623	1,319	1,066	580	250	54	8,165
Int'l Falls	70	111	353	743	1,184	1,714	1,878	1,530	1,283	772	419	164	10,222

STATION	JUL	AUG	SEP	OCT	NOV	DEC	JAN	FEB	MAR	APR	MAY	JUN	ANNUAL
Itasca	51	82	323	716	1,164	1,654	1,824	1,502	1,274	775	403	142	9,910
Leech Lake Dam	30	53	256	654	1,101	1,599	1,778	1,436	1,192	699	341	106	9,246
Luverne	8	23	157	507	963	1,416	1,506	1,211	984	539	211	40	7,566
Marshall	9	25	180	544	990	1,449	1,572	1,285	1,073	592	243	48	8,008
Milan	8	24	166	520	981	1,461	1,578	1,275	1,025	533	206	43	7,819
Montevideo	10	28	194	572	1,022	1,508	1,640	1,344	1,116	614	261	55	8,364
Mora	26	47	249	625	1,043	1,541	1,699	1,382	1,136	651	314	92	8,804
Morris	18	40	219	610	1,062	1,551	1,706	1,401	1,152	637	280	70	8,748
MSP	5	14	154	507	939	1,404	1,531	1,236	998	530	218	44	7,580
New Ulm	7	19	164	514	951	1,443	1,559	1,267	1,023	546	214	40	7,747
Olivia	13	34	205	566	1,015	1,497	1,634	1,338	1,102	609	260	55	8,328
Park Rapids	39	70	289	693	1,161	1,654	1,803	1,473	1,218	713	360	128	9,600
Pine River Dam	26	45	250	637	1,086	1,590	1,753	1,429	1,192	703	334	98	9,144
Pipestone	17	35	206	577	1,020	1,486	1,601	1,302	1,071	604	268	67	8,252
Pokegama Dam	36	57	264	673	1,121	1,634	1,810	1,462	1,201	705	349	116	9,427
Red Wing	8	18	161	519	932	1,410	1,553	1,278	1,041	558	237	54	7,769
Redwood Falls	1	14	148	520	970	1,456	1,614	1,299	1,079	582	217	18	7,919
Rochester	11	27	177	521	936	1,406	1,530	1,253	1,011	553	245	52	7,720
Roseau	42	73	300	721	1,208	1,730	1,920	1,582	1,302	752	387	125	10,142
St. Cloud	17	41	228	599	1,040	1,522	1,655	1,344	1,104	617	287	78	8,531
St. James	10	27	177	542	971	1,441	1,567	1,278	1,056	584	238	47	7,940
St. Peter	6	16	167	514	947	1,435	1,562	1,281	1,025	569	229	43	7,792
Sandy Lake Dam	34	53	248	636	1,080	1,595	1,756	1,432	1,198	699	338	112	9,182
Stillwater	6	19	177	537	982	1,470	1,601	1,330	1,071	566	263	43	8,064
Tower	102	144	372	754	1,172	1,669	1,858	1,534	1,299	803	463	199	10,368
Two Harbors	80	62	239	594	960	1,386	1,528	1,267	1,116	746	510	265	8,754
Wadena	40	68	278	666	1,110	1,598	1,768	1,461	1,221	707	352	114	9,383
Warroad	44	79	309	710	1,168	1,696	1,877	1,540	1,290	773	405	136	10,027
Waseca	10	24	173	528	969	1,463	1,607	1,302	1,048	571	236	48	7,978
Waskish	59	90	324	738	1,161	1,706	1,885	1,536	1,293	753	399	141	10,085
Wheaton	12	31	210	601	1,065	1,567	1,705	1,394	1,165	644	270	62	8,726
Willmar	10	25	187	557	1,010	1,505	1,648	1,344	1,097	594	249	55	8,279
Windom	7	19	151	498	941	1,389	1,489	1,208	977	538	216	39	7,472
Winnebago	9	23	164	513	948	1,429	1,559	1,270	1,026	563	234	45	7,783
Winona	3	8	113	433	848	1,313	1,435	1,152	921	464	170	28	6,888
Worthington	13	35	191	545	989	1,449	1,565	1,278	1,068	609	266	56	8,064
Wright	52	76	272	642	1,077	1,562	1,697	1,384	1,161	697	371	136	9,128
Zumbrota	18	39	213	557	996	1,472	1,615	1,327	1,076	603	298	71	8,286

of 7,580 HDD, the historical data for Minneapolis–St. Paul equate to departures of plus or minus 20 percent for most years except for the extreme cases. These figures can produce variability in annual heating costs approaching several hundred dollars for small homes and several thousand dollars for large homes and businesses. For this reason, many Minnesotans choose an equal payment option from their utility company, contributing roughly the same amount each month throughout the year rather than trying to meet very high costs during unusually cold winter months.

Tyndall flowers

Occasionally seen through clear lake ice on bright sunny days in Minnesota, Tyndall flowers are small, water-filled, hexagonal cavities that appear inside ice masses bathed in sunlight. Named for John Tyndall, a nineteenth-century English physicist who studied the scattering of light as it passed through smoke, mist, fog, and ice, Tyndall flowers form when ice melts due to radiative absorption at its weak points.

The character of Minnesota's meteorological winter (December–February) is changing. Many areas are showing greater snowfall amounts and higher temperatures. For example, the long-term average seasonal snowfall in the Twin Cities area since 1885 is 47.0 inches, while the current 30-year normal (1981–2010) shows 54.4 inches. Examining the average statewide temperatures for the period from December through February reveals that seven of the ten warmest winters since 1895 have occurred since 1986. In fact, winter's warming trend is more evident statistically than that of any other of the year's seasons.

DID YOU KNOW?

Lolly ice is a needlelike or plate-shaped ice crystal formed in turbulent supercooled water.

Despite this tendency toward milder winters, Minnesota's reputation for cold weather is solidified by the National Weather Service's National Weather Summary, which highlights the nation's daily warmest and coldest temperatures. Outside of Alaska, Minnesota claims the nation's coldest temperature more often than any other state (in the winter season of 2013–14 this happened 44 times). In reporting the National Weather Summary to the American public, newspapers, radio and television stations, and Internet websites continually reinforce the association of cold temperatures with Minnesota. The state's reputation has endured for well over 150 years thanks to such reporting stations as Babbitt, Cook, Crane Lake, Embarrass, Flag Island (Lake of the Woods), Fosston, International Falls, Roseau, and Tower, many of which frequently report low temperatures colder than -30, -40, and even -50°F. Frankly, most Minnesotans seem proud of this reputation and are not ashamed to embellish the temperature readings with "chilling" personal tales to awed outsiders.

Coldest Places

I would be a wealthy man if I had a dollar for every time a Minnesotan has asked me where the coldest place in the state is. This question is not unexpected in a state that takes pride in leading the nation in chilling temperatures. In the meteorological winter (December-February) of 2013–14, Minnesota reported the coldest temperature in the 48 contiguous states on 44 dates, far more than any other state.

In recent years new climate stations located in somewhat unique positions have gained national reputations for unusually cold temperature readings. One is the small St. Louis County town of Embarrass, on the Iron Range south of Tower and west of Babbitt. The station is located in a relatively low-lying area where cold air can pool at night and which sees little wind compared to the surrounding forested landscapes. Since becoming an official Cooperative Weather Station in 1994, it has reported the state's lowest temperature more frequently than any other location and has established 21 all-time statewide daily low temperature records (see Table 25). In fact, it achieved national recognition during the winter of 1996–97 when the observer reported six nights with lows colder than -50°F, bottoming out at -57°F on January 20. The last remarkable cold reading at Embarrass was -54°F on January 17, 2005.

Another relatively new station is Flag Island, located on the Northwest Angle in Lake of the Woods, making it the northernmost climate station in the contiguous 48 states. Established as an Automated Weather

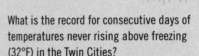

What is the record for consecutive days of temperatures never rising above freezing (32°F) in the Twin Cities?

a. 18

b. 42

c. 83

(answer on page 329)

Observation Station (AWOS) by the Minnesota Department of Transportation in 1998, Flag Island is positioned to provide surface weather data to pilots flying between Minnesota and Manitoba. The climate station's location is unique: it is surrounded by Lake of the Woods. When the lake is completely frozen over in late winter and early spring, it is as though Flag Island is packed in ice. On numerous occasions the state's lowest temperature has been reported there, including readings as cold as -27°F in March and just 1°F in April. As time goes by the climate data at both Embarrass and Flag Island will likely include a greater number of state low temperature records, perhaps surpassing all other Minnesota locations.

Cold soak

Cold soak, *a term used by climatologists and engineers, describes equipment exposure to cold, even polar, temperatures, when metal becomes more brittle, lubricants thicken, and operational tolerances diminish. Preheating is often prescribed for equipment that has been cold soaked—stored or left idle in freezing conditions—for an extended period.*

Of course the state's latitude as well as its varied landscapes contribute to the temperatures its residents experience. Outbreaks of cold, arctic air descending from higher latitudes can encounter a dry landscape that, when combined with clear skies, produces nighttime heat losses sufficient to bring temperatures down below freezing in any month of the year. That's right: places in northern Minnesota have recorded freezing temperatures even in the summer: temperatures in June as cold as 15°F (June 1, 1964, at Bigfork, Itasca County), in July as cold as 24°F (July 7, 1997, at Tower, St. Louis County, and July 13, 2009, at Kelliher, Beltrami County), and in August as cold as 21°F (August 21, 2002, at Kelliher and August 28, 1986, at Tower).

The snow accumulation season starts early and ends late in Minnesota's northern counties, producing average snow cover durations ranging from 130 to 170 days. The presence of snow cover increases the reflection of winter's sunlight, preventing its absorption and gradual warming of the landscape. In addition, snow's radiative properties magnify the effect of overnight winter temperature inversions, when cold air settles near the surface under calm and clear skies. Because of these attributes, communities with persistent snow cover also happen to hold many of the state's record low temperature readings.

One snowy community, International Falls (Koochiching County), touted itself as the nation's icebox for many years during the twentieth century, yet more than 20 other Minnesota locations hold more statewide daily low temperature records. Nearly all such records come from the northern half of the state. These communities are often used as cold-weather testing grounds for manufactured products such as car batteries. The community with the largest number of records is northern St. Louis County's Tower, which was established as a climate station in cooperation with the state forest ranger station near Lake Vermilion in January 1895. Observers at Tower and at the climate station opened later three miles south of town (Tower 3S) have recorded more than 40 all-time state low temperature readings, including the coldest ever in the state: -60°F on February 2, 1996. Pokegama Dam in Itasca County holds the next-largest number of records with more than 30. This station, established in April 1887, has been operated mostly by the Corps of Engineers, as Pokegama is one of the Mississippi headwaters reservoirs used to control flow on the river. Tower and Pokegama Dam share several characteristics: shallow soil layers over bedrock, a rolling landscape that allows cold, dense air to pool in the lower elevations at night, and prolonged seasonal snow cover. These characteristics foster rapid overnight heat loss, allowing temperatures to cool remarkably fast.

Some communities in the drier, prairie-dominated landscapes of the Red River Valley also hold statewide low temperature records in the winter months. Here overnight heat loss is uninhibited by topography and usually enhanced by a drier overlying atmosphere. Water vapor— the ultimate greenhouse gas—effectively absorbs long-wave radiation as it escapes from the surface, thereby keeping overnight temperatures high. But drier air often dominates prairie communities like St. Vincent and Hallock in Kittson County, Angus in northern Polk County, Ada in Norman County, and Beardsley in Big Stone County, which hold more than 20 state low temperature records among them. The southernmost location to hold a state low temperature record is Canby in Yellow Medicine County, where it was just zero degrees on April 22, 1952.

MAP 5: Climate Stations with Four or More Statewide Daily Minimum Temperature Records

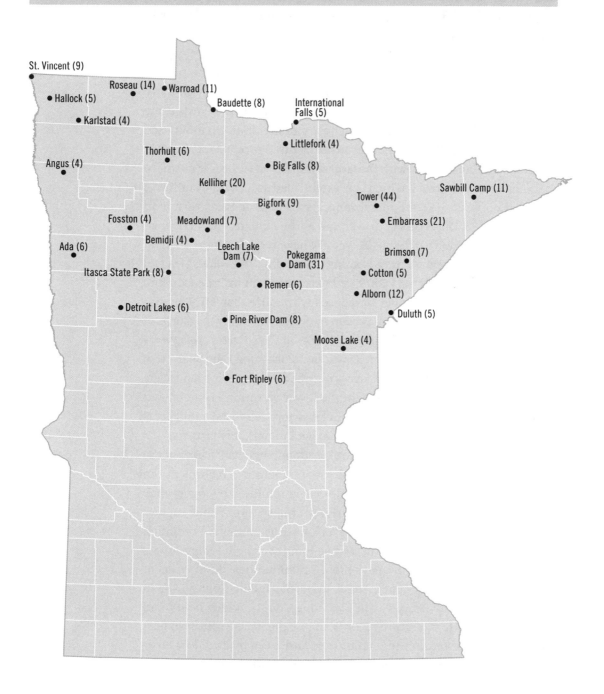

St. Vincent (9)

Roseau (14) Warroad (11)

Hallock (5)

Karlstad (4)

Baudette (8)

International Falls (5)

Thorhult (6)

Littlefork (4)

Angus (4)

Big Falls (8)

Kelliher (20)

Sawbill Camp (11)

Tower (44)

Bigfork (9)

Embarrass (21)

Fosston (4)

Meadowland (7)

Bemidji (4)

Ada (6)

Leech Lake Dam (7)

Brimson (7)

Pokegama Dam (31)

Itasca State Park (8)

Cotton (5)

Remer (6)

Alborn (12)

Detroit Lakes (6)

Duluth (5)

Pine River Dam (8)

Moose Lake (4)

Fort Ripley (6)

Characteristics surrounding Minnesota's late fall, winter, and early spring low temperature records include fresh snowfall; a dry high-pressure system bringing nearly cloudless skies and calm winds; and an arctic (high-latitude) or maritime polar air mass, usually coming directly from the north or northwest and anchored by an immense high-pressure system with a central barometric pressure of 1040 mb (30.71 inches) or higher. On rare occasions, usually in the months of January, February, or March, arctic high-pressure systems can equal or exceed 1050 mb (31 inches). In northeastern Minnesota's forests, where underlying snow cover is shielded from the sun, snow lingers well into spring, amplifying the effects of polar air. Consequently, many of the state's spring low temperature records are set in these areas.

Recently the State Climatology Office discovered a previously unrecognized seasonal snowfall record in what Minnesota county?

a. Beltrami

b. Cook

c. Itasca

(answer on page 329)

Biofog

Biofog is a steam fog that appears when a very cold air mass encounters the warm, moist air that surrounds humans or animals. Usually local in nature, biofog can sometimes be seen around livestock feedlots or near the doorways of a health club as people exit and the warm, moist inside air meets cold night air. Sports fans may have seen biofog when Viking or Packer football players emerge from the locker room at halftime. More historically, massive steam clouds lingered near buffalo herds on the Great Plains during harsh nineteenth-century winters.

Summertime low temperature records primarily coincide with very dry high-pressure systems that produce few or no clouds and little wind. Statewide record summertime lows have occurred in drought years like the 1930s, 1976, and 1988. Such lows have been in the twenties, sufficient to kill most garden vegetation. Consequently, frost-free growing seasons in northern Minnesota can be as short as 35 to 40 days, presenting challenges for the home gardener.

Minnesota has a national reputation to uphold, and the state's coldest places help maintain expectations. Temperature readings, whether sub-zero in January or in the twenties in July, often surprise residents and outsiders alike.

Winter Storms

The National Weather Service defines a winter storm as hazardous conditions resulting from significant quantities of freezing or frozen precipitation—freezing rain or drizzle, sleet, or snow—sometimes combined with strong winds and cold temperatures. In Minnesota *winter storm* also implies 6 or more inches of snowfall within 12 hours or 8 or more inches within 24 hours. The NWS may issue a winter storm watch to indicate that conditions are favorable for such a storm to develop, usually over a period of 12 to 48 hours. A winter storm warning is issued when such a storm is expected to pose a threat to life and property over a period of 12 to 24 hours. On average, Minnesota sees between three and four winter storms each year. On occasion a severe winter storm, sometimes called a midlatitude cyclone, will exhibit a central pressure equivalent to that of a hurricane (996 mb or 29.40 inches on a barometer), along with hurricane-force winds (74 or more mph). Though rare, such storms have inflicted serious damage across Minnesota and the Great Lakes over the past 200 years.

DID YOU KNOW?

Pioneer settlers on the plains melted so-called cooking snow—heavy, dense snow harvested for its water content—for use in the kitchen or bathtub. *Snawbroo* is the Scottish term for cooking snow.

A blizzard is an especially dangerous type of winter storm that lasts for three or more hours and is characterized by low or falling temperatures, strong winds of 35 or more miles per hour, and falling and/or blowing snow that reduces visibility to 0.25 mile or less. When such a storm is expected, the NWS will issue a blizzard warning, which indicates a life-threatening risk for those caught outside. Though rare, ground blizzard warnings are announced in the absence of snowfall but when wind threatens to move large quantities of snow cover across the landscape, drastically reducing visibility. According to research by Robert Schwartz and Thomas Schmidlin, the annual probability of a blizzard ranges from 50 to 76 percent across all Minnesota counties. It is especially high in the Red River Valley, where the NWS Grand Forks Office, with forecast jurisdiction over 17 North Dakota and 18 Minnesota counties, shows an 85 percent annual probability for at least one blizzard.

The term *blizzard,* also used by the Canadian and United Kingdom Meteorological Services, dates back to the nineteenth century. Although winter storms and blizzards can occur in Minnesota any time between

October and April, the vast majority appear from November through March. Their occurrence is less well documented than that of flash floods and tornadoes; nevertheless, some particularly severe storms stand out in the historical record.

The first observed thunder snow occurred on February 15, 1820, at cantonment New Hope, precursor to Fort Snelling. It was followed the next day by high wind with rain and sleet and a steep drop in temperature, by 25 degrees. Such readings indicate a winter storm's passage, but no precipitation measurements were taken.

October 11–14, 1820, brought 11 inches of snowfall to Camp Coldwater, later Fort Snelling, situated above the junction of the Minnesota and Mississippi rivers. It is still, nearly 200 years later, one of the earliest winter storms of the season for the Twin Cities area. Such October storms are so rare that similar ones can be found only for October 22, 1835, when strong winds accompanied 6 inches of snow, and October 16–18, 1880, when strong winds brought a mixture of snow, sleet, and rain to Fort Snelling and a blizzard to southwestern parts of the state. The 1820 storm followed a difficult year: many soldiers had died from pneumonia and scurvy during that first winter of 1819–20, an unremarkable one except for the exceptionally cold January, when the average minimum temperature was -9°F and there were 20 mornings of subzero temperatures. Consequently, this mid-October storm inspired a good deal of apprehension about the coming winter. Indeed, it was a signal of things to come, as the fort recorded a very snowy November—more than 20 inches—with blizzard-like conditions on November 25, which saw 9 inches of snow and high winds. December's total snowfall is unclear, but the month was very cold—on average 9.4°F—the temperatures undoubtedly affected by the already significant snow cover. Having survived the first winter on the site, the soldiers were better prepared for the second.

On December 20–23, 1825, a blizzard struck western and northwestern parts of Minnesota with devastating suddenness, resulting

Q & A

Which of these Minnesota communities has never recorded a minimum temperature of -50°F or colder?

a. St. Cloud

b. Two Harbors

c. Grand Rapids

(answer on page 329)

in 33 recorded deaths and leaving many settlers and traders stranded between Pembina, North Dakota, across the river from St. Vincent in Kittson County, and Selkirk, near present-day Winnipeg, Canada. The Red River Valley was plagued by a series of blizzards through the middle of February, and snow depths in excess of 5 feet were reported in areas of undisturbed prairie—both conditions setting the stage for a disastrous Red River flood in May 1826. Alexander Ross, an eyewitness to this winter, wrote in 1856,

> The winter had been unusually severe, having begun earlier and continued later than usual. The snows averaged three feet (0.9m) deep, and in the woods, from four to five feet (1.2–1.5m). The cold was intense, being often -45 degrees F (-43 degrees C); the ice (on the Red River) measured five feet seven inches (1.7m) in thickness.

This pattern followed two years of spring snowmelt flooding on the Red, prompting many settlers to give up and move elsewhere.

March 19–20, 1826, brought one of the most severe late-winter storms on record: up to 18 inches of snow fell in southwestern Minnesota and snowdrifts from 6 to 15 feet deep were reported. The Fort Snelling observer noted snow on March 19, but no measurements were taken; the record also indicates great loss of life among the Sisseton and Dakota tribes. March brought about 40 inches of snow to parts of Minnesota, and April also saw above-normal snowfall and then, as the snowpack melted, significant flooding on both the Minnesota and the Mississippi rivers. An American Indian settlement near present-day South St. Paul was swept away, and Fort Snelling landing, on the Minnesota River below the garrison, was inundated by a flood crest estimated at 20 feet above the previous year's low-water mark. Fort Crawford, another outpost along the Mississippi River near Prairie du Chien, Wisconsin, had to be evacuated in April as the flood crest moved downriver.

An all-day snow with temperatures consistently in the low twenties was recorded on October 22, 1835, another early start to winter. The Fort Snelling observer noted a sudden shift as the wind swung around to come from the northwest, movement suggesting a storm system passing to the south. Then, November 10, 1835, brought a deep low-pressure system that deposited 6 inches of snow at Fort Snelling overnight. Winds associated with this sizeable and intense winter storm were very strong, producing large waves on the Great Lakes and contributing to 19 ship-wrecks and as many as 254 sailors' deaths.

■ The first week of April in 1837 at Fort Snelling was quite cold, with temperatures as low as 6°F (normal lows are in the thirties). A very severe winter storm with high winds was noted over April 4–6, 1837. Temperatures fluctuated from the mid-twenties to low forties over those three days, yielding mixed precipitation that totaled a significant 0.90 inches. The total quantity of snow was not recorded.

■ The Fort Snelling records note a "violent snowstorm" on February 25–26, 1843, one that produced well over 5 inches of snow, dropping temperatures, and winds in excess of 35 miles per hour. It capped a month that had deposited 22 inches of snow on top of an already deep snowpack that had been accumulating since November. An arctic cold front held temperatures consistently below zero on February 27 and 28. This storm and weather pattern contributed to the coldest March in Minnesota history—the average monthly temperature was 3.9°F, compared to the modern monthly normal of 32.8°F—and one of the state's longest-ever winter seasons. Lake Pepin was not ice free until May 20 that year.

DID YOU KNOW?

The Scottish yowdendrift, perhaps a derivative of *yowt*, meaning "to scream or howl," is a term for snow driven by wind, such as in a blizzard.

■ Two heavy snows were noted at Fort Snelling during December 1849. One occurred the week before Christmas, when 10 or more inches blanketed an already snowy landscape on December 22, a day when temperatures barely climbed above the single digits. A second snowstorm came the week between Christmas and New Year's, delivering 6 or more inches as temperatures fell 32 degrees over December 28–29. Although wind records are incomplete, both weather systems were undoubtedly winter storms. Over January 20–22, 1850, a "very severe snowstorm" with "high winds," presumably a blizzard, dumped at least a foot of snow on Fort Snelling. Winds were estimated at 35 to 45 miles per hour, producing zero visibility. This storm added to a snowy winter season that yielded the greatest late-January snow depth the fort had ever seen. A

10-inch snowstorm on St. Patrick's Day (March 17) increased the threat for spring snowmelt flooding, and sure enough the floods arrived in April, affecting both the Minnesota and the Mississippi rivers.

December 27–30, 1852, brought strong northeast winds and snowfall of a foot to downtown St. Paul. This strong winter storm capped a snowy December, which saw 20 inches of snow in all, producing fine terrain for sleigh rides. Temperatures dipped from -20° to -28°F, and snow depths were up to 3 feet for the start of the New Year.

March 12–14, 1855, produced 14 to 16 inches of snow, one of the heaviest-ever falls for that month, with easterly winds up to 45 miles per hour at Fort Snelling. The month's snowfall total was more than 22 inches, with snow depths as great as 20 inches. Fortunately, a dry April and May helped alleviate the threat of spring snowmelt flooding.

Friday and Saturday, December 15–16, 1855, brought a very heavy snowfall, "with flakes as big as a featherbed," according to the weather observer, who also noted that "sleighing was excellent." Snow fell continuously for several hours as temperatures hovered in the mid-twenties. Total snowfall was probably between 8 and 12 inches, contributing to a monthly total of 22 inches. This heavy blanket of snow reinforced an arctic air mass that arrived for the holiday, yielding a Christmas Eve temperature reading of -33°F in the St. Paul area.

. . . .

Mercury on the chute

During the National Weather Service's first 50 years, the phrase mercury on the chute, *in reference to thermometer readings, was used to report the onset of a cold wave, when temperatures fall rapidly and reach levels that pose a threat to agriculture and commerce.*

. . . .

December 28–30, 1856, saw back-to-back storms drop 13 to 14 inches of snow, followed by bitter cold temperatures in the -20°F range. These storms resulted in a monthly snowfall total of 30 inches and produced conditions in St. Paul that greatly inhibited pedestrian traffic. Two heavy

January snowstorms continued this cold, snowy winter in 1857. The first dropped 7 inches on January 13, followed by cold arctic air that brought many readings below -20°F. The second, on the last day of the month, deposited 10 inches of snow and brought high winds. Both strong winter storms contributed to the highest snowfall total observed in January to that date: 25 inches. Low temperatures in January reached -40°F on the eighteenth. Then, back-to-back storms over February 5–7 brought 11 additional inches. The *St. Paul Daily Pioneer and Democrat* described the latter storm as "one of the most violent in recollection of [our] oldest residents." The winds were undoubtedly fierce, producing blizzard-like conditions and ushering in a strong arctic air mass that dropped temperature readings to -50°F or colder in the Fort Ripley (Crow Wing County) area.

When living in the high arctic, it is considered good manners to "degomble" before coming inside. What does this mean?

a. blow your nose

b. comb your hair

c. brush all the ice and snow off your clothing

(answer on page 329)

March and April 1859 ushered in heavy snows. Over March 10–12, a massive snowstorm dumped 12 slushy inches on St. Paul as temperatures hovered on either side of freezing. Then, on April 2–3, another 7 to 8 inches of snow came to Fort Snelling while 12 inches were reported in St. Paul. With temperatures mostly in the twenties, the snow from this unusually heavy April storm drifted and blew, affecting travel and visibility.

On February 13–15, 1866, arguably one of the worst blizzards in Minnesota history struck. In St. Paul, when the storm began the temperature was about 17°F; by the time it ended it was -31°F. In western Minnesota, snow cover and snowfall were blown around violently by strong winds that produced 20-foot-high drifts that buried barns. Though a fierce and long-lived blizzard, the loss of life was somewhat limited because the storm struck at night.

January 16–17, 1870, brought a blizzard to eastern sections of the state, with downtown St. Paul recording 9 inches of snow, strong winds, and low visibility. The *St. Paul Pioneer* reported that the wind blew "great guns," driving and whirling snow in every direction. Two months later, on March 14–16, a blizzard struck northern Iowa and southwestern

Minnesota with an estimated 16 inches of snowfall. This storm inspired the first use of the term *blizzard,* which appeared in the Estherville, Iowa, *Vindacator* newspaper; the U.S. Signal Corps Weather Service would not adopt the term until 1876. This March storm also dropped 8 inches of snow in St. Paul and 11 inches in Minneapolis. Railroads were blocked for two days. With more than 40 inches of snow accumulating during the first three months of the year, many residents wondered what spring would hold, but April proved to be a dry month, with some 80°F daytime temperatures.

A storm on November 7–8, 1870, produced strong winds and heavy rainfall in the Twin Cities area: 1.38 inches of precipitation were recorded in St. Paul, a large amount indeed for the month of November. The U.S. Army Signal Corps was inspired to issue its first winter storm warning, an alert that proved more relevant to areas east of Minnesota. Regardless, the NWS's mandate to issue such warnings dates from this storm, and that remains a significant bookmark in history.

January 7, 1873, began as a mild day that encouraged outdoor activities, but then a blizzard struck, bringing with it a drastic temperature drop that produced severe wind-chill conditions. The Smithsonian weather observer at New Ulm remarked that it was a "most violent snow storm," suddenly reducing visibility to fewer than 20 yards with winds from the northwest up to 45 miles per hour that persisted from late on January 7 to early on January 9. Total snowfall over the course of the storm ranged from 9 to 16 inches. Trains were stuck in high drifts for days, 70 deaths were reported, and hundreds of cattle were lost.

DID YOU KNOW?

A barber is a Great Lakes blizzard that produces wind-driven ice particles that can cut the face.

Second in earliness only to the October 11, 1820, blizzard, the October 16, 1880, blizzard struck Minnesota's southwestern and west-central counties particularly hard. Huge drifts exceeded 20 feet in the Canby (Yellow Medicine County) area, leading people to use their houses' second-floor windows as exits. An observer at New Ulm (Brown County) reported 15 inches of snow, while at Breckenridge (Wilkin County) visibility was reported to be zero. The snow deposited by this storm lasted until spring, setting up one of the longest and snowiest winters in Minnesota history, written about by Laura Ingalls Wilder in *The Long Winter*. In fact, observations in St. Paul suggest that the winter's total snowfall exceeded 110 inches. Not unexpectedly, as the landscape shed excess snow in April and May, all the major watersheds saw spring flooding—in some cases the worst since 1844.

During the "big snow" of 1880–81, manpower and snowplows were needed to keep trains on track.

November 16–18, 1886, brought more than 15 inches of snow to the Twin Cities area as well as Spring Valley and Rochester in what was likely a heavy winter storm rather than a blizzard. Temperatures generally remained in the twenties and thirties, and winds were moderate.

January 12, 1888, began as a relatively mild day, with temperatures in the thirties. Following a bitterly cold December, when temperatures had plummeted into the thirties below zero, this warm spell offered a respite for citizens of the prairie. Children had walked or sledded to school; adults were working outside or attending to business in town. Suddenly at midday, a dark squall line approached from the west, bringing high winds, blinding snow, and rapidly falling temperatures, which bottomed out at -37°F the next day. Children were sent home early from school, but many lost their way and died. Deaths from this storm, perhaps Minnesota's worst, totaled more than 200. It later became known as the Children's Blizzard, documented in a book of that name by David Laskin.

Duluth saw one of its worst blizzards on March 9–10, 1892. Sixty-mile-per-hour winds off Lake Superior not only provided ample water vapor to the storm but also dislodged and damaged the NWS rain gauge. The estimated snowfall total was 13 inches, though the wind piled drifts more than 10 feet high, blocking some second-story windows. Over the course of the storm, the temperature plummeted from above the freezing mark to -5°F and colder.

January 31, 1893, brought a blizzard to Park Rapids in sparsely populated Hubbard County. The temperature dropped 40 degrees in fewer than five hours and an incredible 56 degrees in 18 hours. In addition, nearly a foot of snow was blown around, reducing visibility to near zero and producing large drifts. This resulted in livestock losses for farmers who had set out their animals for winter grazing. A second heavy snowstorm over April 19–21, 1893, delayed spring field work for farmers. This storm delivered 3 feet of snow to St. Cloud (Sherburne County) and more than a foot and a half to Cambridge (Isanti County), Maple Plain (Hennepin County), Park Rapids (Hubbard County), and Wabasha (Wabasha County). The Twin Cities received 10 inches. The storm brought the railroads to a halt as well.

What does the term *pagophobia* mean?

a. fear of ice or frost

b. fear of strong wind

c. fear of typhoons

(answer on page 329)

November 26, 1896, became a memorable Thanksgiving Day as a storm with rain, lightning, thunder, hail, wind, and snow encompassed the western Great Lakes region. There were reports of rain and thunderstorms in southern Minnesota, snow and blizzard conditions in North Dakota, and a mixture of the two in central and northern Minnesota. Seven climate stations in Minnesota reported more than 3 inches of precipitation, and four northern stations reported a foot-plus of snow. A severe cold wave followed the storm: Pokegama Dam in Itasca County recorded a temperature of -45°F.

On November 27–28, 1905, another blizzard hit Duluth as winds blew from 50 to 60 miles per hour for an extended period, gusting to 68 miles per hour and producing zero visibility. Seven communities reported more than a foot of snow, with huge drifts. This strong storm yielded enormous waves of 20 to 30 feet that wrecked 30 ships on the Great Lakes and washed away the lighthouse at the end of the harbor entrance pier. The ship *Mataafa,* caught in the storm as it left the harbor, dashed against the rocks and was wrecked with the loss of all nine sailors aboard. In all, 39 people lost their lives in this storm. Consequences involving the Great Lakes shipping industry helped to spark federal funding for the construction of Split Rock Lighthouse.

November 7–11, 1913, brought one of the worst-ever November storms to the Great Lakes. The Duluth Weather Service Office recorded wind speeds up to 62 miles per hour, but the storm hit harder to the east, producing hurricane-force winds (74 miles per hour and greater) and causing a significant loss of property, ships, and crews in Lakes Michigan, Huron, and Erie. The strong winds and low visibility inspired the press to term this storm the "White Hurricane" or "Freshwater Fury." Two ships were lost on Lake Superior: the *Leafield,* with 18 victims, and the *Henry B. Smith,* with 23 victims.

October 18–20, 1916, brought one of the season's earliest blizzards. Beltrami, Kittson, Roseau, and Lake of the Woods counties were hit with blinding snows and accumulations of 5 to 16 inches. The snow and a 50-degree temperature drop abruptly halted farmer's fall field work. Like the early blizzards of 1820 and 1880, this one set up one of Minnesota's longest and snowiest winters. In fact, Minnesota's southern counties were hit by another blizzard on January 21–22, 1917, as temperatures ranged through the single digits and teens and strong winds blew the snow into large drifts. Tracy in Lyon County reported 24 inches of snow; Redwood Falls in Redwood County, 22 inches; Stillwater in Washington County received 19 inches; Glencoe in McLeod County reported 16 inches; and the Minneapolis downtown Weather Service Office, 15.8 inches. This very strong low-pressure system produced a barometer reading of just 29.19 inches at La Crosse, Wisconsin, and wind speeds well over 40 miles per hour across southern and central Minnesota communities—conditions that yielded wind-chill values of -25 to -35°F and caused enough blowing and drifting snow to block roads and highways for days.

DID YOU KNOW?

Intense motion of snow or soil particles, carried by strong gusts of wind, is known as saltation, from the Latin *saltare,* "to dance."

Two more heavy snowstorms struck Minnesota in 1917: one in the northeast on March 13–14 and another across the south and central sections on March 16–17. The earlier storm, which produced blizzard-like conditions in a few places, dumped 22 inches of snow at Duluth and Grand Marais, 21 inches at Two Harbors, and 13.5 inches at Meadowlands. The second storm was very much a blizzard, with winds from 30 to 50 miles per hour and blinding snow that started during the night and continued through most of the next day. The storm brought thunder and lightning to some places as well. Lynd in Lyon County reported 24 inches of snow; Canby (Yellow Medicine County), 20 inches; Willmar (Kandi-yohi County) and Tyler (Lincoln County), 18 inches; and Montevideo (Chippewa County), 12.8 inches. Duluth tallied 48.2 inches of snow for the month, and many other places reported more than 30 inches, concluding one of the coldest and snowiest winters in Minnesota history. Fortunately, the last few days of March brought temperatures in the fifties and sixties, rapidly melting the abundant snow cover.

■ January 16, 1921, brought a blizzard to many northern counties, including Duluth, where winds up to 59 miles per hour were recorded. Snowfall was significant but not extreme: Baudette in Lake of the Woods County reported 10 inches; International Falls in Koochiching County, 6 inches; Cloquet in Carlton County, 4 inches. Winds above 40 miles per hour blew around the abundant snow cover in places that received only light amounts of new snowfall. In the southern counties, which lacked snow, the wind produced a significant dust storm, reducing visibility and filling ditches with soil. Wind-chill conditions ranged from -20 to -30°F.

• •

Snoweaters

Weather conditions that accelerate the melting of snow cover, such as wind and fog, are known as snoweaters. Warm air descending on a mountain's leeward side or strong southerly winds coming from warmer latitudes can rapidly melt snow in the western or central plains states. Fog resulting from warm air advection can release heat through condensation, thereby accelerating melting as well.

• •

■ The worst February winter storm to strike Minnesota since 1866 arrived on February 21, 1922, bringing thunder, lightning, rain showers, sleet, freezing rain, and snow as it developed over a three-day period. The ice storm hit mostly southeastern counties, coating power lines and trees with thick layers that caused breakage and much damage and left communities without power for days. The observer at Grand Meadow in Mower County called it "the worst ice storm" to ever hit that community. Winds were strong at 30 to 40 miles per hour, and though the Twin

Cities recorded 6.3 inches of snow, communities to the west and north set new February snowfall records. Fergus Falls in Otter Tail County reported 13.2 inches; Willmar in Kandiyohi County, 14 inches; Morris in Stevens County, 15 inches; Montevideo in Chippewa County, 19 inches; Milaca in Mille Lacs County, 22 inches; and Detroit Lakes in Becker County, 25 inches.

Yet another unusually strong winter storm struck the state with blizzard conditions on February 12–14, 1923. Snowfall amounts were generally less than a foot, but winds of 40 or more miles per hour blew the snow into large drifts that blocked roads. Wind-chill values ranged from -35 to -45°F, capable of causing frostbite in only ten minutes of exposure. Detroit Lakes in Becker County reported a record 12 inches of snowfall; Waseca in Waseca County, 10.5 inches; Bemidji in Beltrami County, 8 inches; and Fergus Falls in Otter Tail County and International Falls in Koochiching County, 7 inches. This storm swept up black dust or soil from North Dakota's snow-barren landscape and deposited it on top of Minnesota's newly fallen snow, leading some citizens to coin the phrase *black blizzard.*

After a March 1924 storm, roads were closed to all but trucks—and pedestrians—in St. Paul.

A rare strong winter storm arrived on April 5, 1933, as temperatures hovered around 30°F and a heavy, wet snow fell along Lake Superior's north shore. Two Harbors in Lake County reported 17 inches of snow, and Pigeon River in Cook County reported 28, an enduring state record 24-hour snowfall for the month of April.

Robin Hood's wind is raw and penetrating and usually occurs in saturated air with temperatures at or below freezing. This wind steals heat from even the well dressed, much like the legendary Robin Hood robbed the rich to give to the poor.

Over November 11 and 12, 1933, a severe dust storm struck in Minnesota's southern and central counties, sending visibility to near zero; the observer at Redwood Falls in Redwood County called it the worst dust storm in the city's history. Farther north this storm produced a blizzard with 40- to 60-mile-per-hour winds. Baudette in Lake of the Woods County reported 12 inches of snow; Pigeon River in Cook County, 18 inches; Orr in St. Louis County, 13.5 inches; and Grand Marais in Cook County, 9 inches.

A significant snow and ice storm occurred in north-central and northeastern Minnesota over March 3–5, 1935. This storm brought 8 inches of snow to Baudette (Lake of the Woods County) and 9 inches to Pigeon River, but the real disruption was caused by accumulation of up to 2 inches of ice on trees, power poles, and streets in Duluth and icy roads from Beaver Bay (Lake County) to Moose Lake (Carlton County), where there were numerous accidents. Four streetcars in Duluth had to be abandoned during the storm as they were frozen to the tracks. Thousands of trees were damaged, with 75 percent of all shade trees in the Moose Lake area reported to be ruined. All power and phone service was out in Duluth by March 5, the only communication with the outside world coming by short-wave radio. Damages in the city were estimated at over half a million dollars.

Dominated by numerous storms as well as by arctic outbreaks that brought record-setting cold temperatures, the winter of 1936 was one of Minnesota's coldest during the twentieth century. An especially strong winter storm passed across the northern counties on February 13–14, bringing wind-chill conditions of -40 to -50°F, some of the most dangerous values recorded in the state. Grand Marais in Cook County reported

22 inches of snow; Pigeon River in Cook County, 20 inches; and Two Harbors in Lake County, 10 inches.

The afternoon of November 11, 1940, brought the devastating Armistice Day Blizzard. A mild morning offered a promising start for the duck hunters who were out in force, but the afternoon deteriorated rapidly with the approach of a strong winter storm that dropped the Twin Cities' barometer to 28.93 inches and the Duluth barometer to a record low of 28.66 inches, indicating hurricane-like low pressure. Gusts to 45 miles per hour were reported from Collegeville in Stearns County, where 26.6 inches of snow fell. The winds at Duluth reached 63 miles per hour, and the temperature dropped 41 degrees over a 24-hour period. In most places traffic came to a standstill; even the Twin Cities streetcar system halted. The snowfall intensity was measured at times between 2 and 3 inches per hour; across parts of the state ice accumulation took down power lines. Drifts up to 20 feet were reported in the Willmar area (Kandiyohi County). Snowfall measurement showed 15 inches at St. Peter in Nicollet County; 16.7 inches at Bird Island in Renville County; 16.8 inches in the Twin Cities; 19.3 inches at Milaca in Mille Lacs County; and 22 inches at Orr and 24 inches at Meadowlands, both

The infamous Armistice Day Blizzard of 1940 stymied vehicles along numerous state roads.

in St. Louis County. Damages were estimated at more than $1.5 million. Forty-nine people lost their lives; included in that number were many duck hunters caught unprepared on Mississippi River islands. This storm system was unusually and extremely slow moving, maintaining its intensity for a long period of time.

The very next spring, on March 15–16, 1941, another terrible blizzard struck Minnesota's western counties: along the Red River Valley, 85-mile-per-hour winds blew at Grand Forks and 74-mile-per-hour winds at Fargo-Moorhead; 75-mile-per-hour winds were also measured at Duluth. The poorly forecasted storm struck suddenly on a Saturday evening, causing 32 deaths. Wind-chill values plummeted as low as -30 to -35°F; a number of people died from exposure as they spent the night in stalled vehicles or out in the open, and many livestock were lost. Snowfall from this storm was relatively light, in most places just a few inches, although Crookston in Polk County reported a foot of snow. One significant outcome from this storm was the opening of a regional forecast office in the Twin Cities. After Governor Harold Stassen and congressman R. T. Buckler of Crookston vehemently protested the lack of warning and inadequate forecasting for the state, the NWS consented to shift the Twin Cities' weather jurisdiction from observational to forecasting. Previous jurisdiction had rested with the Chicago regional office, but after its poor performance with both this storm and the previous fall's Armistice Day Blizzard, the change seemed long overdue.

Q & A

The Gunflint Trail's snow cover season is the longest of any Minnesota climate station. What is the average duration of snow cover there?

a. 110–120 days

b. 140–150 days

c. 170–180 days

(answer on page 329)

January 1943 is remembered for blizzard conditions that occurred in the latter part of the month, but the heaviest snowfall came with a winter storm on January 3. Dry and fluffy snow fell for hours in temperatures that ranged from the single digits into the teens. Accumulations in western communities were significant: Beardsley in Big Stone County reported 14.5 inches and Willmar in Kandiyohi County 15.5 inches, both records at those locations. In addition, Wheaton (Traverse County) reported 9.5 inches and Milan (Chippewa County), 9 inches.

December 1945 brought the snowiest-ever Christmas to southeastern Minnesota. All-day snow on December 25 carried over into the next day, producing a record total of 15.5 inches at Theilman (Wabasha County), 15 inches at Bricelyn (Faribault County) and Albert Lea (Freeborn County), 13 inches at Waseca (Waseca County) and Stillwater (Washington County), and 12 inches at Wells (Faribault County). Though no record amounts were tallied, Winona reported 8.2 inches and Grand Meadow (Mower County) received 11 inches. This strong winter storm is one of only two to arrive with heavy snow on Christmas Day, the other coming in 2009.

▨ A very significant snow and ice storm stretched from New Ulm in Brown County to Duluth and the Iron Range over November 6–7, 1947. The storm produced 1 to 4 inches of ice on power lines and tree branches, causing a great deal of tree damage and many power outages. More than 2,400 power poles were knocked down, and thousands of wire breakages were reported. Estimated damages: $1.2 million. Babbitt and Virginia (St. Louis County) and Litchfield (Meeker County) all reported more than 10 inches of snow.

▨ March 26–27, 1950, brought a severe ice and wind storm mostly to northern counties. The Duluth National Weather Service Office called it a "phenomenal ice storm," which produced ice up to 2 inches thick on power and light poles, as well as trees. Wind gusts peaked at 83 miles per hour, causing many of the weighted-down power lines and trees to break. Many roads and highways were blocked. Most schools from Duluth west to Crookston (Polk County) were closed for days until the roads improved.

▨ One of Minnesota's snowiest-ever Decembers was that of 1950. Many eastern communities reported measurable snowfall on 15 consecutive days, essentially the first half of the month. A strong and slow-moving winter storm crossed the state from December 5 to 8, bringing record snowfall to some places. Duluth reported 35.2 inches of snow during one of its longest-lived snowstorms; 23.2 of those inches arrived on December 6, but temperatures remained in the twenties throughout. Other notable snowfall amounts from this four-day storm were 13.7 inches at Rochester in Olmsted County, 13 inches at Waseca in Waseca County and Cambridge in Isanti County, 11.8 inches at New Ulm in Brown County, and 11 inches at Bird Island in Renville County.

▨ A mixture of thunderstorms, sleet, freezing rain, and snow was produced by a deep low-pressure system on January 14, 1952. Campbell

(Wilkin County), Aitkin (Aitkin County), and Cloquet (Carlton County) observers all reported more than 10 inches of snow, but across central and southern areas of the state icy roads generated hundreds of accidents. Power and telephone lines were taken down by the weight of the ice, and Minneapolis General Hospital treated 81 patients for falls on icy sidewalks and streets.

On May 2–3, 1954, a rare winter storm moved across northern Minnesota, its heavy snowfall closing rural schools. Though other such May snows occurred during 1890, 1935, 1938, and 2013, they were geographically small in comparison. This unusual storm brought 12 inches to Leonard in Clearwater County, 11.5 inches to Walker in Cass County, 11 inches to Winnibigoshish Dam in Itasca County and to Indus in Koochiching County, and 10 inches to Tower and Babbitt, both in St. Louis County—all totals that still stand as records for the month. More than 25 state climate stations reported 5 or more inches of snow from this storm.

On November 17–18, 1958, a huge winter storm tracked over Minnesota, bringing 60-mile-per-hour winds to the Twin Cities and 67-mile-per-hour winds to Duluth. Snowfalls from 5 to 9 inches were reported in the Red River Valley, and rainfalls of 1 to 2 inches came to many southeastern communities, topped by a record 2.63 inches at La Crescent in Houston County.

On November 28–29, 1960, a severe winter storm and blizzard, dubbed a nor'easter, hammered the Lake Superior shoreline, producing 20- to 40-foot waves that destroyed a good deal of lakeside property. Three feet of water flooded the streets of Grand Marais in Cook County; thousands of cords of pulpwood washed into Lake Superior; winds gusted to 73 miles per hour. Duluth recorded nearly a foot of snowfall, while Grand Portage in Cook County measured 14 inches and Grand Marais and Isabella (Lake County) each saw 10 inches. Across the western part of the state, Ada in Norman County reported 12 inches and Beardsley in Big Stone County, 10. Many southern communities received only modest rainfall.

Two major blizzards struck the state during March 1966. From March

2 to 5, heavy bands of snow slowly moved across Minnesota's central and northern sections, bringing record-setting snowfall totals: High Landing in Pennington County, Isabella in Lake County, Itasca State Park in Clearwater County, and Park Rapids in Hubbard County all reported more than 30 inches of snow; 22 other communities measured at least 20 inches. The storm caused four deaths, many road closures, and considerable building damage, usually roofs giving way due to heavy snow loads. The Crookston (Polk County) police department used ten snowmobiles for emergency runs during and after the storm, showcasing for the first time the maneuverability of these machines when other forms of transportation were unavailable. The second blizzard struck on March 22–23, this time across east-central, south-central, and southeastern Minnesota. Lightning, thunder, and freezing rain preceded the snow. Travel was hampered by both icy roads and blowing and drifting snow, and many power lines came down. The Twin Cities reported 13.6 inches of snow; Forest Lake in Washington County received 11 inches; Mankato in Blue Earth County and Rosemount in Dakota County, 13 inches; Hastings in Dakota County, 14 inches; Farmington in Dakota County, 15 inches; and Montgomery in Dakota County, 18 inches. For the first time, the Twin Cities campus of the University of Minnesota closed due to weather.

Derived from the Norwegian *sweel,* "to whirl around," sweevil is occasionally used by Scottish meteorologists to describe a gust of wind.

Stacking

When the geographic center of low or high pressure at the Earth's surface tends to be the same for low or high pressure aloft—meaning there is little tilt to the pressure field—this condition is known as stacking. Stacking, which occurs with large-scale, slow-moving low-pressure systems, appears in satellite water vapor imagery as a large rotating white blob.

On January 16, 1967, a short-lived, fast-moving blizzard and ice storm resulted in seven deaths statewide, some from snow shoveling. In the southern counties, accumulated ice plus very strong winds, with gusts over 75 miles per hour that brought visibility to near zero, made driving hazardous and sent many cars into roadside ditches. Snowfall ranged from 3 to 6 inches, but Meadowlands in St. Louis County, Pokegama Dam in Itasca County, and Two Harbors in Lake County reported 10 inches of new snow. A rescue team on snowmobiles saved a Le Sueur girl who had become lost in the blizzard.

■   December 1968 brought three blizzards to Minnesota. The first one, on December 12 and 13, blanketed west-central and northeastern counties with snowfall totaling around 5 to 10 inches, although up to 14 inches fell in the Walker (Cass County) and Bemidji (Beltrami County) areas. Winds blew at 40 to 50 miles per hour, with gusts unofficially estimated near 70 miles per hour. Roads closed, and power outages lasting up to 12 hours were common. During what was one of the worst storms to hit Mille Lacs, those strong easterly winds piled ice floes as high as 20 feet onto State Highway 169 and destroyed many ice houses and shoreline buildings along the lake's west end. A second blizzard struck southern counties on December 19, bringing up to 10 inches of snow to the Marshall (Lyon County) and Springfield (Brown County) areas. Dozens of schools closed before rural roads became blocked. A third blizzard struck on December 21 and 22, bringing 5 to 12 inches of snow and winds up to 40 miles per hour. Scores of highways closed all day on December 22, and many rural roads closed for an entire week, curtailing holiday travel. The result of these three blizzards: one of the snowiest Decembers in Minnesota history. More than 30 communities reported a monthly snowfall of 30 or more inches.

That snowy December was followed by two more blizzards in January 1969. The first struck on January 8 and 9. Though snowfall amounts ranged from only 2 to 6 inches, blowing of the already abundant snow cover by the 50-mile-per-hour winds blocked hundreds of roads. Nearly all schools except those in larger cities closed. Another blizzard arrived on January 23–24, bringing 6 to 12 inches of snow to some areas and freezing rain to others. Many motorists were stranded as hundreds of roads remained closed from continuous drifting in the 30- to 40-mile-per-hour winds.

DID YOU KNOW?

Niphablepsia is more commonly known as snow blindness, a condition caused by the high reflection of sunlight from snow cover. Occurring on sunny days in areas where snow has drifted into a relatively uniform surface, this reflection is so intense that it can cause impaired vision or even temporary blindness. The medical term is derived from the Greek *nipha*, "snow," and *blepsia*, "affliction of the eye."

On January 24, 1972, a fierce blizzard struck most of the state with heavy snow and high winds. Snowfall totaling 10 inches spanned from Park Rapids in Hubbard County to Duluth in St. Louis County, while many other areas received 4 to 6 inches. Winds of 40 to 50 miles per hour were standard, though gusts to 72 miles per hour were reported from Worthington in Nobles County and 70 miles per hour from Mankato in Blue Earth County. Nearly all schools closed by noon. When school buses from the districts of Redwood Falls (Redwood County), Mankato, and Worthington became stranded in deep snow coupled with poor visibility, hundreds of students weathered the storm in nearby farm homes. The NWS and local television and radio meteorologists received widespread praise for their forecasting and for the warnings they issued about this storm.

What Minnesotan hasn't gotten his or her car stuck in a snow bank? These men make valiant efforts following a blizzard in 1948.

The year 1972 concluded with a blizzard on New Year's Eve. In some places, rain started falling on December 29, but as the storm intensified the winds grew stronger and the precipitation turned to freezing rain and then to snow on December 30–31. An ice storm caused damage across an area from Morris in Stevens County and Appleton in Swift County east through St. Cloud in Sherburne County to Pine City (Pine County). Utility damages were estimated to be $1.5 million; power and phone service was out to thousands of farms and many smaller communities for days. Across the north from Fargo-Moorhead to Duluth, 6 to 14 inches of snow was blown around by 30- to 40-mile-per-hour winds, bringing all vehicle and air traffic to a halt and subduing residents' New Year's celebrations.

One of the worst blizzards of the twentieth century struck Minnesota on January 10–12, 1975. This storm brought hurricane-like—and record-setting—low-pressure readings to Rochester (28.63 inches), the Twin

Cities (28.62 inches), and Duluth (28.55 inches). During what was labeled the "Storm of the Century" by many meteorologists and the media, constant winds blew at 30 to 50 miles per hour, with gusts ranging from 50 to 80 miles per hour, producing snowdrifts up to 20 feet high. Snowfall amounts varied from just a few inches in southeastern communities to 20 or more in a number of central and northern communities, topped by 27 inches at Riverton in Crow Wing County. Near Willmar in Kandiyohi County, 168 passengers were trapped for hours in a stalled train because wind-chill values of -25 to -35°F made it too dangerous to walk to shelter. Thirty-five storm-related deaths were reported, 14 from the blizzard and 21 from heart attacks. Assessed damages to structures and costs of snow removal and restoration of power and phone service totaled $6 to 8 million. The American Red Cross provided food and shelter to nearly 17,000 people. Again, the NWS was widely praised for its forecasting and warning efforts, which together likely saved many lives.

■ Two more blizzards came in March 1975. The first, on March 23 and 24, was preceded by freezing rain across southwestern and central counties. As the southeast largely escaped the storm's impact, elsewhere winds gusted to 60 miles per hour and snowfall was quite heavy, ranging from 5 to 17 inches in central and northern counties. A coating of 1 to 3 inches of ice brought down power and telephone lines across southwestern Minnesota. Interstate 35 was closed from Forest Lake in Washington County to Duluth, and Interstate 94 was closed from Alexandria in Douglas County to Sauk Centre in Stearns County. Waves up to 20 feet high pounded the Lake Superior shoreline, causing property damage at Duluth and Two Harbors (Lake County). Then, a second blizzard struck over March 26–29, bringing snow mixed with intermittent freezing rain and wind gusts to 70 miles per hour. The total snowfall from this slow-moving storm was 4 to 10 inches across the south and 5 to 17 inches in the north, where air, bus, and train travel was curtailed. The Duluth airport closed, as did Interstate 90 near Austin in Mower County, stranding more than 100 travelers.

Q & A

In meteorology, what is meant by the term *dog teeth*?

a. a frightening forecast

b. the symbols on a blue cold front that show direction

c. a weather pattern that won't let go

(answer on page 329)

■ A most memorable winter storm struck the northeastern area of the state on November 10–11, 1975, bringing strong wind, rain, and snow. Southern counties reported 0.5 to 1 inch of rainfall, while northern counties saw a few inches of snow. This mild-mannered storm intensified over Lake Superior, however, producing winds up to 71 miles per hour and creating waves of 12 to 15 feet. The ore carrier *Edmund Fitzgerald* was caught in the storm and sank rapidly on November 10, 1975, taking with it all 29 crew members. Immortalized by the Gordon Lightfoot song "The Wreck of the *Edmund Fitzgerald*," this tragic storm is commemorated each November 10 in a ceremony at Split Rock Lighthouse, along the shores of Lake Superior.

Q. What is the greatest amount of snowfall recorded in one month in the Twin Cities?

(answer on page 329)

■ On November 18–19, 1981, a winter storm brought heavy snow in a narrow band across central Minnesota: Gaylord in Sibley County, Minneapolis, and St. Paul all reported 10 inches or more. The heavy, wet snow brought down power lines and caused the inflated roof of the relatively new Hubert H. Humphrey Metrodome to tear and collapse.

■ In January 1982, back-to-back winter storms delivered very heavy snowfalls across much of the state. Over January 20 and 21, many communities reported totals ranging from 10 to 20 inches, with a record 20 inches at Itasca State Park in Clearwater County and 17.4 inches in the Twin Cities. Then, after a brief respite, a second winter storm produced more snowfall starting late on January 22 and continuing the next two days. The second storm brought another 10 to 20 inches to even more communities than the first. Strong winds produced blizzard conditions in parts of the state, closing Interstates 90 and 35 for a time on Friday, January 22. The accumulated snow collapsed the roofs of several buildings; as one result, a tool for removing snow loads—the roof rake—gained in popularity.

■ On February 4, 1984, a fast-moving blizzard traveled north to south across western Minnesota, bringing wind gusts up to 80 miles per hour as arctic air spilled down the Red River Valley and through the Minnesota River Valley. Snowfall was highly variable and very light, but strong winds blew the accumulated snow cover, causing near-zero visibility. Temperatures plummeted by 20 to 30 degrees from afternoon through evening, and extreme wind-chill values, ranging from -25 to -35°F, compounded the danger. The storm's rapid approach—it moved at 50 miles per hour—and deteriorating conditions left many people stranded in their vehicles or ice houses; 16 perished in the storm.

▨ Late on March 2, 1985, a blizzard formed in southwestern Minnesota and spread east with high winds up to 40 miles per hour, producing freezing rain, thunder, lightning, sleet, and heavy, wet snow over the following two days. Many areas saw more than a foot of accumulation; in the west and north, storm totals exceeded 20 inches, topped by 24.4 inches at Brainerd in Crow Wing County. Interstate 94 was closed briefly near Fergus Falls in Otter Tail County, and other roads were blocked for longer periods. The wind gusted to 68 miles per hour at Rochester and 71 miles per hour at Duluth, creating drifts as high as 20 feet at the latter location during one of the strongest and longest-lasting March storms the state has seen.

DID YOU KNOW?

A snow garland is a rare and beautiful snow rope formed between fences or trees during a heavy snowfall with nearly calm winds.

▨ One of the most memorable weather systems to visit Minnesota during the extreme drought year of 1988 was a blizzard that progressed from southwest to northeast on November 26–27. Striking on the Saturday and Sunday following Thanksgiving, the storm stranded many travelers: Interstate 94 was closed near Fergus Falls (Otter Tail County), as was Interstate 90 from Sioux Falls, South Dakota, to Fairmont, Minnesota (Martin County). Whiteout conditions prevailed in western counties; Windom in Cottonwood County recorded a wind gust to 63 miles per hour. Snowfall amounts were generally 3 to 9 inches, but some central and northern communities saw more than 10, including Wright in Carlton County, which measured 14 inches.

▨ The Red River Valley experienced one of its worst blizzards on January 6–8, 1989, as a heavy dose of snow and wind affected Traverse County north to Kittson County. Winds of 35 to 50 miles per hour accompanied an invasion of arctic air, yielding dangerous wind-chill values of -30 to -40°F. Snowfalls of 18 to 24 inches were common, and drifts up to 10 feet blocked many roads, including Interstates 90 and 29. The storm contributed to an extremely large amount of snow cover that winter, eventually leading to spring flooding on the Red.

Over March 22–23, 1991, heavy, wet snow accumulated across the Minnesota landscape, from southwestern counties to northeastern counties along Lake Superior. Tracy (Lyon County) reported more than 10 inches of snow, while Montevideo (Chippewa County) saw more than 16 inches. Along the North Shore, a mixture of sleet, snow, and freezing rain caused a number of traffic accidents, and the ice buildup on the 850-foot WDIO TV tower in Duluth was so great that it caused the structure to topple.

One of the state's most remarkable winter storms occurred from October 31 to November 3, 1991. During the aptly named Halloween Blizzard, at least 30 communities reported 20 or more inches of snowfall, including record amounts in the Twin Cities, 28.4 inches, Two Harbors, 36.0 inches, and Duluth, 36.9 inches. Snowfall intensity ranged up to 2 inches per hour, and a 180-mile stretch of Interstate 90 closed, stranding many motorists. In response to widespread power outages, the National Guard provided generators to rural areas. Winds up to 60 miles per hour produced 10-foot snowdrifts, closing roads for several days. Perhaps Minnesota's largest and longest-lasting blizzard, this storm inaugurated record-setting November snowfall totals for many communities, including the Twin Cities (46.9 inches), Chaska in Carver County (48.6 inches), Duluth (50.1 inches), Two Harbors in Lake County (51.5 inches), Cedar in Anoka County (52.2 inches), Forest Lake in Washington County (47.2 inches), and Bruno in Pine County (58.6 inches).

What is cooking snow?

a. snow that the sun is melting

b. heavy, dense snowpack harvested for water

c. snow falling in large, cookie-shaped flakes

(answer on page 329)

Five blizzards struck during the winter of 1995–96, repeatedly creating harsh and dangerous conditions across Minnesota. The first, hitting west-central counties on December 8, 1995, produced snowfall totals up to 8 inches and brought strong winds of 30 to 60 miles per hour and dangerous wind chills of -20 to -30°F, which justified school cancellations.

On January 17–18, 1996, a blizzard hit central and southwestern counties with wind gusts to 60 miles per hour and snowfall of 6 to 12 inches. Wind chills of -25 to -35°F along with severe blowing snow prompted school and business closures in western sections of the state on January 18 and shut down Highway 14 between New Ulm and Sleepy Eye, both in Brown County. In the east, this storm brought freezing rain and, in the Twin Cities, ice buildup that led to power outages for 180,000 customers. The month's second blizzard, striking on January 26–29, dropped snow most heavily across south-central and southeastern counties, including 28 inches in Hokah, Houston County. Interstate 90 closed, becoming in essence a parking lot for more than 200 abandoned vehicles.

February 10–11 brought blizzard number four, this time to northwestern counties, where snowfall was light but strong winds of 35 to 40 miles per hour swirled snow cover and reduced visibility to near zero. Highways 10 and 2 were closed, clogged with abandoned vehicles; Interstate 94 was similarly affected. As winter abated, yet another blizzard struck western counties on March 24 and 25, bringing strong winds and dangerous wind-chill conditions that prompted schools and businesses, as well as Interstate 94 and Highways 10 and 2, to close. Many areas reported 6 to 10 inches of snowfall; Moorhead in Clay County measured 11 inches. Not surprisingly, many communities recorded 1995–96 as one of their snowiest-ever winters. Seventeen cooperative weather observers reported more than 90 inches for the snow season, topped by 153.9 inches at Lutsen (Cook County) and 147.5 at Isabella (Lake County).

That harsh winter was followed by another in 1996–97, during which many blizzards blew through the state. The first struck northwestern counties on November 16–17, 1996, with winds up to 45 miles per hour and 8 to 13 inches of snow. Highway 2 was shut down between Polk County's Crookston and East Grand Forks. Three more blizzards came the next month—on December 17–19, December 20–23, and December 31—mostly affecting western counties with heavy snow and dangerous wind chills. The last one spoiled many a New Year's Eve celebration in Clay, Norman, and Wilkin counties: though little snow fell, 50-mile-per-hour

Q & A

Where is the official Twin Cities snowfall measurement taken?

a. Fort Snelling

b. Minneapolis–St. Paul International Airport

c. National Weather Service–Chanhassen

(answer on page 329)

winds moved the abundant snow cover with such force that huge drifts formed and visibility dropped to zero, making travel impossible.

Then, the first month of 1997 brought four more blizzards, which reigned on January 4–5, January 9–10, January 15–16, and January 21–22 and affected all parts of the state except the northeastern-most counties. Each brought heavy snow; dangerous wind-chill conditions; road, school, and business closures; and long days and nights for snowplow operators. Though February was blizzard free, additional and significant snow fell in northern, southwestern, and southeastern counties. Another blizzard struck the northwest on March 4, bringing 10 to 15 inches of snow and closing many roads. The season's final blizzard socked the northwest on April 5–6, arriving just as the worst spring snowmelt flooding in a century was starting along the Red River Valley. Many volunteers continued sandbagging along the banks of the Red during this raging blizzard, which brought 65-mile-per-hour winds and an additional 4 to 7 inches of snow. At least eight cooperative observers in the Red River Valley area reported more than 100 inches of snow for the season, topped by Fargo-Moorhead, with 117 inches. Little wonder that the worst spring snowmelt flooding of the twentieth century for many valley communities ensued that spring.

A strong winter storm on November 10, 1998, set new barometric pressure records across the state, including in the Twin Cities, which measured 28.55 inches; at Duluth, 28.47 inches; and at Austin in Mower County and Albert Lea in Freeborn County, 28.43 inches, a new state record low. These hurricane-like low-pressure readings produced strong winds; five communities reported wind speeds greater than 55 miles per hour, and St. Cloud State University in Stearns County measured 64 miles per hour—enough force to blow vehicles off roads. Significant snow came with this winter storm as well, particularly in the west: 13.5 inches at Canby in Yellow Medicine County, 10 inches at Madison in Lac qui Parle County, and 8 inches at Montevideo in Chippewa County.

Hibernal is the Latin term for winter.

Heavy snow descended upon eastern sections of the state on March 8–9, 1999. The Twin Cities airport reported 16 inches; Forest Lake in

Washington County saw 15 inches; and Fridley (Anoka County), New Prague (Le Sueur/Scott counties), Richfield (Hennepin County), Spring Lake Park (Anoka/Ramsey counties), and Waconia (Carver County) all reported 14 inches—for some, their heaviest-ever March snowfall. Some roads were closed until plowing was under way.

A rare early-season blizzard struck northwestern Minnesota on October 24–25, 2001, bringing 50- to 55-mile-per-hour winds and a heavy band of snow that ran from Clay County north through Kittson County. Argyle in Marshall County reported 14 inches; Thief River Falls in Pennington County, 11 inches; Grand Forks, North Dakota, 10.9 inches; and Hallock in Kittson County, 10 inches. Following this early, well-forecast blizzard, a powerful winter storm struck on November 26–27, bringing heavy snows to central Minnesota and a "freshwater fury" to the Lake Superior shoreline. Willmar in Kandiyohi County received a near–state record snowfall of 30.4 inches, while New London in Kandiyohi County, Granite Falls in Chippewa/Yellow Medicine counties, and Collegeville in Stearns County all tallied more than 20 inches. This storm caused power outages and more than a thousand traffic accidents. Winds greater than 50 miles per hour whipped huge waves on Lake Superior near the Duluth harbor entrance and caused more than $1 million in damages to the Lakewalk area.

Q & A

On August 14, 2001, the National Weather Service announced what major change in its weather advisory procedures?

a. no more NOAA weather radio

b. a new wind-chill index

c. new criteria for blizzard warnings

(answer on page 329)

A winter storm crossed the region on March 14–15, 2002, bringing record-setting heavy snowfalls of 10 to 20 inches: Dawson in Lac qui Parle County, for example, reported a record 21 inches. Thunder, lightning, freezing rain, and sleet were also prevalent during this storm, which effectively halted travel for a time.

A fast-moving blizzard that brought little snow but considerable wind crossed Minnesota's west-central and south-central counties on February 11, 2003. Snowfall totals were only 2 to 3 inches, but wind speeds ranged up to 60 miles per hour, creating whiteout conditions. Virtually all schools in these areas closed early, and law enforcement officials discouraged any travel, precautions that, along with good fore-casting by the NWS, helped prevent any deaths or injuries as a result of this storm. Then, an unusual winter storm closed schools and busi-nesses in southern Minnesota on April 7, 2003. During one of the state's heaviest April snowstorms, 6 to 14 inches of snow fell across a confined geographic area, including Blue Earth, Faribault, Freeborn, Martin, and Waseca counties.

A large winter storm passed over the state on January 24–26, 2004, bringing between 10 and 20 inches of snow that was blown into drifts by 25-mile-per-hour winds. Two Harbors in Lake County measured 30.5 inches; Duluth received 27.1 inches, the city's third-highest storm total; and Finland, along the Lake County highlands, reported 26 inches.

A heavy snowstorm and ground blizzard struck many of the state's western and southern counties on January 21–22, 2005. Freezing rain and sleet mixed with snow to yield accumulations of 6 to 8 inches, and sustained winds of 30 to 40 miles per hour caused zero visibility at times. Wind gusts of 60 or more miles per hour were reported from Blue Earth in Faribault County, New Ulm in Brown County, and St. James in Watonwan County. Interstate 94 closed between Alexandria in Douglas County and Moorhead in Clay County late on Janu-ary 21, and in western counties snowplow operators were pulled off the roads between midnight and 4:00 AM on January 22 because of dangerous blizzard conditions. The storm caused numerous power outages and traffic accidents.

A snowflake's shape primarily depends on

a. how windy it is

b. the air temperature

c. whether it is night or day

(answer on page 329)

Much of eastern and southern Minnesota was affected by a winter storm on March 17–18, 2005, during the state boys' high school basket-ball tournament. The storm produced a good deal of snow: 21 inches at Kiester in Faribault County, 18 at Geneva in Freeborn County, 17 at Ormsby in Watonwan County, 15 at Fairmont in Martin County, and 13.5 at Mankato in Blue Earth County. Blowing and drifting of this heavy snowfall closed Interstate 90 between Worthington in Nobles County and Albert Lea in Freeborn County and Interstate 35 between Owatonna

(Steele County) and the Iowa border, creating travel problems for those returning from the basketball tournament and rekindling interest in the tournament snowstorm legend, established during the 1950s and 1960s. A personal note about this storm is that National Weather Service colleague Bill Togstad, lead forecaster at the Chanhassen office, predicted record-breaking snowfall amounts and blizzard conditions for places like Kiester and Geneva using a new forecast numerical and graphical tool (Bufkit3). Despite objections from other meteorologists, he stuck with his forecast, and it was spot-on, earning the coveted NOAA Isaac Cline Award for expert forecasting for Bill, one of the few Minnesota forecasters to ever win this award.

A winter storm over December 13–14, 2005, yielded very heavy snowfall in northeastern counties. Called a "double-barreled" storm by the NWS, this weather system was fueled by low-pressure systems over Nebraska and the North Dakota–Manitoba border. It brought 6 to 8 inches of snow to many areas, but the north shore of Lake Superior was hit especially hard. Lutsen (Cook County) received a foot of snow, Brimson (St. Louis County) and Silver Bay (Lake County) reported more than 14 inches, and Duluth 15.8 inches. An observer near Two Harbors (Lake County) reported 26.5 inches. Schools were closed in Duluth and all along the North Shore as well.

March 12–13, 2006, brought a winter storm with heavy snow to south-central portions of the state. Most observers reported 6 to 12 inches of snowfall with wind gusts up to 35 to 40 miles per hour. Hastings (Dakota County) reported 19 inches of snow, and Highway 52 in southeastern Minnesota was closed to all traffic for a period of time. The Department of Transportation reported more than 250 vehicles involved in accidents, and flights into and out of MSP airport were cancelled or delayed all day on the thirteenth.

February 23–26, 2007, brought a slow-moving winter storm that produced ice, heavy snow, and even blizzard conditions in some areas of the state. In southwestern counties, the storm dropped nearly a quarter inch of ice to coat the roads, then snow fell upon the ice to make roads and highways all but impassable. Weighted with ice, many power lines and trees came down. Other areas of the state saw snow accumulations of 10 to 15 inches. Southeastern counties noted some record-setting amounts for February, including 23 inches at Rushford, 25 inches at Lanesboro (both Fillmore County), and 26.8 inches at Winona (Winona County).

■ Back-to-back winter storms dropped heavy snow on many parts of northern Minnesota at the end of 2007. On December 1, nearly every county received significant snowfall, with the greatest amounts in the northeast. Lake County's Two Harbors and Silver Bay and St. Louis County's Babbitt, Duluth, and Orr reported more than a foot of snow, while Grand Marais (Cook County) recorded 20 inches. Then December 4 brought another significant winter storm, with widespread snowfall amounts of 4 to 6 inches. Duluth received 15 more inches from this storm.

■ Heavy snows and blizzards dominated the Minnesota landscape in April 2008. A stationary front draped across the state April 5–7 and brought many long hours of snow, yielding some of the heaviest total snowfalls ever measured in April. Virginia (St. Louis County) reported 32 inches, Cass Lake (Cass County), 29 inches, Babbitt and Chisholm (St. Louis County) and Bemidji (Beltrami County), 26 inches, Grand Rapids (Itasca County), 25 inches, and Park Rapids (Hubbard County), 24 inches. The heavy snow shut down Interstate 94 between Alexandria (Douglas County) and Fergus Falls (Otter Tail County). Then another storm over April 10–11 brought heavy snow and blizzard warnings to many parts of northeastern Minnesota, where winds gusted to more than 60 miles per hour. Snowfall ranged from 9 to 14 inches in many places, with Askov (Pine County) reporting nearly 17 inches and huge drifts. Near the end of the month, another winter storm brought a mixture of rain, sleet, and snow, with blizzard warnings issued in west-central and north-central counties over the twenty-fifth and twenty-sixth. Many observers reported 6 to 10 inches of snow, while Hawley (Clay County) and Fergus Falls received more than 15 inches. With the frequent storms and heavy snowfalls, many observers reported new April records for total snowfall, including five communities that received more than 40 inches.

Q & A

In the Twin Cities climate record, which three months show all-time snowfall records of 40 or more inches?

a. January, February, and March

b. November, January, and March

c. December, January, and February

(answer on page 329)

■ The year 2008 concluded with a very snowy December. Four significant winter storms arrived during that month, the first coming over the thirteenth to the fifteenth and affecting primarily central and northern counties with 5 to 10 inches of snow. International Falls (Koochiching County) and St. Louis County's Babbitt, Embarrass, and Orr reported closer to 15 inches. Blizzard conditions closed down Interstate 94 between Alexandria (Douglas County) and North Dakota on the fourteenth, and wind-chill warnings were posted by the NWS as readings fell to -43°F at Thief River Falls in Pennington County. A second winter storm over the eighteenth and nineteenth brought significant snowfall to southern counties, in amounts ranging from 4 to 8 inches. The third storm came December 20–21 and delivered a good deal of snowfall to every county, primarily ranging from 4 to 8 inches, although Wolf Ridge climate station at elevation above Lake Superior on the North Shore reported more than 16 inches. Yet one final snowstorm for 2008 occurred on December 30 and delivered 8 to 16 inches across central counties. More than 25 Minnesota communities received 30-plus inches of snow during December 2008, while observers at Two Harbors (Lake County) and Wolf Ridge reported more than 50 inches.

DID YOU KNOW?

Frigophobia is fear of cold.

■ Thunder snow came to central and southern Minnesota on February 26, 2009. A short-lived storm brought 4 to 10 inches of snow, some of which arrived at the rate of 3 inches per hour. The snowfall intensity was so great that evening classes were cancelled at several colleges in the Twin Cities area, including the University of Minnesota.

■ A winter storm brought a mixture of rain, sleet, and snow to the state over March 10–11, 2009. At least a dozen observers reported more than a foot of snow, with peak accumulation of 18.8 inches at International Falls (Koochiching County). Cold northwest winds drove the wind-chill values to -35°F, and a new statewide low temperature record was set at St. Louis County's Embarrass and Babbitt with a reading of -35°F on March 12.

■ One of the most devastating ice storms to ever hit the North Shore coastal communities struck over March 23–24, 2009, bringing a half inch to as much as 2 inches of ice to areas between Two Harbors (Lake County) and Grand Marais (Cook County). There were widespread power outages and the ice damaged thousands of trees. To the west, heavy rains fell over the Red River Valley, worsening the spring snow-melt flood threat.

A late-season blizzard crossed northern Minnesota on March 31 to April 1, 2009, bringing heavy snowfall to many areas. Campbell (Wilkin County) reported 27 inches of snow, while nearby Breckenridge received 24 inches. Many other areas reported more than a foot of snow. More than 10 inches of snow at International Falls (Koochiching County) pushed its seasonal total to a new record 125.6 inches. This storm set up a difficult and prolonged Red River Valley snowmelt flood season (lasting 66 days at Moorhead) in the spring of 2009.

After an extraordinarily mild November, a blizzard hit southern Minnesota counties over December 8–9, 2009. Snowfall amounts of 6 to 10 inches were common, though Minnesota City in Winona County reported a record 16.3 inches. Winds gusted to 54 miles per hour during the storm, bringing visibility along portions of Interstates 90 and 35 to near zero. Temperatures dropped to well below zero following this storm. A long winter storm, with reports of freezing rain, sleet, and heavy snow, reigned over the state from the twenty-third to the twenty-sixth, making the Christmas travel season difficult. Every area of the state received significant snowfall, mostly ranging from 6 to 12 inches. Seven communities reported more than 20 inches, topped by Windom (Cottonwood County) with 28.5 inches. Windom also ended up setting an all-time monthly snowfall record with 48.4 inches.

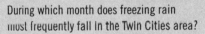

Q & A

During which month does freezing rain most frequently fall in the Twin Cities area?

a. March

b. November

c. December

(answer on page 329)

A double-barreled low-pressure system brought a challenging winter storm to the state over January 23–26, 2010. The early portion of the storm yielded freezing rain and sleet, causing icy roads and numerous accidents in the Brainerd Lakes region (Crow Wing County). Then snow fell across most areas, accompanied by winds of 40 to 50 miles per hour.

Blizzard warnings were issued in northwestern counties, and wind-chill warnings in the southeast pointed to temperatures dropping below zero. Windom (Cottonwood County) reported 10 inches of snow, while Duluth airport recorded nearly 14 inches.

◼ A rare late-season snowstorm occurred over May 7–9, 2010, in north-central and northeastern Minnesota. Grand Rapids (Itasca County) reported 6 inches of snow, and St. Louis County's Hibbing, 7. Following the storm, the temperature dropped to just 17°F at Embarrass, also in St. Louis County.

◼ A classic winter storm brought a mixture of freezing rain and snow to swaths of the state over November 13–14, 2010. Many areas reported 6 to 11 inches of snow; Maple Grove (Hennepin County) saw a foot of snow. In southwestern Minnesota, ice accumulation on roads caused numerous accidents. The water content of the snow was very high, bringing more than 2 inches to many areas. A second significant snowstorm came at the end of the month, November 29–30, dropping 7 to 9 inches of snow on many central and northern communities. Blizzard warnings were issued for portions of the Red River Valley as winds gusted from 30 to 50 miles per hour, reducing visibility to less than a quarter of a mile. Bruno (Pine County) reported nearly 14 inches of snow, while Cook (St. Louis County) received 11 inches and Morgan (Redwood County) and Olivia (Renville County) reported 10 inches.

DID YOU KNOW?

A ram penetrometer is a cone-shaped device forced into snow or ice to measure its density.

◼ A strong winter storm brought blizzard conditions to many parts of the state over December 10–12, 2010. Central and southeastern counties were affected most, as heavier snow amounts were blown around by 30- to 50-mile-per-hour wind gusts. Along the Mississippi River, Winona Dam reported 26 inches of snow, La Crescent Dam, 21.5 inches, and Read's Landing, 17. Areas around the Twin Cities saw 15 to 20 inches of snow, some of the largest December amounts ever measured. The heavy snow caused the Metrodome in the Twin Cities to rip and deflate due to excessive weight on the Teflon roof. Strong winds and blizzard conditions closed Interstate 90 between Albert Lea (Freeborn County) and the South Dakota border for several hours.

Back-to-back blizzards visited the state over December 30–31, 2010, ending a snowy month. Western counties saw blizzard conditions persist over two days, with very strong winds and snowfall amounts ranging from 9 to 14 inches. Interstate 94 along with Highways 10 and 2 were all closed for a time due to large drifts and low visibility. Many residents in the Red River Valley stayed home for New Year's Eve. Heavy snows also occurred in north-central and northeastern counties. Bigfork and Grand Rapids in Itasca County reported more than 17 inches, while Cook (St. Louis County) recorded 15.

A winter storm brought thunder snow to Minnesota over February 20–21, 2011, especially in southern and central portions of the state. The storm system moved from south to north, starting out with freezing rain and sleet along the Iowa border. Winds were strongest (30–40 mph) in west-central counties, where a blizzard warning was in effect. Many roads were closed and events cancelled on both Sunday and Monday. Record-setting amounts of snow were reported at Ortonville (Big Stone County) and Madison in Lac qui Parle County (more than 19 inches), MSP airport (13.8 inches), and Springfield in Brown County (17.5 inches).

Yet another spring sports tournament snowstorm occurred over March 22–23, 2011. The snow was preceded by thunderstorms and rain before falling temperatures brought freezing precipitation. Scores of traffic accidents occurred during the morning commute in the Twin Cities on the twenty-third, with the snow tapering off later that day. Many central Minnesota communities received from 4 to 8 inches, with Redwood Falls (Redwood County), Little Falls (Morrison County), and Rush City (Chisago County) reporting more than 10 inches. Portions of the Red River Valley were under a blizzard warning as winds gusted up to 60 miles per hour in that area. The added moisture from this storm aggravated the spring snowmelt flooding already under way along the Minnesota River Valley. Many roads and highways were closed because of flooding.

The most significant winter storm of the 2011–12 season occurred over Leap Day, February 28–29, 2012. Much of the south received rain, sleet, and freezing rain, while northern counties saw full-fledged blizzard conditions develop, with wind gusts up to 55 miles per hour. There were many school cancellations on Wednesday the twenty-ninth, including at the University of Minnesota, Morris campus. Many observers reported 6 to 9 inches of snowfall, and some reported more than 10 inches. The snow's moisture content ranged from 1 to 2 inches, high for the month of February. Numerous communities reported all-time record amounts of precipitation for the day, including 0.65 inches at MSP airport, 0.78 inches at Duluth, 0.51 inches at Rochester, and an all-time statewide precipitation record for Leap Day of 2.23 inches at Faribault.

Q & A

On January 21, 2005, the Twin Cities recorded a snowfall of 5.5 inches, establishing what new climate record?

a. new record total for January 21

b. most snowfall ever in a one-hour period

c. latest date for the first winter snowfall of 1 or more inches

(answer on page 329)

A strong winter storm brought blizzard conditions to western counties and heavy snowfall elsewhere over December 8–9, 2012. Many places received more than a foot of snow, while Chisago City (Chisago County), Forest Lake and Lake Elmo (Washington County), Marshall (Lyon County), Sacred Heart (Renville County), and St. Francis (Anoka County) reported 17-plus inches. The Minnesota State Patrol responded to more than 600 crashes, and as the snow melted and refroze, many highways took on a washboard effect, making for hazardous driving conditions that lasted for a few days. Some businesses in southwestern Minnesota closed due to low visibilities and dangerous wind-chill conditions (-35°F or colder). Another blizzard struck southern Minnesota over December 19–20, dropping up to 10 inches of snow on Minnesota City (Winona County) and causing whiteout conditions along Interstate 35, which was closed between the Iowa border and Albert Lea (Freeborn County).

A winter storm brought blizzard conditions to far northwestern communities and an ice storm to many north-central counties over January 10–11, 2013. Up to a quarter inch of ice coated trees and power lines near Bagley (Clearwater County), Bigfork (Itasca County), Littlefork (Koochiching County), and McGregor (Aitkin County). Many schools cancelled classes on Friday as a result of the icy roads. Up to 7 inches of snow with winds of 50 miles per hour caused blizzard conditions for a time farther west in Kittson, Marshall, and Polk counties. Later in the month, the twenty-seventh to the twenty-ninth, a wintry mix of precipitation greeted both northern and southern Minnesota residents. In southeastern counties, an ice storm brought down power lines and tree limbs and closed or delayed most schools in the area on Monday the twenty-eighth. Many flights into and out of the Rochester airport were cancelled. Farther to the north, 6 to 12 inches of snow fell across northwestern and north-central counties. For the month of January, International Falls (Koochiching County) reported snow on every day except for the fourth and the tenth.

February 9–11, 2013 brought heavy snow and a blizzard to some parts of the state. Many western Minnesota communities reported a foot or more of snow with winds of 30 to 50 miles per hour. Rothsay (Wilkin County), reported 21 inches of snow, Wheaton (Traverse County), 17.1 inches, Bemidji (Beltrami County), 16 inches, and Breckenridge (Wilkin County), 15.7 inches. Many schools were closed on Monday the eleventh. Another blizzard visited the northwestern counties over February 18–19, bringing 4 to 8 inches of snow with 40- to 50-mile-per-hour winds. Several highways and country roads were closed.

On February 4, 2005, Pipestone in Pipestone County reported a daytime high temperature of 60°F, which was how many degrees above normal for that date?

a. 20

b. 30

c. 40

(answer on page 329)

Over March 17–19, 2013, a strong low-pressure system brought a blizzard to northern Minnesota. Snowfall amounts ranged from 6 to 11 inches, with wind gusts from 50 to 60 miles per hour. Interstate 94 was closed between Alexandria (Douglas County) and the North Dakota border for several hours. Over the course of the storm, the temperature at Kabetogama (St. Louis County) fell from 15 to -25°F.

One of the worst ice storms ever recorded in southwestern counties occurred over April 9–10, 2013. The heaviest coating was between the cities of Luverne (Rock County) and Worthington (Nobles County), where ice thickness ranged between a half inch and an inch. Combined with strong winds, the result was extensive tree damage, widespread power outages, and stoppage of all traffic on some roads and highways. The power outages lasted for days, and Governor Mark Dayton activated the State Emergency Operations Center to help citizens of various communities find shelter and see their power restored. Around Worthington, the storm clean-up effort took more than two months. Damages caused by this ice storm were estimated to exceed $70 million. In southeastern counties, the storm brought record-setting heavy rains. Harmony in Fillmore County established a new statewide precipitation record, with 3.14 inches reported on the tenth. A little over a week later, one of the heaviest April snowstorms of all time struck northeastern sections of Minnesota over the eighteenth and nineteenth. Many observers reported more than 10 inches, with snowfall rates of up to 3 inches per hour during the storm's peak. Some of the heaviest amounts included 29 inches at Isabella (Lake County), 23 inches at Babbitt (St. Louis County), 19 inches at Floodwood (St. Louis County), and 16 inches at Cloquet (Carlton County).

The last significant snowstorm of spring 2013 occurred over May 1–3 and was record setting, especially in southeastern counties. A number of observers reported three-day snowfall totals of 10 inches or more. Dodge Center (Dodge County) set a new statewide record May snowstorm total of 17.2 inches, as well as a new all-time May daily snowfall record of 15.4 inches on the second. Rochester airport also reported a new record snowfall total of 14.5 inches. The heavy snow damaged trees and power lines and caused numerous school cancellations in southeastern communities. Some rural buildings and barns collapsed as a result of the heavy, wet snow.

Significant Days

Christmas Eve and Christmas Day

Those who dream of a white Christmas in Minnesota are likely to see their wish come true: the chances of snow cover on this holiday vary from 60 percent in the south to 100 percent in the north. While virtually all Christmases are white in International Falls (Koochiching County) and Hibbing (St. Louis County), three-fourths meet that standard in the Twin Cities, and even the less-snowy landscapes of Worthington and Redwood Falls show a 60 percent or better chance of a white Christmas Day. Additionally, snowfall of some sort occurs between Christmas Eve and New Year's Day at most Minnesota locations about 85 percent of the time. And, in the Twin Cities area, snow is observed on Christmas Eve and/or Christmas Day approximately 43 percent of the time. Gifts like sleds, skis, snowshoes, snowmobiles, caps, and gloves thus have a high probability of immediate use.

Not every Christmas holiday is white, however. Lack of snow—defined as the absence of both snow cover and snowfall on Christmas Day—has been a reality in the Twin Cities 11 times since 1950: 1958, 1965, 1967, 1977, 1979, 1986, 1988, 1997, 2002, 2006, and 2011. Described as "brown Christmases," they have often been followed by 20 or more inches of snow, redeeming the balance of the season for those who crave a snow-covered landscape.

The state record snowfalls for Christmas Eve and Christmas Day are sizable. The Christmas Eve record amount is 15.5 inches at Isabella (Lake County) in 1959 and at Windom (Cottonwood County) in 2009. The Christmas Day record snowfall is 14.9 inches at Two Harbors (Lake County) in 2009. All-time state temperature records for Christmas Eve originate from the same decade of the nineteenth century: a high of 57°F at Northfield in Dakota and Rice counties in 1888, and a low of -43°F at St. Vincent in Kittson County in 1884. Christmas Day records include a high of 62°F at Faribault in Rice County in 1923 and a low of -50°F at Orr in St. Louis County in 1933.

Many holidays are memorable for somewhat intolerable weather. For example, Christmas Eve and Christmas Day 1903, 1933, and 1934 each brought cold wave warnings and wind-chill readings of -25 to -45°F; consequently, most Minnesotans celebrated quietly at home. On Christmas Eve and Christmas Day 1945, 20 hours of continuous snowfall blocked roads and required snowplow operators to work the holiday in southern Minnesota. In 1982 two separate winter storms paralyzed many communities, dumping 10 to 18 inches of snow between December 24 and 31. Both storms produced thunder snow, and the second brought a coating of glaze ice up to 1 inch thick, causing extended power outages in the southeast.

More recently, in 2009 a winter storm brought heavy snow and a blizzard to parts of the state over December 23–26, causing real travel

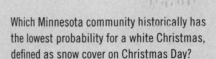

Q & A

Which Minnesota community historically has the lowest probability for a white Christmas, defined as snow cover on Christmas Day?

a. Albert Lea

b. Springfield

c. Winona

(answer on page 329)

problems for the holiday season. Full-fledged blizzard conditions pre-vailed from the Red River Valley to Duluth, where strong winds pro-duced 10-foot waves on Lake Superior. Many cities reported 15 to 25 inches of total snowfall. More than 100 flights at MSP airport were delayed or cancelled on Thursday the twenty-fourth, and travel by car was slow and treacherous in many cases. At the Mall of America, some shoppers had entire stores to themselves.

Perhaps the most amazing Christmas weather story is one related by St. Paul weather historian Tom St. Martin. In 1879, following snow-falls totaling several inches just before the holiday, a strong high-pressure system ushered in an arctic air mass. Christmas morning brought record cold tem-peratures to the Twin Cities: -38°F at the signal corps offices in St. Paul and Minneapolis and -39°F at Fort Snelling. The *St. Paul Dispatch* reported that the in-tense cold was "very discouraging to outdoor amuse-ments and even [interfered] with visits of friendship, courtesy, charity or religion." It added, "yesterday will be long remembered as the cold Christmas." The coldest Twin Cities Christmas morning since then was in 1996, with a temperature of -22°F.

Three of the Twin Cities' coldest Christmas Days in the twentieth century—with average wind-chill values of -33°F—occurred in which consecutive years?

a. 1933, 1934, 1935

b. 1950, 1951, 1952

c. 1974, 1975, 1976

(answer on page 329)

In the Twin Cities area, extreme wind-chill factors have occurred on Christmas Eve 1983, -54°F, and on Christmas Day 1903 and 1934, -40°F. At these temperatures, exposed human skin can freeze in as few as ten minutes. There have been streaks of three consecutive Christmas Eves and Days when wind-chill factors exceeded advisory criteria. Perhaps the worst stretch occurred from 1933 to 1936: these four holiday seasons were plagued by wind-chill conditions ranging from -20 to -40°F. These readings, on top of statewide agricultural and economic failures, must have truly tested the Christmas spirit.

The chilliest week preceding Christmas occurred in 1983, when most Minnesota locations reported average daily temperatures that were 30 to 35 degrees below normal. Even in the Twin Cities' urban heat island the average daily high was -5°F and the average low -21°F. On Christmas Eve, Olivia (Renville County) and Willmar (Kandiyohi County) reported a daytime high of only -20°F. Ten to 30 inches of snow cover around the state amplified the effects of every polar air mass during the month, making December 1983 the coldest of the century. In fact, only the pioneer era records for 1822, 1831, and 1872 showed colder Decembers.

The warmest week preceding Christmas in the modern Twin Cities record is a virtual tie between 1923 and 1931, both of which show an average temperature of 35°F, about 17 degrees above normal. In 1923 there was no snow cover until a storm on December 27, while in 1931 the ground was brown until 1.6 inches of snow fell on Christmas Eve. These readings cannot compare, however, to the warmest-ever Christmas week in the pioneer record. In 1877 the average daily temperature for December 18–24 was 42°F at Fort Snelling and 40°F at Minneapolis. The Mississippi River was ice free to the north of St. Anthony Falls, and some flowers bloomed as late as December 28. This climate singularity is best remembered as the "muddy Christmas," when abundant rainfall made travel difficult for other than the usual reasons.

New Year's Day

In Minnesota, snowfalls occur on New Year's Eve or New Year's Day about 35 to 40 percent of the time and, with fewer people traveling for this holiday, present fewer challenges than Christmas storms. Several years are remembered for heavy snows, including 1911, 1916, 1921, 1960, 2005, and most recently 2011. Each New Year's Day storm was memorable in its own way.

In 1911 it was actually dangerous to venture outside. A "severe snowstorm raged all day," reported Martin Hovde in the official Twin Cities National Weather Service diary. "Raged," indeed: with temperatures ranging in the single digits to the teens and howling winds of 30 to 40 miles per hour, falling snow was blown into drifts of several feet and visibility was nil for hours. Wind-chill values ranged from -20 to -30°F, and snowfall across the southern half of Minnesota totaled 3 to 6 inches.

On New Year's 1916—during one of the state's hardest winters—the National Weather Service issued a "cold wave and heavy snow warning" early in the day, and central and western sections of the state received plenty of winter weather. Beardsley (Big Stone County), Collegeville (Stearns County), and Bird Island (Renville County) all reported 5 or more inches of snow, but the real problem was the wind. Persistent winds of 20 to 30 miles per hour blew around the already deep snowpack, which ranged from 1 to 3 feet in many places.

Fun in the snow: twin Macalester students ski during the school's 1936 winter carnival.

New Year's 1921 brought snow to many Minnesota communities, including 3.8 inches in the Twin Cities. The northwest received an especially heavy load: travelers in the Red River Valley were paralyzed by 10 to 18 inches, the latter reported from Ada of Norman County, which holds the statewide record for the date. Fergus Falls in Otter Tail County saw just 4.9 inches of snow, but Baudette of Lake of the Woods County reported 8. Interestingly enough, January 1921 ended up being a relatively dry month for most of Minnesota.

New Year's Day 1960 was dominated by widespread and persistent snow, thanks to a storm that covered most of the state. Although total amounts were not record setting, many communities reported several inches of snow: the Twin Cities, 3.7; St. Cloud (Sherburne County), 5.3; Milan (Chippewa County), 6; Cambridge (Isanti County), 7. A low, lingering cloud ceiling kept temperatures nearly isothermal, ranging from only 27 to 29°F in the Twin Cities and 23 to 27°F at St. Cloud.

In 2005 a classic large-scale winter storm brought a variety of conditions to the state, including rare displays of thunder and lightning in the Twin Cities, St. Cloud, and Duluth on New Year's Day. An ice storm deposited up to half an inch across southeastern Minnesota, while sleet and ice pellets with diameters of up to one-quarter inch fell in central counties, including Ramsey and Hennepin. As colder air swept in, all precipitation turned to snow, 1 to 3 inches across the south and 6 to 10 in the north. Duluth reported 9.5 inches and Babbitt in St. Louis County 10.2 inches—new daily records for both sites. For Duluth this storm was only the beginning: its total monthly snowfall was 45.7 inches, the city's second-highest January reading.

New Year's Eve (2010) and New Year's Day (2011) saw back-to-back blizzards visit the state, separated in time by a lull of only 14 hours. Both blizzards brought fierce winds, exceeding 40 miles per hour in many areas, and snowfall amounts that generally ranged from 4 to 16 inches.

Fergus Falls (Otter Tail County) reported nearly 17 inches of snow. Visibility was so poor in western sections of the state that Interstate 94 along with Highways 10 and 2 were closed down for periods of time. Some people away from home to celebrate the holiday had to book an extra night's stay.

Whether snow falls or not, a Minnesota year can begin with very cold weather. Northern communities typically record below-zero temperatures on New Year's Day with a frequency of about 40 to 60 percent, while those in southern counties record such values about 20 to 40 percent of the time. Four of the coldest starts to the year occurred in 1885, 1899, 1974, and 2014, all under very dry high-pressure systems with mostly ample snow cover. St. Vincent in Kittson County holds the state record cold temperature for New Year's Day: -46°F in 1885, with 10 inches of snow cover. Pokegama Dam in Itasca County recorded -44°F in 1899 with 15 inches of snow on the ground; Tower in St. Louis County and Wannaska in Roseau County reported -44°F in 1974 with 8 inches of snow cover; and Embarrass reached a low of -46°F in 2014, surprisingly with no snow cover.

In contrast to these chilling temperatures, highly unusual New Year's Day weather came to the state in 1897 and 1998. Fog, rain, and warm temperatures opened 1897 as record rainfalls of greater than 1 inch occurred at New London in Kandiyohi County and Campbell in Wilkin County. Lawrence in Mille Lacs County reported 2 inches, a statewide record level of precipitation for the date. Afternoon temperatures reached the fifties at Caledonia in Houston County (55°F), St. Charles in Winona County (54°F), and Albert Lea in Freeborn County (53°F), marking one of the mildest starts to the year in the state's history. New Year's Day 1998 is remembered in southern Minnesota as a day for golfing. With an El Niño–inspired mild winter and no snow cover in the south, many golf courses had already opened for a few days in December. On New Year's Day temptation was difficult to resist: temperatures hit 50°F or higher at such places as Pipestone in Pipestone County, Canby in Yellow Medicine County, Madison in Lac qui Parle County, Marshall in Lyon County, Lamberton in Redwood County, and Luverne in Rock County. In fact, the Luverne reading of 56°F remains the state high temperature record for this date.

Groundhog Day

If groundhogs see their shadows on February 2
Men can take for granted
six more wintry weeks are due

Groundhog Day, or Candlemas, the halfway point of winter, has been recognized in many midlatitude cultures for centuries. Inspired by anxiety concerning how long winter will last, it is also founded on the belief that some of Mother Nature's creatures are better at anticipating weather than humans are.

Popularized by German communities in Pennsylvania, Groundhog Day festivals flourish on the premise that this hibernating animal can predict the weather. When the groundhog or woodchuck emerges from its burrow on a bright and clear February 2, it may be frightened by its shadow and retreat back into the earth for another six weeks. If the day is cloudy and dull, no shadow is projected, and the groundhog may stay out for a while, indicating to observers that spring—with its increased cloudiness and rain—is just around the corner.

In early Christian times, this date was known as the Feast of the Purification of the Virgin Mary or the Feast of the Presentation of Christ in the Temple. Later, in pre-Reformation England, it became Candlemas, when a church's annual stock of candles was blessed. For centuries there were public ceremonies on February 2, and folklore evolved from historical weather observations:

If Candlemas Day be fair and bright
Winter will have another fight;
But if Candlemas Day be clouds and rain
Winter is gone, and will not come again.

In Europe, Canada, and the United States, the behavior of hibernating animals such as badgers, bears, and woodchucks has long been observed as a means to predict the weather that follows this date.

A look at the historical record for cloud cover on Groundhog Day in Minnesota suggests that between 70 and 80 percent of the time an animal will cast a shadow on this date and thus forecast continued winter weather. About five percent of the time it is foggy on Groundhog Day, as it was in 2012, when a dense fog prevailed for most of the day, especially in the Twin Cities area. Average daytime temperatures around the state range from the teens into the twenties, though the Twin Cities tend to record temperatures above freezing about one Groundhog Day in five.

Two of the most unusual Groundhog Days occurred during the 1990s. The warmest ever—by a significant margin—was in 1991, when several locations reported daytime highs in the sixties, the statewide record for the day being 66°F at Wheaton in Traverse County. The Twin Cities' record high of 48°F on February 2 was also reported that year. Groundhog Day 1996 brought the coldest temperature ever measured in the state of Minnesota: -60°F at Tower in St. Louis County. In fact, that

Did he see his shadow? These boys await the groundhog's appearance on February 2, 1945.

morning's weather established many local records, as 11 communities reported low temperatures of -50°F or colder.

Sometimes Groundhog Day weather leaves little doubt that winter will continue. Such was the case in 1915, when February 2 was plagued by an all-day storm that dumped 15 to 20 inches of snow in southern Minnesota, closing schools and obstructing transportation. In 2004 a similar storm dumped 5 to 12 inches on the state; some locations reported 45 hours of snowfall from January 31 to February 2. The 10.7 inches of snow the Twin Cities saw on just February 1–2 was the most ever for the month. In both 1915 and 2004, Mother Nature's message was unambiguous.

Valentine's Day

February 14 is a day for romance, indoors if you please, though some Minnesotans prefer to take their sweetheart for a sleigh ride.

In the Twin Cities, Valentine's Day has a reputation for being snowy: there is snow cover nearly 80 percent of the time, and it actually snows on February 14 about one year in four—6.4 inches in 1950, for example. Statewide, snow typically covers the ground, often to great depths in the north. On Valentine's Day 1969, many northern locations reported more than 3 feet of snow on the ground. Northeastern sections of the state are prone to significant snows on Valentine's Day, as in 1936 when Grand Marais in Cook County received 22 inches—yielding good sleighing conditions.

February 14 can be quite cold as well. Though the Twin Cities' average high and low temperatures are 25°F and 9°F, respectively, on Valentine's Day the mercury has read below zero 31 times since 1891, bottoming out at -21°F in 1936. For many the cold is a good excuse to cozy up to the fireplace and share some spirits or a cup of tea, activities surely justified in 1906 when temperatures as low as -40°F kept even the hardiest citizens indoors. Conversely, outside activity must have been the rule for Valentine's Day 1954, as much of the state enjoyed temperatures in the fifties and sixties with abundant sunshine and no snow cover. One could have hand-delivered fresh-cut flowers that day without any risk of spoilage.

Minnesota's Spring

Minnesota's spring weather pattern is perhaps best described as erratic. As the days grow perceptibly longer and the landscape sheds winter's snow cover, edgy residents begin to anticipate a warmer season. March days may offer beautiful clear skies, warm southerly winds, and increased temperatures and humidity, prompting some citizens to wear short-sleeved shirts. After months sealed against winter's cold, windows and doors are thrown open to admit fresh, spring-like air. All too often, however, these warm spells are simply a tease, awakening Minnesotans from their late winter lethargy and inciting enthusiasm for assorted outdoor activities.

With or without such teasers, winter can extend deep into spring, testing residents' endurance and their patience. March blizzards have been well documented, and in northern counties snow cover, frozen soil, and lake ice have on occasion lingered into May. Backward-looking Aprils bring precipitation in the form of snow more so than rain; damaging frosts can occur as late as Memorial Day weekend. Spring snowmelt floods have consumed entire communities, inspiring stories of trauma, heroics, and despair. Perhaps *erratic* is too mild a word to describe Minnesota's spring.

March sometimes sets the tone for the spring season, bringing above-normal warmth and more rain than snow or favoring winter-like temperatures and heavy snowfalls. The range of extreme temperatures—from 88°F at Montevideo (Chippewa County) to -50°F at Pokegama Dam (Itasca County)—is larger in March than in any other month. March not only brings the vernal equinox but also the first phenological signs of spring. Fishing houses are removed from lakes as the ice begins to crack and rot; a chorus of frogs chirps from thawing ponds; American robins return to the state and visit backyard bird feeders; snow cover melts away and soils begin to thaw, exposing muddy fields; cumuliform clouds become more visible and occasional thunder is heard. In most communities, average daytime high temperatures are below freezing to start the month but above 40°F by its end. For many areas, especially in western counties, March brings the year's first 60-degree days. High temperatures can even reach the eighties: the month's all-time high of 88°F was reported in Montevideo, Chippewa County, on March 23, 1910. For such temperatures to occur the landscape must be free of snow, a condition far more prevalent in southern and western counties than in the north.

· ·

Gustiness factor

The gustiness factor is the ratio of the range in wind gusts (maximum minus minimum) divided by the mean wind speed for short periods. For example, if the wind gusts during a particular hour ranged from 30 miles per hour to six miles per hour and the mean wind speed was 24 miles per hour, the ratio would be 24/24, a gustiness factor of 100 percent. This measurement used to be reported from airports to alert pilots to significant wind variation that might affect their approach and takeoff.

· ·

A snow-free landscape was just the case in 1878, 1910, and 2012, the three warmest Marches in state history. Early spring wheat planting was the rule in 1878 and 1910—in fact, by the end of March 1878 all spring grains had been sown, a state agricultural record. That March was exceptionally warm because of persistent cloudiness and high minimum temperatures—on many nights the temperature never fell below 40°F. After 13 consecutive days with temperatures in the forties and fifties, Duluth harbor was opened to shipping on March 17. Lake Pepin was ice free by March 9, Lake Minnetonka by March 11, and Lake Osakis by March 13, all the earliest recorded dates for these conditions. Similarly, March 1910 saw dry weather, clear skies, and very warm daytime temperatures, with many communities reporting seventies and eighties nearly every date from the nineteenth to the closing of the month. More than 50 climate observers reported a daytime high of 80°F or greater. Measurable precipitation was extremely light, occurring on only two days in the Twin Cities, and comparable rates were recorded elsewhere.

What weather record was set in the Twin Cities on March 5, 2000?

a. record snowfall

b. record dew point

c. record high temperature

(answer on page 329)

Most recently, March 2012 broke many of the daily temperature records set during March 1910 and further established a new statewide record for the warmest March in history (41.8°F statewide mean). That month began colder than average, with snow on the ground, but all the snow disappeared as temperatures warmed by the tenth. The last 20 days of March 2012 were the warmest of any comparable period in history, with at least 50 climate observers reporting at least one daytime high in the eighties and several overnight lows in the fifties. At La Crescent, four nights never dropped below 60°F, and on St. Patrick's Day the mean temperature was 72°F, 38 degrees warmer than normal! During March 2012, seven new statewide daily highest maximum temperature records were set, six statewide daily highest low temperature records were set, and more than 700 daily station temperature records were set within the Cooperative Climate Station Network.

Conversely, in many years March has simply been a continuation of winter: in 1875, 1888, 1899, 1923, 1951, 1960, 1965, 2002, and 2014, the month was dominated by persistent snow cover, frequent and heavy snows, and polar air masses that brought minimum temperatures below zero degrees. But by far the coldest March occurred in 1843, when average temperatures were around 27 degrees below normal and an estimated 12 inches of snow fell at Fort Snelling. Few locations other than Fort Snelling were equipped to record the weather: the soldiers there measured below-zero temperatures on 19 days, compared to an average frequency of just two days in the Twin Cities' modern climate record. Eight daily record low temperatures from March 1843 still exist, having withstood more than 170 subsequent years of observations.

In northern counties, where snow cover is generally deeper and longer lasting, March has brought polar air that matches the coldest midwinter temperatures. On March 2, 1897, following snowfalls totaling more than 18 inches, Pokegama Dam in Itasca County reported a low of -50°F. On March 24, 1974—a date well past the vernal equinox—with 11 inches of snow on the ground, Thorhult in Beltrami County reported a low of -41°F, the latest such spring reading; and as recently as March 3, 2014, Embarrass in St. Louis County reported a low of -44°F with 30 inches of snow on the ground.

March was once the snowiest month of the year for many communities. In fact, from 1951 to 1980 a legend developed concerning snowy state boys' high school basketball tournaments. Much of this reputation was built on heavy snows in the Twin Cities area (home to the tournament) that arrived on March 22, 1952 (13.7 inches) and March 23, 1966 (11.4 inches). Today, however, for the vast majority of the state March ranks lower than January in terms of average total snowfall. In recent years, rainfall has made up a larger fraction of the monthly precipitation, as illustrated by the historically wettest Marches statewide. These were 1977 (average 2.75 inches), 1979 (average 2.63 inches), 2009 (average 2.60 inches), 1951 and 1965 (both average 2.49 inches). In 1951 the statewide average March snowfall was 31 inches, while in 1965 it was 27 inches. In both years numerous communities recorded more than 40 inches of snow for the month, topped by a remarkable 66.4 inches at Collegeville in Stearns County in 1965. For the other wet Marches, the average monthly snowfalls were 9 inches in 1977, 11 inches in 1979, and 7 inches in 2009. No individual climate station reported any snowfall totals close to 40 inches in any of those three years. Despite this trend toward rain rather than snow, March can still produce heavy snowfalls and blizzard-like conditions. More than a dozen Minnesota communities have reported 24-hour snowfalls of 20 or more inches in March. During the blizzard of March 17–18, 2005, for example, portions of southeastern Minnesota were hit with up to 21 inches of snow—as at Kiester in Faribault County—and zero visibility. More recently, the blizzard of March 31–April 1, 2009, brought 27 inches to Campbell and 24 inches to nearby Breckenridge (both in Wilkin County).

On March 6, 2003, what were the weather conditions on Lake Superior?

a. massive waves due to a winter storm

b. near-complete ice coverage

c. unusually warm water temperatures

(answer on page 329)

March can also be quite dry, as it was during the record-setting warmth of 1910, when the statewide average total monthly precipitation was only 0.28 inches and many communities reported no measurable precipitation at all. Curiously, the normally drier Red River Valley was the wettest area of the state, though most communities there still reported less than an inch of rain. Other dry Marches include those of 1883, 1887, 1895, 1909, 1912, 1930, 1958, and 1959, all of which recorded average precipitation values of 0.5 inches or less.

• •

Abraham's tree

Abraham's tree—named for the father of all Hebrews—is a cirrus cloud with feather- and plume-like appendages rising from a point on the distant horizon. In midlatitudes, it is most often seen on the western or southern horizon with the approach of a warm front or as cirrus blows out from the top of a remote thunderstorm cloud. Thus, from the observer's perspective, rain is not far away. Examples in Minnesota abound in spring, as cloud ceilings rise and convective precipitation becomes more frequent.

• •

March may wreak havoc with its uncertain weather, but usually by the time April arrives soils have thawed and root systems are active, swelling perennial buds. Many mammals come out of hibernation, and Minnesotans remove the mulch from their gardens. April is the core spring month, when familiar scents from blooming lilac, elderberry, and apple trees begin to emerge. Showers and thunderstorms are more frequent and winds more intense. In fact, average wind speeds show April to be the windiest month over most of the state. The polar jet stream begins to migrate to higher, Canadian latitudes and frequently steers weather fronts across Minnesota, bringing mostly rain, sometimes snow, and occasionally even severe thunderstorms. The NOAA–National Weather Service routinely conducts tests of the storm warning systems (including sirens) and offers special educational broadcasts during the second week of April, which is labeled "severe weather awareness week."

Many northern communities like Duluth and International Falls (Koochiching County) report at least one or two days with snow, on rare occasions heavy snow. St. Cloud (Sherburne County) recorded 24 inches on April 19, 1893, on the way to a monthly total of 49 inches, the second-highest in its station record. Pigeon River of Cook County recorded 28 inches on April 5, 1933, leading up to a monthly total of 30 inches. Several southern communities like Albert Lea (Freeborn County) and Worthington (Nobles County) see measurable snowfall about one of every two Aprils. While the month's precipitation usually falls as rain, sometimes it comes from intense thunderstorms, as at Morris (Stevens County) and Milaca (Mille Lacs County) on April 26, 1954, when it rained 6.90 inches and 5.05 inches, respectively.

Minnesota's driest Aprils occurred in 1926, 1980, and 1987, when the average monthly precipitation was just 0.51 inches or less statewide. April 1926 was a cool, dry month, especially in the northern counties, where measurable precipitation was recorded only once. Similarly, both 1980 and 1987 brought very dry Aprils to the north, this time coupled with warm temperatures. In April 1980, two communities reported no measurable precipitation, more than 100 set records with temperatures in the nineties, and six reported daytime highs of 100°F or greater, the only time such warmth has been observed during this month. In April 1987 eight Minnesota communities reported no measurable precipitation and 16 saw daytime highs in the nineties. Prior to twentieth-century agricultural practices, such dry, warm Aprils often led to widespread prairie fires, like those observed by the Fort Snelling soldiers in 1853 and 1856.

Virtually all of April's lowest temperatures occur with abundant snow. Many below-zero readings have been made in northern communities, the latest ever at St. Vincent in Kittson County on April 28, 1892: -2°F. More dispiriting than extreme April cold, extraordinary temperature swings can result from the presence or absence of snow cover combined with increasing day length and alternating passages of warm and cold fronts. For example, on April 2, 1982, strong southerly winds brought convective thunderstorms and even a tornado watch to southwestern Minnesota. Lamberton (Redwood County) recorded an afternoon high of 78°F just before a thunderstorm poured 0.90 inches of rain on the town. That night a cold front passed through, bringing strong northwesterly winds and dropping the temperature a remarkable 71 degrees to a low of just 7°F.

April is also a key month for agriculture, with producers typically out in force for tilling, fertilizing, and sowing. Most of the state's small grain crop is planted during this month, and in very warm springs the corn crop may be planted as well. In fact, the University of Minnesota Extension recommends planting corn shortly after April 20. In a number of recent years, over half of the state's corn crop (7–8 million acres) has been planted by the end of April.

Though May brings the occasional snowstorm, its weather pattern usually whets residents' appetites for the summer months to come. Several days may see temperatures in the eighties, perhaps one or two in the nineties. More than half the calendar dates show a statewide record high of 100°F or greater, topped by 112°F at Maple Plain

Minnesota's spring is predictably unpredictable: these golfers hope the rain will move on so they can resume their game.

(Hennepin County) on May 31, 1934. For many areas of the state, the daily range in temperature increases as the energy from the sun is amplified by longer day length and a higher sun elevation angle. Southerly winds can usher in very warm air from the southern plains, where the spring season is already well advanced, while easterly winds off the cold waters of Lake Superior or northwesterly winds from the prairie landscapes of Canada can greatly suppress temperature values in Minnesota. Such was the case on May 19, 2009, when at 4:00 PM bright sunshine and southerly winds brought the temperature to a record-setting 100°F at Milan (Chippewa County) and Madison (Lac qui Parle County), while at the same moment easterly winds off Lake Superior were holding temperatures at Grand Marais harbor to just 34°F!

Soil temperatures generally warm in May, and germination and rapid growth become visible in gardens and agricultural fields. Plants started indoors earlier in the spring are transplanted into the garden or yard as the threat of frost passes. May's freezing temperatures are short lived, typically lasting only an early-morning hour or two. Many southern Minnesota locations, including the Twin Cities, show freezing temperatures in May less than one year out of two. May rainfall is usually abundant. Thunderstorm frequency increases, and hail and tornadoes may bedevil

Minnesotans as well. Routine testing of severe weather warning sirens is done across the Twin Cities metro counties throughout the month and for the rest of the summer at 1:00 PM every Wednesday.

Many stations have recorded daily rainfall amounts exceeding 3 inches, the greatest being Thief River Falls in Pennington County: 7.50 inches on May 29, 1949. Strong thunderstorms over May 23–24, 2012, brought rainfall that exceeded 4 inches to Andover, Buffalo, Chanhassen, Montrose, and St. Francis in central Minnesota. May 2012 was one of the wettest in state history, with a statewide average monthly rainfall of nearly 6 inches. The very next year, on May 19–20, 2013, severe thunderstorms delivered 4.47 inches of rain to Spring Valley (Fillmore County), flooding Highway 63, and then May 29–30 saw severe thunderstorms dump 4 to 5 inches of rain in the Fargo-Moorhead area, causing widespread street and basement flooding there. In fact, May 2013 was the wettest in history for many Minnesota cities, including Grand Meadow in Mower County (14.64 inches), Spring Valley in Fillmore County (13.75 inches), Rochester (12.26 inches), and Austin (10.98 inches).

Q. What is the latest summer date when a measurable snowfall has occurred in Minnesota?

(answer on page 329)

The University of Minnesota Southern Agricultural Research Station at Waseca reported 21 days with measurable rain in May 1915 as well as nine consecutive such days in May 1974 and 1996. Rochester concluded May 2013 with 13 consecutive days of rain. Very wet Mays often herald an early and abundant mosquito population, while very dry Mays, such as those of 1934 and 1976, can produce significant drought, leading to problems with seed germination or stunted crops and what will likely be a disappointing agricultural season.

Minnesota's state parks open in May, and by then most of the golf courses are busy as well. The fishing opener occurs on the second weekend, and all of May is designated Arbor Month to promote tree planting. The pace of outdoor activity intensifies as increased day length fosters early evening recreation. Unpredictable human behaviors inspired by early spring's erratic weather patterns give way to a boundless, ambitious, and focused energy. Many Minnesotans have but one priority: spend as much time outdoors as possible.

Snowmelt Floods

Early History

The spring season has brought frequent flooding to Minnesota for centuries. Paleo records of deposited sediments along with tree-ring analysis and other documentation indicate that such floods occurred prior to landscape disturbances by European settlers. For example, the *Geoscientific Insights into Red River Flood Hazards in Manitoba*, published in 2003, shows that since 1648 there have been three periods of multiple, high-magnitude flooding, two of which predate any significant settlement: the mid-1700s, the early 1800s, and the latter half of the twentieth century. Evidence further suggests that significant floods on the Red occurred in 1776, 1790, 1809, and 1815. Fur trader Alexander Henry's journal notes ice-dam flooding on the Red near Pembina, in present-day North Dakota, in April 1808.

As more of the Minnesota landscape was settled in the early part of the nineteenth century, observations, diaries, and other documents provide accounts of spring snowmelt floods on the major watersheds, including those of the Red, the Minnesota (formerly the St. Peter), and the Mississippi rivers. Three successive spring floods on the Red in 1824, 1825, and 1826 motivated early settlers to reconsider their choice of homesteads, and many packed up and relocated to St. Peter (in today's Nicollet County) and Galena, Illinois. Though no climate measurements are available to quantify or characterize these flood years, documentation shows that they followed very snowy winters and rapid spring thaws compounded by above-normal rainfall. The Red River's 1826 flood is considered one of the highest of all time, at least since 1648: evidence suggests the crest was fully 66 feet at Pembina (current flood state there is 42 feet). Only a few estimates from the Red River flood of 1852 rival the 1826 flood. Since 1900, the highest-ever water level at Pembina is 54.94 feet, measured during the devastating flood of April 26, 1997. The fact that the 1826 flood exceeded the 1997 flood by more than 10 feet is sobering indeed for those who live on the floodplain.

Beginning in the 1820s, when Fort Snelling was established at the juncture of the Minnesota and Mississippi rivers, documentation of spring snowmelt floods on those watersheds became routine. In fact, during the spring of 1826, when the all-time historical flood occurred on the Red River, observers at Fort Snelling also noted a huge flood on the Minnesota and Mississippi rivers. March 19–20 brought "one and a half feet of snow [that] drifted into heaps of six to fifteen feet" at Fort Snelling, followed on April 5 by another 8 inches and arctic temperatures that prolonged ice cover on the rivers. On April 21, 1826, the river ice began to break up and rapid melting of the deep winter snowpack brought flooding. A 20-foot rise in the Mississippi River wiped out all the low-lying buildings along the bank, including those of Jacques Baptiste Faribault, a French fur trader and Indian liaison living on Pike Island. Dakota leader Little Crow saw his village, located near present-day South St. Paul in Dakota County, swept away as well.

• •

Precipitable water

Measurement of precipitable water shows the liquid equivalent of water vapor in a vertical column of air, that is, the depth of liquid that would result if all the water vapor were wrung out of the atmosphere through condensation. Atmospheric profiles are taken twice daily by instrumented balloons, or radiosondes, to characterize the vertical patterns in temperature, humidity, pressure, and wind away from the Earth's surface. Calculations of precipitable water made from these measurements guide precipitation forecasts. Less than 0.1 percent of the total planetary water supply is found in the atmosphere, and the average residence time for water vapor is about ten days—though it may vary from hours to weeks—before it condenses as liquid in the form of precipitation.

• •

Heavy April rains combined with abundant snowmelt produced another episode of moderate flooding on the Minnesota and Mississippi rivers in April 1844. The Fort Snelling observers noted that spring rains kept the river flows very high well into the first week of May.

April 1850 brought a Mississippi River flood to rival that of 1826. In this case, heavy snowfall was compounded by above-normal rainfall. On April 9, 1850, the *Minnesota Pioneer* reported, "we could hear the noise of the masses of ice tumbling over the Falls of St. Anthony, 8 miles distant." In St. Paul, several low-lying warehouses were submerged and some log booms destroyed.

Tracks along the riverbank can be hazardous during flooding years. This locomotive was inundated in 1916.

Deep snow cover and a wet spring combined to produce spring snowmelt flooding on both the Minnesota and the Mississippi rivers in April 1861. Nearly 2 inches of rain fell between April 4 and 7, accelerating the rivers' thaw. At Mankato in Blue Earth County, the Minnesota River was said to have reached record levels. By April 15 the Mississippi River was three miles wide south of St. Paul. The river level was so high that the steamboat *Itasca* tore down Winona's overhead telegraph line with its tall smokestack.

A second consecutive spring snowmelt flood occurred the following April. Many settlers anticipated the rivers' rise because in the Minnesota River Valley the snow depth was reportedly 4 feet. Even in Minneapolis, observers measured 3 feet of snow with high water content. Heavy rains on April 13 and 17, 1862, thawed the river ice, and the waters rapidly began to rise. At Mankato in Blue Earth County, the Minnesota River exceeded the previous year's record flood stage. The St. Croix River at Stillwater in Washington County exceeded flood stage as well, inundating several warehouses.

Other notable and damaging nineteenth-century spring snowmelt floods occurred in 1851, 1852, 1853, 1871, 1873, 1882, 1893, and 1897 on the Red River and in 1849, 1873, 1875, and 1881 on the Minnesota and Mississippi rivers. The winters preceding the majority of these floods were quite snowy; indeed, 1880–81 was probably the century's snowiest. However, not every flood came as a result of abundant snowmelt: some were compounded by ice dams that backed up the river flow, causing it to inundate low-lying areas, and some stemmed from significant spring rainfall that arrived during peak snowmelt. These latter causes hold true along the Red River, which drains from south to north, from relatively warmer landscape toward relatively colder landscape, where frozen conditions exist later into the spring season. Flooding in this flat area of northwestern Minnesota is further exacerbated by the shallow slope of the river channel, which in places drops only 1 or 2 feet per mile and drains an area of nearly 40,000 square miles between Lake Traverse in Traverse County and St. Vincent in Kittson County. During the famous 1897 Red River flood, ice gorges blocked the floodwaters, forcing them to spread as wide as 12 to 20 miles across the plain.

Coming to Understand Snowmelt Floods

Much of Minnesota's pioneer settlement took place along its major rivers, which not only provided drinking water for settlers and their livestock but also fishing, irrigation for crops, a transportation channel for trade goods, a power source for mills, and ice for refrigeration. Despite these advantages, settling on any floodplain poses a serious risk to life and property. The frequency of spring snowmelt floods (nearly 45 percent of all years produce flooding along the Red River of the North) and the damages and disruption they brought to pioneer endeavors during much of the nineteenth century prompted greater attention to the measurement of seasonal snowfall and analysis of its liquid water content. In addition, some rivers' hydrological dimensions were taken to help establish a relationship between stage—a relative measure indicating height above a fixed elevation on the landscape—and volume of flow—usually estimated in cubic feet per second. These data provided a better understanding of the variable effect of seasonal climate patterns and prolonged or excessive wet and dry periods on watershed flow. By the 1860s many Smithsonian weather observers began to report daily snowfall and snow depth in

DID YOU KNOW?

Flood stage is an arbitrarily set local measurement of stream flow that designates damage potential from high water.

inches, and several began to convert snowfall into liquid water equivalent. Some early efforts to predict floods using these measurements were marginally successful; two examples are anticipation of the 1881 flood on the Minnesota River and the 1897 flood on the Red River.

Aside from spring flooding, the weather of April 2001 set what new record for scores of Minnesota communities?

a. wettest April

b. sunniest April

c. coldest April

(answer on page 329)

Developing a conceptual model for spring snowmelt flood forecasting involved a long and slow process over several generations as new data were gathered and lessons learned. Various National Weather Service section directors analyzed spring snowmelt floods, seeking to describe and quantify possible climate-related causes. Typical of these efforts is a report by Martin Hovde, section director from the mid-1930s to the mid-1950s, that examines flooding on the Red River:

> The April 1952 flood in the Red River of the North Valley was mainly an upper valley flood with relatively minor flooding downstream from Fargo-Moorhead to the Canadian Border. In the Breckenridge-Wahpeton and Moorhead-Fargo areas the flood assumed major proportions. The crest at Moorhead, Minnesota, 34.65 feet on April 16, 1952, has been exceeded only once in the past 60 years, by 40.1 foot crest in the flood of 1897. The cause of the flood was the rapid melting during the second week in April of the heavy snow accumulation during the winter in the area to the south of Fargo. Other contributing factors were (1) unusually cold weather in December resulting in rather deep frost in the spring, and (2) a late runoff with consequent quick change to warm temperatures. Flooding began in the upper valley on the 8th. Many buildings were inundated and many others were affected by seepage. At Grand Forks, the Red River of the North was above flood stage from the 12th to the 30th. The loss on the Minnesota side of the Red River of the North was estimated at $985,600.

Researchers recognized that abundant seasonal snowfall followed by a wet spring greatly elevated the risk for flooding, as evidenced in 1881 and 1897. But widespread and extreme flooding did not always result from these two conditions—sometimes flooding was relatively minor or did not occur at all. Other important factors surely existed. Joe Strub, an NWS meteorologist and hydrologist in the Twin Cities during the 1950s, 1960s, and 1970s, is credited with developing the conceptual model for spring snowmelt floods. Study of historical floods in Minnesota, including those of the 1880s and 1890s, along with experience in forecasting floods provided a framework for his research and analysis, out of which he developed a five-factor model:

1. High levels of fall soil-moisture recharge prior to freeze-up inflates the risk of spring flooding because the soil's storage space is filled, preventing it from absorbing much spring meltwater.

2. Deep soil frost exacerbates the soil's inability to hold much more water because the frozen surface remains impermeable to meltwater or additional liquid precipitation.

3. Abundant seasonal snowfall that produces snow cover with high water content, on the order of 3 to 6 inches of water, is also a significant factor. Snow's water content can be highly variable, but high water content, especially when measured across the entire drainage basin of a major river like the Red, can translate into a tremendous volume of runoff.

4. Above-normal spring rainfall contributes because as rain falls on a snow-laden landscape it adds greatly to the snow cover's water content and accelerates the melting process. A 2-inch rainfall on top of a snowpack that already contains 3 inches of liquid water may double the expected flood crest.

5. Warmer-than-normal spring temperatures, even over just a few days, can greatly accelerate snowmelt, creating large volumes of runoff over a short period of time. This outcome holds particularly true when a warm front stalls over the state, bringing clouds that keep nighttime temperatures warm so that melting occurs continuously, day and night.

Though the NWS's modernization has produced heavily numeric hydrologic models that forecast probabilities for spring floods on most Minnesota watersheds, the conceptual model of these five critical factors still applies, embedded in the measurements and equations used in computer-based calculations. When at least three of the five critical factors exist across any given Minnesota watershed, there is a risk for spring snowmelt flooding. When four or five are present, a significant flood invariably ensues, sometimes near or exceeding record levels.

Many of Minnesota's major twentieth-century spring floods can be linked to Strub's conceptual framework. In both 1897 and 1997, all five factors pointed to a great Red River flood: fall soil moisture recharge was abundant, soil frost was deep, snowfall was significant (117 inches at Fargo-Moorhead in 1997), the spring was unusually wet, and above-normal temperatures accelerated a delayed melt. Other noteworthy floods on the Red occurred in 1950, 1952, 1966, 1969, 1978, 1979, and 1989. In addition, the Red River has reached flood stage at Fargo-Moorhead every year since 1993 (except for 2012), and in some of those recent years not only did floodwaters reach a very high stage of flow but the flood persisted for many weeks. Most of these years show that at least three of the five climate-based flood factors played a significant role. In 2009 the Red River at Fargo-Moorhead set a new record flood crest (40.84 feet) and also remained at flood stage for 66 consecutive days, a new record for longevity. Partially as a result of the long (66 days), stressful, and economically disastrous 2009 flood on the Red River, the Army Corps of Engineers designed and engineered a nearly $2 billion flood diversion channel project to help alleviate the threat in the Fargo-Moorhead area. The new diversion, scheduled to take several years to build, will be able to handle a flow volume of 20,000 cubic feet per second over a 36-mile-long diversion channel.

The Red River snowmelt flood of April 1997 was the worst in a hundred years. Only farmsteads on higher ground or those with dikes constructed around them managed to stay dry.

Climate elements are also important factors related to spring flood crests on the Mississippi River at St. Paul. In the years of 1965 and 2001, all five of the climate factors related to spring snowmelt floods were in play. In fact, in April 2001, the Mississippi River at St. Paul crested twice: heavy rainfall during the first half of the month stimulated a rapid melt and flood peak on April 18 followed by a brief recession; another heavy dose of rainfall, 2 to 3 inches in many places, during the fourth week of the month caused a second flood peak on April 30. Flooding along the

Minnesota River at Mankato in Blue Earth County was also attributed to the five climate-based factors during the springs of 1965 and 2001. Other, more recent floods on the Minnesota River in 2010, 2011, and 2014 were primarily due to some but not all of the climate factors being in play. Perhaps the most consistent factor contributing to spring flooding in the Minnesota River basin has been the frequent occurrence of wetter than normal spring months.

∙∙∙

Depression storage

Hydrologists refer to water stored in puddles, ditches, and other small indentations as depression storage. In agricultural landscapes where moldboard plowing or ridge tilling has been done, this small-scale storage can add up to quite a bit of water during the winter; such moisture must be either absorbed or drained before soils can be worked in the spring.

∙∙∙

Today the North Central River Forecast Center (RFC) of the National Oceanic and Atmospheric Administration (NOAA), located in Chanhassen, has a mandate to provide flood outlooks for all of Minnesota's major watersheds (along with those of surrounding states as well). These outlooks are typically released to the public in late February and updated monthly until the threat of spring flooding has passed. The RFC employs an Advanced Hydrologic Prediction System (AHPS) that uses a probabilistic framework. Flood crest levels—stage readings on various river gauges—are assigned a probability based on ensemble numerical models. The outlooks are derived from climatic measurements made during the fall and winter of stored soil moisture levels, frost depth, accumulated snow cover and its water content, and watershed flow volumes. Such data are combined with details from the 90-day seasonal climate outlooks for temperature and precipitation. Periodically each watershed gauging point is calibrated based on local historical measurements of flow and probability distributions. Current conditions of stream flow and soil moisture within specific watersheds initialize the outlook models, which run into

Q & A

Rochester reported what new weather record on Memorial Day (May 29), 2006?

a. rainfall of 4.50 inches

b. low of 28°F

c. high of 95°F

(answer on page 329)

the future with multiple scenarios using more than 30 years of historical climatological data. The output gives a complete range of probable values of stream stage and discharge for each watershed gauging point and consequently offers a risk assessment tool that can be used in long-range decisions involving both flooding and low-flow concerns. The resulting outlook for April 1 might show a ten percent probability of a crest near flood stage and a 50 percent probability of a crest 10 feet below flood stage. Thus, rather than a single forecast, residents and city managers consider an assigned risk, or probability, for potential flood crests that could affect their community. The AHPS system has been fully deployed since 2000 and, with some training for users provided by the NWS, has gained wide acceptance.

Because spring snowmelt floods can cause property damages in the billions of dollars—losses from the 1997 floods were estimated at more than $2 billion—the NWS prioritizes flood forecasting. Fortunately for the Minnesota public, the North-Central River Forecast Center is staffed by experienced hydrologists with comprehensive knowledge of the state and its watersheds. Through their efforts, communities are typically given enough lead time to prepare for and even to develop mitigation strategies for a spring snowmelt flood: evacuating at-risk citizens, opening shelters, sandbagging stream banks, constructing or reinforcing dikes to protect vulnerable property, opening gates to diversion channels and storage reservoirs—all of which may save time, property, and lives. Additionally, recent concerted efforts by the Minnesota Association of Floodplain Managers have produced improvements to community-based mitigation of potential flood damage by, for example, purchasing properties near or along the watershed channel or moving homes and other structures off the floodplain.

> **DID YOU KNOW?**
>
> Hydrologists use a rating curve graph to depict discharge (flow rate) versus stage (height) at fixed points along a stream or river.

History demonstrates that Minnesota's climate is inclined to produce large-scale spring snowmelt flooding; in response, a partnership among meteorologists, climatologists, and hydrologists has evolved over the past century to provide citizens with a highly accurate and valued prediction service, as well as an analysis of adaptation strategies and mitigation measures to reduce the amplitude and impacts of flooding events.

May Rituals

As might be expected of a region that suffers from long, hard winters, spring is a most welcome season, especially the month of May, after the frequent snowmelt floods of March and April have subsided. Soils, frozen hard and deep during winter, release nutrients and moisture for newly planted seeds or perennials just coming out of dormancy. Rivers and streams drain the landscape of excess moisture. Farmers plant their fields, and the land of 10,000 lakes begins to green up.

Pleasant temperatures make the increasing number of outdoor celebrations and rituals more enjoyable. May first is May Day, a celebration of spring with parades, floral demonstrations, and folk dances. Ceremonies, speeches, and outdoor activities commemorate V-E (Victory in Europe) Day on May 8 and Statehood Day on May 11. The month concludes with Memorial Day, a national holiday with parades, speeches, and cemetery decorations to honor and remember those who have served in the armed forces.

The most perplexing of Minnesota's May rituals—the annual fishing opener—usually falls on the second weekend. The start of Minnesota's fishing season regularly coincides with Mother's Day, when mothers across the nation are showered with flowers, cards, and gifts and enjoy the time-honored tradition of a nice meal in a restaurant of their choosing. Minnesota may be unique among states in that fishermen (or -persons) outnumber mothers by a considerable margin. In fact, across America Minnesota ranks first in the number of fishing licenses issued per capita. It goes without saying that the state also claims the top spot for the number of fishing houses dotting its frozen lakes, offering an attractive alternative winter lodging for the man occasionally thrown out of the bedroom by his wife.

Starting about a week before the fishing opener, local meteorologists like myself begin to get calls for a weather forecast. As any fisherman will attest, the weather can dictate the success or failure of a fishing expedition; however, this reality is not the common motivation for requesting a forecast. Minnesota fishermen are going to the opener *regardless* of the weather—unless they die first. The forecast simply provides guidance on what to wear and what type of spirits to bring along. In contrast, I hardly ever receive requests for a Mother's Day forecast. Probably most mothers are either in the boat fishing or at home enjoying some quality time with a hobby, a good book or movie, or some friends. There are stories of the guilt-ridden fisherman calculating how to spend the entire opener in a boat and somehow simultaneously honor his wife or mother on her special day. Such agonizing may lead to extravagant gifts, such as diamond earrings or a health spa gift certificate, making the true cost of participating in the fishing opener rather substantial.

It is interesting to note that in Minnesota, fishing, and to a lesser extent hunting, is the preferred and affordable form of weekend recreation for the everyday citizen, while theater or concerts are reserved for the few with the means to pursue such pastimes. In England, the opposite seems to have evolved: people there treasure music and theater, which are available to most citizens at affordable prices, while the few—the elite—fish and hunt. At least the English do not confound their traditional Mothering Sunday, the fourth Sunday of Lent, with other activities or celebrations, the way Minnesotans do.

Minnesota governor Karl Rolvaag and Iowa governor Harold E. Hughes at Gunflint Lake, 1965. Judging by their faces and their catch, this fishing opener wasn't as miserable as some.

Meteorological Lamentations

Here before the dying embers
I sit and weigh my last regrets;
When I'm right no one remembers;
When I'm wrong no one forgets.
A WEATHERMAN'S LAMENT

For meteorologists living in Minnesota, stress is just another sign of spring.

In May, suddenly it seems that everyone's chief priority is to spend some time outside. Any psychologist will describe this desire as quite logical and perfectly natural. Temperatures are getting warmer, days growing longer. Sight and smell are stimulated by renewed colors and fragrances, dormant since the last growing season. Fresh air pours in through the screen doors and open windows. Crickets methodically serenade people to sleep at night; singing birds merrily wake them in the morning.

All of this encouragement is enough to energize even the most lethargic Minnesota hibernator. Time to act on the plans and procrastinations of winter, to set expectations for the coming outdoor season: improve the golf game, redesign the garden, finish the home remodel, fish every other weekend, travel to the Boundary Waters, train for the Twin Cities Marathon, stain the deck, camp in the state parks. Spirits and hopes rise as one contemplates the possibilities.

While most Minnesotans feel joyous at this time of year, meteorologists are a notable exception to the rule. Instead of planning for the coming months, they steel themselves to endure the slings and arrows of the disappointed, the disgruntled, the dissatisfied. "Crunch time" for professional weather forecasters in this part of the world begins in May. Hours, days, weeks, even months of activities are planned on their advice: the pressure to make accurate forecasts is never greater than at this time of year. Good weather complements outdoor activity as good wine complements fine dining. But lousy weather can ruin an outdoor event, and the devastation and disappointment is especially magnified if inclement weather was not anticipated.

As if the immense pressure was not enough, the weather can be difficult to forecast in May. After all, temperatures as high as 112°F and as low as 4°F have been recorded around the state during this month. Rain has fallen on as many as 20 days, and the arrival of several inches of snow is not unusual. In some years, northern lakes have been too frozen for a successful fishing opener, traditionally held the second weekend of May. It is also tornado and hail season. Contrasting air masses battle for dominance over the territory, and the polar jet stream can meander several hundred miles in a day or two.

Accuracy is expected day in and day out; reputations rise and fall like the stock market. Meteorologists handle these somewhat frightening circumstances in various ways; some choose this month for their vacation. Those who face the public may become edgy after repeated servings of humble pie. Visible symptoms include slurred speech, loosened ties, mussed hair, shaky hands, nervous eye ticks, and increased use of the third person in referring to forecast models, as in "the models show that . . ." or "the models are calling for a change to . . ." Yet another sign of stress is the inabbility to coreckly spell; and punctuate a ritten weather forcast or advizory.

- -

Dumbbelling

When a low-pressure system splits into two circulating lobes that remain close together, this behavior is called dumbbelling. The lobes act as a single system but appear as two distinct rotating cloud masses on satellite imagery. In these cases, dumbbelling refers to the shape of the pressure pattern rather than to the character of the forecaster.

- -

This stress was reduced somewhat for National Weather Service forecasters with a new technology, introduced in 2000, called the Console Replacement System or NOAA weather radio, which used a computer-driven synthesized voice broadcaster affectionately known as "Igor." Igor did not stutter or slur his speech, nor did he use weasel words like "should," "maybe," or "possibly." Igor gave the forecast without emotion: no inflections to indicate hope, doubt, panic, or drama, just the pertinent details. Some said he sounded like an Arnold Schwarzenegger clone. Fortunately, in more recent years friendlier synthesized voices, known as "Donna" and "Tom," have been introduced by NOAA weather radio; they, too, offer stress reduction for forecasters.

Significant Days

St. Patrick's Day

St. Patrick's Day, March 17, is a time for festivals, feasts, and parades in many Minnesota communities. In the capital city's parade, a tradition since 1967, colorfully costumed people and bedecked vehicles make their way from Union Depot through St. Paul's downtown to Rice Park. Food vendors offer traditional selections such as corned beef and cabbage and shamrock-green beer. And what about the weather? All joking about the Irish gift for gab aside, it happens that March 17 is arguably the windiest celebration day on the Minnesota calendar. The average wind speed in the Twin Cities on St. Patrick's Day is 11 to 12 miles per hour, but gusts of 40 or more miles per hour have been recorded. Revelers commonly have to hold on to their Irish hats while watching or marching in parades.

March, especially St. Patrick's Day, is known for wind: pedestrians fight "a light breeze" in St. Paul, 1948.

In fact, the high wind gusts recorded for 1987, 1990, 1999, and 2011 prompted the National Weather Service to issue wind advisories for the St. Paul parades in those years. In the Twin Cities area, average daily wind speeds in March are higher than those of any other month except April; thus, the brisk breezes of St. Patrick's Day should come as no surprise.

Minnesota's average daily high temperatures for March 17 range from the upper twenties to the upper thirties, but highs of 50°F or warmer are not without precedence. The Twin Cities have enjoyed such unseasonable warmth on this holiday 21 times since 1891, topping out with 80°F in 2012. That same year, St. Patrick's Day was 80°F or higher in more than 40 Minnesota communities, setting a new statewide high temperature record with a reading of 83°F at Browns Valley (Traverse County), Redwood Falls (Redwood County), St. James (Watonwan County), and Winona (Winona County).

Perhaps the worst St. Patrick's Day weather occurred in 1965. As temperatures hovered below freezing all day, heavy snow and blizzard conditions prevailed over the entire state but for the far northwestern counties. Starting at about 5:00 AM and continuing for most of the day, wind-driven snow minimized visibility. Drifts blocked scores of highways and roads, and hundreds of schools around the state were closed. Total snowfall ranged from a few inches to more than a foot. The Twin Cities received a record 11.2 inches, while St. Cloud (Sherburne County) reported 13 inches and Duluth 17.2 inches. Collegeville in Stearns County set the state record for the date—23.6 inches—contributing to a monthly total of 66.4 inches there. The festive spirit appropriate to St. Patrick's Day must have been difficult to muster under those extreme circumstances.

TABLE 14: Average Wind Speed Reported from Minneapolis–St. Paul International Airport for March 17, 1945–2014

YEAR	AVERAGE WIND SPEED miles per hour (maximum gust noted in parenthesis)	YEAR	AVERAGE WIND SPEED miles per hour (maximum gust noted in parenthesis)
1945	15.8	1981	15.5
1946	12.4	1982	5.6
1947	9.7	1983	10.3
1948	5.9	1984	14.5
1949	11.7	1985	9.2
1950	16.6	1986	9.0
1951	8.3	1987	20.5 (39)
1952	13.4	1988	9.8
1953	10.9	1989	11.3
1954	14.5	1990	18.3 (37)
1955	16.5	1991	7.2
1956	8.9	1992	12.3
1957	9.7	1993	6.8
1958	10.7	1994	14.1
1959	8.4	1995	10.9
1960	9.0	1996	10.3
1961	10.9	1997	10.3
1962	6.9	1998	12.7
1963	13.4	1999	19.8 (46)
1964	13.8	2000	12.9 (30)
1965	20.1 (34)	2001	5.0
1966	16.4 (33)	2002	12.0
1967	11.3	2003	3.0
1968	17.4 (40)	2004	6.0
1969	7.6	2005	10.0
1970	7.5	2006	3.5
1971	4.6	2007	2.7
1972	8.7	2008	10.4
1973	11.8	2009	9.2
1974	6.7	2010	6.5
1975	12.0	2011	10.3 (35)
1976	8.4	2012	12.0 (33)
1977	13.9	2013	12.6
1978	11.2	2014	13.3
1979	15.2		
1980	12.3	**70-year average equals 11.0**	

Easter

The date of Easter varies from March 23 to April 25, opening the way for substantial differences in weather, from warm to cold and snow to rain. On average a March date for Easter Sunday brings highs in the thirties and forties and lows in the twenties. Since the Minnesota National Weather Service Offices opened in 1891, Easter Sunday has occurred 28 times in March. In the Twin Cities, nine of those have been wet and nine have brought snowfall (even if only a trace), the most being 2.5 inches on March 31, 1929. Average daytime highs for Easter Sundays in April range in the fifties and sixties, with lows in the forties. Since 1891, Easter has occurred in April 96 times. In the Twin Cities, 28 of those have been wet and eight have brought snowfall of at least a trace. In just two years, 1941 and 1998, thunderstorms occurred on April Easters.

Statewide climate extremes for Easter Sunday present quite a range—from 93°F to -28°F in terms of temperature. The warmest Easters occurred in 1977 and 1987. On April 10, 1977, Browns Valley in Traverse County and Madison in Lac qui Parle County registered a high temperature of 92°F. On April 19, 1987, warm weather was more widespread: Argyle in Marshall County, Marshall in Lyon County, Milan and Montevideo in Chippewa County, Redwood Falls in Redwood County, Tyler in Lincoln County, Warroad in Roseau County, and Wheaton in Traverse County all reported 90°F or higher. Madison trumped them all with 93°F, the state's highest-ever Easter Sunday reading.

In contrast, the coldest Easter Sundays occurred in 1939 and 1975. On April 9, 1939, Sawbill Camp in Cook County reported the day's statewide record low, -19°F. On March 30, 1975, most areas of the state saw temperatures dip far below zero degrees. After a hard winter, abundant snow cover remained, with 18 to 20 inches still on the ground in the northwest. Argyle in Marshall County, Itasca State Park in Clearwater County, and Thorhult and Waskish in Beltrami County reported lows of -22°F or colder. Thorhult's reading of -28°F is a statewide record low for Easter Sunday.

Major Easter storms are historically rare. On April 13, 1941, three separate thunderstorms dropped 1.54 inches of rain on Carver County's Chaska. On April 18, 1976, an all-day rain poured 2.35 inches on La Crescent and 2.72 inches on Caledonia, both in Houston County. On April 12, 1998, thunderstorms brought little rain but immense winds to southwestern Minnesota: downbursts from 48 to 55 miles per hour damaged buildings in Jackson, Lincoln, Pipestone, and Rock counties. Among these structures was a potato chip storage shed, inspiring visions of bite-sized snacks whirling across the landscape. Without doubt the worst Easter Sunday storm was the heavy snow and blizzard that hit northwestern Minnesota on April 6, 1947. Red Lake Falls in Red Lake County reported 14 inches of snow; Fosston in Polk County set a state record for the date and for Easter Sunday with 18 inches. The most recent winter storm to have an impact on Easter Sunday occurred over March 21–23, 2008. This storm brought 4 to 8 inches of snow to many areas of the state, including nearly 7 inches in the Twin Cities. Portions of western Minnesota reported up to 17 inches of snow with very difficult travel conditions.

Daily precipitation and maximum and minimum temperatures for the Twin Cities on Easter Sunday from 1891 to 2014 are recorded in Table 15. Temperatures of 80°F or higher have been observed only twice, in 1977 and 1987. Temperatures below zero have similarly been observed twice, in 1894 and 1940. In the category of unusual weather, the last dense fog on Easter was in 1993, and a beautiful display of aurora borealis was visible on Easter Sunday evening, March 27, 1910.

Memorial Day

Originally known as Decoration Day, a time set aside to remember Civil War soldiers by adorning their graves with flowers and ribbons, Memorial Day was for generations celebrated on a fixed date, May 30. After World War I, Memorial Day more broadly honored those who died in their nation's service. The holiday continued to be observed on May 30 until 1967, when Congress passed the Uniform Holidays Bill to create three-day weekends for Memorial Day and other historically significant days. Beginning in 1971, Memorial Day was assigned to the last Monday in May; consequently, its date varies from May 25 to May 31.

Memorial Day weekend marks the end of spring for most Minnesotans, whose thoughts now turn to summer vacations or weekends at the lake. Indeed, by late May daytime temperatures typically reach the middle sixties to low seventies, sometimes even the summer-like eighties. The Twin Cities climate record for Memorial Day shows that it rains about one-third of the time. The heaviest Memorial Day rainfall occurred at Red Lake in Beltrami County on May 30, 1949, as a thunderstorm

TABLE 15: Easter Sunday Weather and Climatology for Minneapolis–St. Paul, derived from National Weather Service data for 1891–2014

DATE	PRECIP (INCHES)	HIGH TEMP (°F)	LOW TEMP (°F)	NOTES
3/29/1891	0.00	49	35	
4/17/1892	0.00	56	31	
4/2/1893	0.00	62	30	
3/25/1894	0.00	15	-2	wind chill -15 F
4/14/1895	0.00	62	41	
4/5/1896	0.00	48	30	
4/18/1897	0.00	62	31	
4/10/1898	0.00	58	35	
4/2/1899	0.00	26	13	3″ snow on ground
4/15/1900	0.00	60	36	
4/7/1901	0.00	49	33	
3/30/1902	0.00	38	31	
4/12/1903	0.01	48	39	
4/3/1904	0.00	43	26	
4/23/1905	0.00	66	41	
4/15/1906	0.00	54	32	
3/31/1907	0.00	32	12	
4/19/1908	0.00	81	50	
4/11/1909	0.00	63	35	
3/27/1910	0.00	66	57	
4/16/1911	0.00	62	35	
4/7/1912	0.00	51	28	
3/23/1913	0.25	34	25	trace of snow, 0.3″ on ground
4/12/1914	0.00	62	35	
4/4/1915	0.00	56	40	
4/23/1916	0.00	60	35	
4/8/1917	0.00	45	25	
3/31/1918	0.00	57	44	
4/20/1919	0.00	65	42	
4/4/1920	0.00	27	9	wind chill -15 F
3/27/1921	0.00	45	5	wind chill -20 F
4/16/1922	0.00	58	38	
4/1/1923	0.00	45	24	

TABLE 15: Easter Sunday Weather and Climatology for Minneapolis–St. Paul, derived from National Weather Service data for 1891–2014 *cont'd*

DATE	PRECIP (INCHES)	HIGH TEMP (°F)	LOW TEMP (°F)	NOTES
4/20/1924	0.10	51	33	
4/12/1925	0.00	67	45	
4/4/1926	0.23	36	20	1.7" snow
4/17/1927	0.00	64	53	
4/8/1928	0.00	30	20	
3/31/1929	0.23	33	23	2.5" snow
4/20/1930	0.00	59	34	
4/5/1931	0.00	55	33	
3/27/1932	0.01	45	29	
4/16/1933	0.00	67	41	
4/1/1934	0.03	39	31	trace snow, 0.3" on ground
4/21/1935	0.00	63	40	
4/12/1936	0.00	50	40	
3/28/1937	0.00	42	24	
4/17/1938	0.23	73	51	
4/9/1939	0.00	44	26	
3/24/1940	0.00	17	-1	5.2" snow on ground
4/13/1941	0.49	76	59	thunderstorm
4/5/1942	0.12	38	27	
4/25/1943	0.00	60	47	
4/9/1944	0.00	49	36	
4/1/1945	0.00	58	37	
4/21/1946	0.00	81	51	
4/6/1947	0.13	36	29	0.9" snow
3/28/1948	0.00	49	24	
4/17/1949	0.02	42	29	
4/9/1950	0.29	33	26	0.3" snow
3/25/1951	0.06	40	19	0.8" snow, 22" on ground
4/13/1952	0.01	51	34	
4/5/1953	0.00	46	32	
4/18/1954	0.00	59	39	
4/10/1955	0.00	78	49	
4/1/1956	0.00	50	35	

TABLE 15: Easter Sunday Weather and Climatology for Minneapolis–St. Paul, derived from National Weather Service data for 1891–2014

cont'd

DATE	PRECIP (INCHES)	HIGH TEMP (°F)	LOW TEMP (°F)	NOTES
4/21/1957	0.00	70	42	
4/6/1958	0.26	42	30	1.6" snow
3/29/1959	0.00	54	28	
4/17/1960	0.00	47	31	
4/2/1961	0.00	43	15	
4/22/1962	0.06	63	42	
4/14/1963	0.00	73	26	
3/29/1964	0.05	26	7	0.7" snow, 1" on ground
4/18/1965	0.10	49	34	trace snow
4/10/1966	0.00	47	27	
3/26/1967	0.28	38	34	0.1" snow
4/14/1968	0.16	41	29	trace snow
4/6/1969	0.00	61	28	
3/29/1970	0.01	37	16	0.4" snow
4/11/1971	0.00	66	42	
4/2/1972	0.00	43	31	
4/22/1973	0.00	49	36	
4/14/1974	0.00	53	35	
3/30/1975	0.03	18	4	1" snow, 10" on ground
4/18/1976	0.15	62	41	
4/10/1977	0.00	88	55	
3/26/1978	0.00	52	24	
4/15/1979	0.00	59	33	
4/6/1980	0.03	64	44	
4/19/1981	0.10	60	38	
4/11/1982	0.00	48	22	
4/3/1983	0.01	42	35	
4/22/1984	0.00	55	44	
4/7/1985	0.00	46	31	
3/30/1986	0.00	62	39	
4/19/1987	0.00	84	58	
4/3/1988	0.11	50	43	
3/26/1989	0.00	59	37	

TABLE 15: Easter Sunday Weather and Climatology for Minneapolis–St. Paul, derived from National Weather Service data for 1891–2014

DATE	PRECIP (INCHES)	HIGH TEMP (°F)	LOW TEMP (°F)	NOTES
4/15/1990	0.00	53	29	
3/31/1991	0.00	52	31	
4/19/1992	0.38	65	52	fog
4/11/1993	0.33	44	34	trace snow
4/3/1994	0.00	50	22	
4/16/1995	0.04	50	40	
4/7/1996	0.00	44	22	trace snow, fog
3/30/1997	0.00	41	23	
4/12/1998	0.09	73	57	thunderstorm
4/4/1999	0.06	55	39	
4/23/2000	0.00	69	48	
4/15/2001	0.28	52	28	
3/31/2002	0.00	35	26	blowing snow, mist
4/20/2003	0.10	48	40	
4/11/2004	0.00	44	25	
3/27/2005	0.00	56	32	
4/16/2006	0.00	70	53	
4/8/2007	0.00	40	19	
3/23/2008	0.06	32	20	1″ snow, 5″ on ground
4/12/2009	0.00	59	38	
4/4/2010	0.00	67	50	
4/24/2011	0.00	62	36	foggy morning
4/8/2012	0.00	61	37	
3/31/2013	0.00	42	24	mist and haze
4/20/2014	0.10	74	45	warmest day of the month

Table Summary Statements

- Longest consecutive streak of dry Easter Sundays, 12 years, 1891–1902
- Warmest Easter, 1977, 88°F
- Coldest Easter, 1894, -2°F
- Coldest daytime wind-chill conditions, 1894 and 1920, -15°F
- Coldest nighttime wind-chill conditions, 1921, -20°F

- Most precipitation on Easter, 1941, 0.49 inches
- More snow events in March than in April
- Record snow amount on Easter, 1929, 2.5 inches
- 22 inches of snow on the ground in 1951; 10 inches in 1975

delivered 4.34 inches there and more than 2 inches in other locations. May 30, 1977, also saw a Memorial Day thunderstorm, which brought 1 to 2.5 inches of rainfall to the state. The Twin Cities recorded 1.09 inches and a high of 80°F that day. With summer hovering just around the corner, it is of course rare to see snow on Memorial Day. However, on May 30, 1897, Bemidji in Beltrami County reported 0.1 inches of snowfall, and, more recently, on May 25, 1992, New Ulm in Brown County reported 1.3 inches, the statewide record for the holiday.

Memorial Day 1992 was arguably one of the coldest as well. Cloudy and damp, with rain and trace snow, May 25 brought daytime high temperatures in the forties and overnight lows in the twenties, bottoming out with 25°F at Tower in St. Louis County. Other notably cold Memorial Days occurred in 1897, 1947, and 1964, with many overnight lows dipping into the twenties. But the coldest temperature measured statewide on this holiday was 20°F at Pokegama Dam, Itasca County, on May 30, 1889.

Although it is not uncommon for northern counties to record frosts on Memorial Day, rather hot weather can also prevail. The Twin Cities record since 1906 shows daytime high temperatures reaching 80°F or higher 33 times and 90°F or higher six

Decoration Day, 1940: fine weather for honoring the war dead

times, topping out at 98°F in 1934. The statewide high for the holiday also occurred in 1934: 108°F at Pipestone in Pipestone County. In fact, 21 Minnesota communities reported a daytime temperature of 100°F or higher on that particular Memorial Day—not the most accommodating weather for planting graveside flowers.

The most recent hot Memorial Day was in 2006, when more than 100 communities reported a daytime high of 90°F or higher. Benson (Swift County), Olivia (Renville County), and Springfield (Brown County) reached 99°F. At Canby (Yellow Medicine County) and Mankato (Blue Earth County), the temperature never cooled below 75°F that day as dew points rose to near 70°F. Later in the day, atmospheric instability combined with the high water vapor content to produce some severe thunderstorms with hail, especially in far northern counties.

Minnesota's Summer

There's an old joke about summertime in Minnesota: sometimes it falls on July Fourth, and sometimes it comes on another day. Indeed, summer can be very short in the North Star State. In some northern communities, snowfalls have been observed in early June; in 1992, widespread frost damaged gardens and crops on June 20, the summer solstice. A number of communities in northern counties have recorded freezing temperatures during every summer month; a smaller number have never seen temperatures as high as 100°F. Indeed, many midwesterners consider northern Minnesota an ideal summer vacationland, offering escape from the oppressive heat so dominant in most other states. The highest reading ever reported from Lake County's Two Harbors, located along cool Lake Superior, is 99°F (recorded most recently on August 14, 1961), while the highest from Beltrami County's Waskish, embedded in north-central Minnesota's wetlands, is 98°F (July 20, 1977). Even the Twin Cities' historical record shows a few summers with no daytime temperatures of 90°F or higher: 1867, 1902, 1915, and 1993.

However, most summers bring some spells of excessive heat, humidity, and thunderstorms. The average number of days with maximum temperatures of 90°F or higher ranges from one day in northeastern counties to as many as 22 days in western counties. High humidity combined with these warm temperatures can produce oppressive Heat Index (HI) values of 105°F or higher. Even northern communities like Duluth and Kabetogama (both St. Louis County), International Falls (Koochiching County), and Roseau (Roseau County) have reported HI values ranging from 105 to 115°F.

Perhaps most Minnesota citizens are unaware that the highest Heat Index measured in the United States was recorded at 6:00 PM on July 19, 2011, at Moorhead (Red River Valley), when the air temperature was 95°F and the dew point was 88°F, giving a Heat Index reading of 134°F. In fact, Moorhead was the warmest place in the nation that day.

Thunderstorms may occur on an average of 20 days in northern counties and 40 or more in southern and western counties. And the average number of days with hail ranges from one to three across the state, the higher frequency being in southwestern counties. Peak occurrence for hail in Minnesota is on or about June 1, but, similar to thunderstorms, hail has been reported in every month of the year.

· ·

Heat Index

The Heat Index, also known as the Comfort or Temperature-Humidity Index, evaluates the combined effects of temperature and humidity on the body's ability to cool itself. According to the Heat Index, an air temperature of 85°F with a relative humidity of 60 percent feels the same as a temperature of 90°F with a humidity of 30 percent. For nighttime combined values of 75°F or above and daytime values of 105°F or more expected for 48 hours or longer, the National Weather Service usually issues an excessive heat advisory to warn about health risks, including fatigue, heat cramps, sunstroke, or heat exhaustion.

· ·

When the landscape is exceptionally dry, summer-like heat can arrive early in the spring, as it did at Montevideo in Chippewa County on March 23, 1910, when the afternoon high climbed to 88°F, a statewide record for the month. Other dry springs like 1931 and 1980 brought excessive heat to western counties in April, with many reported temperatures of 90°F or higher. In fact, persistent dry conditions from March to May sometimes lead to summer droughts, as they did in 1910, 1926, 1931, 1934, 1976, and 1988. There are relatively few cases of lasting drought in Minnesota; nevertheless, in response to historical conditions, water flow on major watersheds, including the Red River of the North and the Minnesota and Mississippi rivers, is controlled using a system of twentieth-century reservoirs. These structures permit better management of low flow that can lead to serious consequences for navigation—particularly barge traffic—for fisheries, for recreation, and even for some municipal water supplies.

Extreme summer heat in Minnesota is rare and usually short lived. Such uncomfortable episodes are most often due to persistent southerly airflow from either the Sonoran Heat Ridge—located in northern Mexico and the desert southwest—or the Bermuda Heat Ridge—situated in the western Atlantic. These dominant summertime high-pressure cells fluctuate in position and intensity, affecting the frequency of tropical-like air masses flowing from the south. The polar jet stream, which steers air masses across North America, is usually displaced slightly north of Minnesota during the summer, but periodic dips allow cooler air from high latitudes to slip down over the state and bring relief from hot temperatures. Another reprieve can come from a slow-moving or stalled Hudson Bay low-pressure system: a counterclockwise flow around the low pressure sends cool and moist air on easterly or northeasterly winds across the Great Lakes. Though rare, these winds tend to bring cloudiness and cooler temperatures to the state.

A dry summer wreaks havoc: a farmer during the dust bowl era despairs of his corn crop, July 1936.

Since 1871 the Twin Cities area has seen temperatures of 100°F or higher in only 30 summers, a frequency of roughly 21 percent. Temperatures of that magnitude tend to occur over just one or two days, the approximate time required for a summer air mass to travel from west to east across the state. Exceptions result from stalled or slow-moving tropical-like air masses, as was the case in 1936 at Beardsley, in Big Stone County, when 13 consecutive July days brought temperatures of 100°F or higher. That same heat spell produced five such consecutive days in the Twin Cities, inspiring many citizens to sleep on porches, in backyards, or in city parks. Despite these examples, spells of very hot temperatures that persist beyond four days are actually quite rare for almost any Minnesota community. Summers dominated by both drought and so-called heat waves—consecutive days of abnormally warm temperatures, often combined with high humidity—cause problems for Minnesota citizens and farm animals alike. Lack of rainfall affects food and water supplies, and high temperatures cause heat-related health problems for humans and stress, even death, for farm animals, especially those confined without proper ventilation or with insufficient feed, forage, or water.

In recent years, Minnesota heat waves have taken on a different form, combining moderately high air temperatures with extraordinarily high dew points—a measure of water vapor in the air, also known as humidity. Higher dew points with temperatures in the nineties have produced HI values exceeding 100°F, usually in July and August. Consecutive days with such high HI values can truly be termed "heat waves," though they occur in the absence of drought. In fact, the tropical-like humidity these conditions require is fed in part by a wetter landscape: many recent heat waves have occurred in the middle of an otherwise moist or wet summer, the former's rainfall described as slightly above normal, the latter's as much above normal. The summers of 1983, 1995, 1999, 2001, 2005, 2010, and 2011 brought HI values ranging from 110 to 120°F—resulting from air temperatures in the nineties and dew points near or even above 80°F. In fact, the state record HI was clearly a case of extreme dew point rather than extreme air temperature: on July 19, 2011, many Minnesota communities reported late-afternoon highs in the low to mid-nineties, but with dew points in the lower to middle eighties. Such conditions produced Heat Index values that ranged from 118 to 130°F. These dangerously high values, more typical of a Persian Gulf country, are a serious threat to health, causing thermoregulation problems for both people and animals. Indeed, recent statistics from the National Oceanic and Atmospheric Administration show that average annual deaths due to heat exceed those attributed to lightning, flooding, and tornadoes combined. Fortunately, Minnesota experiences such weather only rarely in comparison to states in the central and southern plains.

The cooling season, when commercial and residential air conditioning systems are put to the test, begins in June. Similar to winter's Heating Degree Days (HDD), Cooling Degree Days (CDD) are used by

Q & A

What are sun kinks?

a. a rock band of the seventies

b. a kind of solar flare

c. railroad tracks bent by excessively hot days

(answer on page 329)

climatologists and energy companies to estimate air conditioning needs. The CDD is calculated by taking the difference between the daily average temperature (maximum plus minimum, divided by two) and a base value of 65°F, a tolerable indoor temperature. For example, the average high temperature in the Twin Cities on June 20 is 79°F and the average low is 57°F, a daily average of 68°. The difference between this value and a base of 65°F yields a CDD of 3 for this date. For most Minnesota locations, 85 to 90 percent of the seasonal CDD totals occur in the months of June, July, and August; however, exceptionally warm spring or fall months can also yield significant CDD values. Applying CDD to estimate energy needs for air conditioning is not as precise as applying HDD for heating. Both are built on the assumption that most

Cool off any way you can: in 1955, Jane Dornfield used a fan in front of an open refrigerator to beat the summer heat.

indoor environments are kept in a temperature range of 65 to 68°F. However, in the summer many energy customers will tolerate higher indoor temperatures, setting air conditioning at 75 or even 78°F on extremely hot days, thus requiring much less energy than if the thermostat were adjusted to 65 or 70°F.

. . . .

Dog days

The dog days of summer are usually associated with the greatest heat of the year, characterized by thunderstorms and high dew points. The phrase's origin is both ancient and astrological: the Greeks and Romans observed that one of the brightest stars, Sirius the Dog Star—located in the constellation Canis Major, Latin for "greater dog"—rose in conjunction with the sun during the six weeks of midsummer. The usual hot and sultry weather, which depleted people's energy and wilted vegetation, was attributed to the evil effects of Sirius. In the United States, the dog days occur between mid-July and early September; in western Europe they run from July 3 to August 11.

. . . .

As with HDD, geography plays a significant role: most locations with seasonal CDD values of 400 or more are located in the warmer western and southern counties. The urban heat island effect of the Twin Cities shows in its seasonal CDD values, which closely approximate those of western and southern Minnesota. Seasonal CDD values vary widely year to year. The Twin Cities area has seen 17 years when CDD totals exceeded 1,000, the most coming in 1936 with 1,178. Five years have yielded fewer than 400 CDD, bottoming out in 1904 with only 319. The range in values produced by hot versus cold summers is substantial: for the Twin Cities, the range of 859 CDD—1,178 (1936) to 319 (1904)—shows a departure of plus or minus 50 to 60 percent of the mean seasonal total. This wide span leads to enormous fluctuation in air conditioning costs. Small wonder that many Minnesotans use an equal monthly payment plan for their electricity, a strategy often coupled with a higher indoor thermostat setting to minimize use of air conditioning.

Minnesota summers are rarely dry enough to cause drought conditions. Because 60 to 70 percent of annual precipitation falls during May through September, and because most of this precipitation is thunderstorm rainfall—highly erratic in frequency and quite spotty geographically—summer rainfall statistics show great range and variation. Statewide, the range in June-through-August rainfall totals shows a low value of less than 5.70 inches in 1936 and a high value of more than 17 inches in 1993. A look at individual station records magnifies this range.

Many communities have reported single-day rainfall totals of 6 or more inches, with five observers seeing a daily total of 10 or more inches, topped by Hokah (Houston County), which measured 15.10 inches on August 19, 2007. Rainfall rates as high as 3 inches per hour have also been measured. Such intense thunderstorms lead to flash floods, but fortunately most storms of this sort are fast moving and short lived.

Wet summers naturally generate a bumper crop of mosquitoes as well as widespread plant diseases. In some years, such as 1993, frequent thunderstorms have produced measurable rainfall on 20 to 22 days during a month, leaving little respite for fishermen, golfers, farmers, and outdoor construction workers. State records show two emerging climate patterns: an increase in summer dew points and an increase in overall precipitation, especially that fraction provided by thunderstorms. In combination, these trends make for a wetter environment and more frequent spells of tropical-like humidity and uncomfortable or even health-threatening Heat Index values.

Though Minnesota has a national reputation for relatively short summers, the state's weather can be quite variable, running the gamut from cool, dry polar air that brings midsummer frosts to moist, hot air that mimics the equatorial rain forests and inspires the skimpiest of wardrobe selections. Not surprisingly, Minnesotans recognize that planning outdoor activities is a risky proposition, one not to be undertaken lightly during the summer season.

Four basic thunderstorm types

Thunderstorms occur in a variety of forms.

*An **isolated cumulonimbus** or anvil-shaped cloud, known as a single-cell storm, is usually a convective cloud containing one updraft and one downdraft segment. Single-cell storms may produce some heavy rain, hail, or even a weak tornado, but they are usually short-lived, lasting 30 or fewer minutes.*

*In a **multicell cluster**, a group of convective clouds moves together as a single unit, bringing multiple updraft and downdraft segments, highly variable rates of rainfall, and moderate hail. These systems may last for hours and can produce flash flooding or weak tornadoes.*

*A **squall line** is a row of convective clouds that share a common gust front along the leading edge, sometimes visible as a wall cloud. They can move at rapid speeds and produce heavy rainfall, moderate hail, and even tornadoes, occasionally leading to flash flooding.*

*A **supercell**, the most damaging type of thunderstorm, is a massive convective system of clouds that rotate as one unit, contain embedded strong updrafts and downdrafts, and produce large hail, frequent lightning, flooding, and moderate to severe tornadoes. Such storms may last for hours and travel across multiple states.*

TABLE 16: Cooling Degree Days by Month (1981–2010)

STATION	JAN	FEB	MAR	APR	MAY	JUN	JUL	AUG	SEP	OCT	NOV	DEC	ANNUAL
Ada	0	0	<0.5	2	20	91	171	145	34	1	<0.5	0	465
Aitkin	0	0	0	1	8	51	116	78	14	<0.5	<0.5	0	268
Albert Lea	0	<0.5	<0.5	3	29	138	225	164	52	5	<0.5	0	616
Alexandria	0	0	0	2	21	93	188	141	33	2	<0.5	0	480
Argyle	0	0	<0.5	1	14	67	126	110	24	1	0	0	344
Austin	0	0	<0.5	3	25	125	190	138	45	4	<0.5	0	531
Baudette	0	0	0	<0.5	5	38	92	64	13	<0.5	0	0	213
Bemidji	0	0	0	<0.5	9	54	115	81	15	<0.5	0	0	275
Big Falls	0	0	0	1	15	56	115	90	21	1	<0.5	0	300
Brainerd	0	0	0	1	15	70	145	102	21	1	<0.5	0	355
Browns Valley	0	0	<0.5	2	25	105	211	160	41	3	<0.5	0	549
Caledonia	0	<0.5	<0.5	3	21	110	192	142	40	4	<0.5	0	513
Cambridge	0	0	<0.5	3	25	108	194	141	32	2	<0.5	0	506
Campbell	0	0	0	3	28	101	193	150	34	2	<0.5	0	510
Canby	0	<0.5	<0.5	4	35	138	247	189	57	4	<0.5	0	675
Cloquet	0	0	0	<0.5	6	38	106	83	19	<0.5	0	0	253
Collegeville	0	0	<0.5	3	33	128	238	178	52	4	<0.5	0	638
Cook	0	0	0	<0.5	7	42	105	73	17	<0.5	0	0	244
Cotton	0	0	0	<0.5	6	31	68	49	10	<0.5	0	0	164
Crookston	0	0	0	2	20	85	155	128	28	1	0	0	420
Detroit Lakes	0	0	0	2	26	109	196	171	44	1	<0.5	0	550
Duluth	0	0	0	<0.5	4	30	90	66	14	<0.5	0	0	205
Ely-Winton	0	0	0	<0.5	6	40	97	69	15	<0.5	0	0	227
Fairmont	0	0	<0.5	4	41	166	265	205	66	6	<0.5	0	753
Faribault	0	0	<0.5	4	27	125	205	147	46	4	<0.5	0	559
Farmington	0	<0.5	<0.5	5	35	141	230	166	55	5	<0.5	0	637
Fergus Falls	0	0	<0.5	2	24	98	185	152	40	3	<0.5	0	504
Fosston	0	0	0	1	15	56	110	83	19	1	<0.5	0	286
Grand Marais	0	0	0	0	<0.5	3	31	39	5	<0.5	0	0	79
Grand Meadow	0	0	<0.5	2	19	101	159	114	34	3	<0.5	0	435
Grand Rapids	0	0	<0.5	1	13	63	131	97	19	<0.5	<0.5	0	326
Gunflint Lake	0	0	0	<0.5	5	28	74	54	9	<0.5	0	0	170
Hallock	0	0	0	2	17	86	136	118	19	<0.5	0	0	377
Hibbing	0	0	0	<0.5	4	25	63	46	9	<0.5	0	0	147
Hutchinson	0	<0.5	<0.5	4	33	135	227	156	49	4	<0.5	0	609
Int'l Falls	0	0	0	1	8	38	76	58	11	<0.5	0	0	193

STATION	JAN	FEB	MAR	APR	MAY	JUN	JUL	AUG	SEP	OCT	NOV	DEC	ANNUAL
Itasca	0	0	0	<0.5	9	50	102	81	14	<0.5	0	0	257
Leech Lake Dam	0	0	0	1	12	66	137	101	20	<0.5	0	0	338
Luverne	0	0	<0.5	5	37	148	243	187	61	5	<0.5	0	686
Marshall	0	0	<0.5	4	37	143	239	183	58	6	<0.5	0	670
Milan	0	0	<0.5	6	46	153	246	189	65	7	<0.5	0	712
Montevideo	0	0	<0.5	3	33	120	215	160	44	3	<0.5	0	578
Mora	0	0	0	1	18	84	158	114	30	2	<0.5	0	407
Morris	0	0	<0.5	2	28	108	184	136	33	3	<0.5	0	498
MSP	0	0	0	5	37	158	276	205	66	6	0	0	753
New Ulm	0	0	<0.5	5	40	156	256	188	60	5	<0.5	0	712
Olivia	0	0	<0.5	3	32	128	207	145	41	3	<0.5	0	560
Park Rapids	0	0	0	1	11	56	119	92	21	1	0	0	300
Pine River Dam	0	0	0	1	15	73	155	110	22	<0.5	<0.5	0	376
Pipestone	0	<0.5	<0.5	4	29	119	215	165	51	3	<0.5	0	586
Pokegama Dam	0	0	0	<0.5	13	60	124	97	21	<0.5	0	0	315
Red Wing	0	0	<0.5	4	32	131	236	177	59	6	<0.5	0	645
Redwood Falls	0	0	0	<0.5	17	112	236	184	35	3	0	0	588
Rochester	0	0	<0.5	4	30	130	207	154	54	6	<0.5	0	585
Roseau	0	0	0	1	12	57	111	87	18	<0.5	0	0	287
St. Cloud	0	0	<0.5	2	23	95	182	127	34	2	<0.5	0	467
St. James	0	<0.5	<0.5	4	38	143	230	165	51	4	<0.5	0	637
St. Peter	0	0	<0.5	5	38	157	260	202	60	5	<0.5	0	727
Sandy Lake Dam	0	0	0	<0.5	11	65	133	104	25	<0.5	0	0	339
Stillwater	0	0	0	<0.5	18	103	199	149	42	4	0	0	514
Tower	0	0	0	<0.5	7	32	69	46	12	<0.5	0	0	167
Two Harbors	0	0	0	<0.5	<0.5	10	69	72	11	<0.5	0	0	163
Wadena	0	0	0	<0.5	12	62	122	89	21	1	0	0	308
Warroad	0	0	0	<0.5	10	54	110	84	15	<0.5	0	0	273
Waseca	0	<0.5	<0.5	5	41	152	228	173	60	7	<0.5	0	667
Waskish	0	0	0	<0.5	7	47	96	72	14	<0.5	0	0	235
Wheaton	0	0	<0.5	2	25	113	216	169	40	2	<0.5	0	568
Willmar	0	0	<0.5	3	35	133	223	159	46	3	<0.5	0	603
Windom	0	0	0	7	43	168	271	197	67	7	<0.5	0	759
Winnebago	0	0	<0.5	5	37	147	233	173	56	6	<0.5	0	658
Winona	0	<0.5	<0.5	2	24	125	219	179	55	6	<0.5	0	609
Worthington	0	<0.5	<0.5	3	30	131	216	152	50	4	<0.5	0	587
Wright	0	0	0	<0.5	8	40	98	78	20	1	0	0	245
Zumbrota	0	<0.5	<0.5	3	21	108	178	126	36	3	<0.5	0	476

TABLE 17: Driest and Wettest Summers (June–August) by Select Location

LOCATION	DRIEST SUMMER (YEAR)	WETTEST SUMMER (YEAR)
Austin	4.07 inches (1979)	26.05 inches (1990)
Mankato	6.29 inches (1988)	24.76 inches (1993)
Moorhead	1.86 inches (1936)	21.69 inches (2005)
Rochester	3.78 inches (1910)	23.34 inches (2000)
St. Cloud	3.18 inches (1921)	22.06 inches (1897)
Twin Cities	1.73 inches (1894)	23.52 inches (1987)
Winona	2.34 inches (2010)	25.86 inches (2007)

Hottest Places

Americans generally do not think of Minnesota as a steamy or an arid place, yet its residents do on occasion experience tropical- or desert-like conditions—hot and moist or hot and dry—brought by differing southern air masses. Latitude seems to have a notable effect: most state high temperature records have been set in southern counties, where the sun angle is relatively higher. Cloudiness may also be a factor: a significant fraction of state high temperature records are found in western communities, where cloud cover is somewhat reduced compared to the east and where drier soils and fewer lakes prevail. Further, western prairie communities are more exposed to the sun and have landscape features that absorb and reradiate solar energy rather than consuming it through evapotranspiration. Dry air masses with strong southerly breezes off the arid high plains have yielded temperatures of 100°F as early as April 21 and as late as September 22, both recorded at Ada in Norman County, along the Red River Valley.

Nearly every climate station with a lengthy record has seen the temperature hit the century mark at least once, including those notorious for cold rather than heat, such as Grand Marais in Cook County (100°F on August 7, 1930), Hibbing in St. Louis County (100°F on July 29, 1999), and Tower, also in St. Louis County (101°F on August 14, 1901). Only a handful of stations have never reported a temperature of 100°F, probably due to local influences like lakes or wetlands. Some of these locations include Grand Portage and Lutsen (Cook County), Crane Lake (St. Louis County), and Two Harbors (Lake County).

Two areas of the state stand out when considering high overnight temperatures: the western prairie communities and urbanized areas like the Twin Cities and St. Cloud, all of which store more heat during the daytime and therefore take longer to cool off at night. Many southern and western communities like Beardsley (Big Stone County), Canby (Yellow Medicine County), and Milan (Chippewa County) have recorded multiple warm overnight low temperatures of 80°F or higher. In fact, Canby has measured overnight lows of 87°F, a value roughly equivalent to that found in the desert southwest. Overnight low temperatures in the eighties have been recorded as far north as Itasca State Park (83°F on July 1, 1911) and Moose Lake (84°F on August 4, 2001). The Twin Cities, however, lead all Minnesota locations in terms of the number of nights (31) when the temperature has remained at 80°F or warmer.

Minnesota's western thumb, at the juncture of Big Stone and Traverse counties, is arguably the hottest spot in the state. During its years of operation, from 1893 to 1973, Beardsley climate station in Big Stone County logged 44 state daily maximum temperature records, the most of any station in Minnesota. In 1973 nearby Browns Valley (Traverse County) began providing daily measurements; since then it has set 14 new statewide high temperature records, continuing Beardsley's legacy of extremely high readings. The thumb area boasts nearly ten percent of the state's high temperature records: 58 between the two communities. In fact, Beardsley (July 29, 1917) holds the state's all-time high temperature of 115°F, measured under very dry atmospheric and landscape conditions. Other western Minnesota communities with a large number of statewide daily maximum temperature records include Canby (Yellow Medicine County) with 36 and Madison (Lac qui Parle County) with 19. Another noteworthy hot spot is the Minnesota River Valley, where communities such as Chaska, Milan, Montevideo, New Ulm, Redwood Falls, and St. Peter together hold 61 state high temperature records. Although nearly all record summertime daily high temperatures have been set in the southern half of the state, St. Vincent in far northwestern Kittson County holds the record for August 25 (102°F in 1886), explained in part by the fact that so few measurements were being taken in the western prairie counties at that time.

TABLE 18: One-Day Maximum Temperature Extremes (°F)

STATION	JAN	FEB	MAR	APR	MAY	JUN	JUL	AUG	SEP	OCT	NOV	DEC
Ada	53 24-1981	65 25-1958	78 30-1967	100 21-1980	107 30-1939	104 18-1933	111 6-1936	104 18-1976	101 22-1936	95 3-1922	74 17-1953	60 6-1939
Aitkin	54 6-2012	57 22-1961	77 20-2012	90 21-1980	39 1-1959	95 28-1961	98 27-1988	100 18-1976	98 7-1976	86 3-1971	72 5-1975	60 1-1962
Albert Lea	64 25-1944	64 18-1981	84 23-1910	94 28-1910	104 31-1934	104 27-1934	104 24-1940	101 17-1988	103 6-1922	94 4-1997	79 9-2006	67 3-1998
Alexandria	58 24-1981	58 17-1981	77 28-1946	95 21-1980	97 1-1959	102 24-1988	101 9-1976	104 7-1983	98 7-1978	90 5-1947	76 8-1999	74 31-1998
Argyle	50 23-1942	62 26-1958	80 27-1946	98 22-1980	103 30-1934	100 18-1995	107 11-1936	104 8-1983	100 3-1983	92 2-1992	75 1-1999	56 6-1939
Austin	55 24-1981	62 17-1981	79 31-1978	91 30-1952	100 15-2013	100 21-1988	99 6-1948	99 23-1948	97 9-1955	92 3-1997	78 9-2006	65 1-1998
Baudette	50 23-1942	59 26-1958	77 19-2012	91 27-1952	96 30-1939	100 29-1931	103 13-1936	101 7-1983	95 2-1983	86 1-1992	75 5-1975	54 1-1962
Bemidji	59 23-1900	60 22-1961	84 28-1946	95 21-1980	96 30-1939	100 29-1931	107 11-1936	101 4-1947	103 8-1931	95 5-1963	73 5-1975	62 20-1899
Big Falls	54 23-1942	62 26-1958	83 28-1946	92 28-1952	97 31-1934	99 17-1995	106 12-1936	99 19-1976	97 7-1976	89 5-1963	75 5-1975	59 6-1939
Brainerd	56 25-1981	60 27-1932	80 28-1946	94 22-1980	101 31-1934	100 19-1988	106 10-1936	102 4-1947	103 10-1931	88 2-1992	73 4-1975	61 6-1939
Browns Valley	63 25-1981	65 3-1991	83 19-2012	100 22-1980	96 10-1987	107 21-1988	109 6-1988	107 18-1976	100 3-1976	93 1-1976	79 3-1978	67 10-2006
Caledonia	57 18-1996	58 24-2000	82 30-1986	93 21-1980	90 2-1901	97 9-1985	103 21-1901	102 1-1988	95 13-1897	91 1-1976	75 4-1978	62 3-1982
Cambridge	53 23-1942	58 16-1981	79 30-1968	92 21-1980	98 17-1972	100 30-1963	109 14-1936	102 15-1936	97 7-1976	87 21-1947	73 1-1944	62 6-1939
Campbell	60 20-1944	62 26-1958	91 23-1910	100 21-1980	108 31-1934	104 30-1921	111 10-1936	105 13-1938	105 11-1931	93 3-1953	77 4-2001	70 6-1939
Canby	68 24-1981	67 16-1981	84 23-1939	98 21-1980	98 1-1959	110 29-1931	111 12-1936	108 1-1988	99 7-1978	94 3-1938	82 9-1999	74 30-1899
Cloquet	52 23-1942	60 21-1950	79 28-1946	88 27-1952	94 20-1939	98 29-1931	105 10-1936	98 5-1947	96 10-1931	86 2-1922	70 4-1975	57 1-1962
Collegeville	56 5-2012	60 23-1951	81 23-1910	95 21-1980	96 27-1969	102 7-2011	103 24-1901	100 4-1896	97 15-1948	89 2-1953	78 8-1999	61 3-1998
Cook	51 26-1973	58 24-1976	75 20-2012	88 10-1977	92 21-1964	96 17-2002	98 27-1988	99 19-1976	96 7-1976	88 5-1963	74 9-1999	57 1-1962
Cotton	52 25-1973	59 24-1976	72 31-1986	87 10-1977	93 28-1986	96 18-1995	100 27-1988	96 19-1976	97 7-1976	82 2-1992	72 4-1975	59 1-1962
Crookston	57 5-1902	63 25-1958	78 27-1946	96 21-1980	101 30-1939	102 12-1893	105 10-1936	104 8-1983	99 11-1931	89 6-2011	73 1-1999	56 7-1939
Detroit Lakes	55 24-1942	59 27-1958	78 23-1910	98 21-1980	97 31-1934	102 30-1921	107 11-1936	101 17-1988	100 10-1931	90 5-1963	75 5-1904	60 7-1939
Duluth	55 22-1942	58 26-1895	81 28-1946	88 27-1952	95 30-1939	97 28-1910	106 13-1936	97 1-1930	95 7-1976	86 2-1953	73 3-1903	56 6-1939
Ely-Winton	49 25-1981	58 17-2011	77 20-2012	86 10-1977	94 28-1969	100 19-1995	101 17-2006	98 19-2003	97 7-1976	85 17-1961	70 5-1975	54 1-1962
Fairmont	64 25-1944	64 14-1954	81 30-1968	90 30-1934	108 31-1934	106 10-1933	108 27-1930	109 3-1930	100 2-1925	91 4-1938	79 13-1999	67 6-1939
Faribault	63 25-1944	65 17-1981	81 26-1986	93 28-1980	108 31-1934	106 27-1934	110 14-1936	105 15-1936	102 10-1931	92 4-1997	79 1-1933	69 20-1923
Farmington	59 24-1981	63 26-1896	82 23-1910	93 21-1980	107 31-1934	102 27-1934	110 14-1936	105 15-1936	103 10-1931	91 3-1997	80 1-1933	67 1-1998
Fergus Falls	56 23-1942	57 27-1932	80 28-1946	94 22-1980	105 30-1934	105 30-1921	110 6-1936	105 16-1922	105 10-1931	93 5-1963	74 1-1999	65 6-1939
Fosston	46 17-1973	61 26-1958	77 31-1963	96 21-1980	97 29-2006	98 25-1980	105 28-1917	103 19-1976	97 6-1976	91 5-1963	72 2-1978	58 1-1962
Grand Marais	52 1-1897	58 25-1976	67 28-1986	83 29-1965	87 8-1953	93 29-1970	96 23-1898	100 7-1930	90 1-1929	79 2-1992	67 5-1975	55 9-1939
Grand Meadow	57 26-1944	63 18-1981	83 26-1907	91 22-1980	107 31-1934	105 27-1934	106 14-1936	102 31-1930	100 1-1913	92 4-1997	78 1-1950	62 2-1998
Grand Rapids	51 23-1942	61 22-1961	80 28-1946	93 21-1980	101 31-1934	100 26-1921	104 11-1936	100 19-1976	99 8-1931	86 5-1963	71 8-1999	59 1-1962
Gunflint Lake	47 25-1981	56 17-2011	76 18-2012	88 30-1977	91 20-1994	94 17-1987	98 31-1975	98 8-1983	94 8-1976	79 1-1963	68 3-1978	52 16-1998
Hallock	50 23-1942	61 26-1958	77 23-1910	98 22-1980	105 30-1934	103 29-1912	98 31-1975	103 2-1989	102 3-1983	92 2-1992	75 6-1975	54 3-1982
Hibbing	48 25-1973	60 16-2011	77 18-2012	89 21-1980	93 22-1965	97 17-1995	109 11-1936	103 2-1989	94 7-1976	87 5-1963	72 5-1975	60 1-1962
Hutchinson	61 24-1981	60 25-1976	83 30-1968	95 21-1980	100 15-2013	102 25-1988	100 29-1999	104 1-1988	98 7-1978	89 1-1963	81 9-1999	65 2-1998
Int'l Falls	49 20-1921	58 22-2000	79 18-2012	93 27-1952	95 21-1964	101 27-1912	103 22-1923	96 9-1920	95 7-1976	88 5-1963	73 5-1975	56 1-1962

STATION	JAN	FEB	MAR	APR	MAY	JUN	JUL	AUG	SEP	OCT	NOV	DEC
Itasca	62 23-1942	63 24-1976	81 28-1946	96 21-1980	98 31-1934	98 29-1912	105 28-1917	101 16-1922	99 8-1931	92 5-1963	74 4-1975	59 6-1939
Leech Lake Dam	53 24-1981	61 26-1958	81 28-1946	94 21-1980	102 31-1934	98 29-1931	104 13-1936	100 19-1976	101 9-1931	88 5-1963	72 4-1972	58 1-1962
Luverne	65 24-1981	66 23-1958	85 30-1978	94 21-1980	100 25-1967	104 21-1988	106 9-1976	103 26-1973	104 6-1976	91 5-1963	79 3-1897	66 2-1998
Marshall	67 24-1981	65 17-1981	83 31-1968	94 26-1962	99 18-1998	106 24-1988	109 31-1988	105 15-1988	100 7-1976	92 6-1963	78 8-1999	64 1-1998
Milan	66 24-1981	64 27-1895	85 23-1910	97 21-1980	106 30-1934	106 28-1931	113 21-1934	107 15-1988	108 10-1931	95 3-1938	80 8-1999	73 6-1939
Montevideo	69 24-1981	64 16-1981	88 24-1910	100 21-1980	105 31-1934	107 28-1931	110 31-1988	106 15-1988	105 10-1931	94 5-1963	82 4-1909	72 6-1939
Mora	54 23-1942	58 28-1932	82 23-1910	98 21-1980	107 31-1934	101 1-1934	108 23-1934	104 16-1988	105 10-1931	91 2-1992	74 3-1978	59 9-1939
Morris	60 25-1981	60 25-1902	83 23-1910	98 22-1980	106 31-1934	104 28-1931	109 18-1940	104 10-1947	106 10-1931	93 5-1947	78 9-1999	69 6-1939
MSP	58 25-1944	64 26-1896	83 23-1910	95 21-1980	106 31-1934	104 27-1934	108 14-1936	103 15-1936	104 10-1931	90 10-1928	77 1-1933	68 1-1998
New Ulm	65 24-1981	68 26-1896	87 30-1968	95 21-1980	103 31-1934	107 28-1931	111 15-1936	106 16-1936	96 23-1936	92 5-1963	83 8-2006	69 7-1939
Olivia	63 24-1981	64 26-1896	82 27-2007	95 21-1980	100 19-1998	105 25-1988	102 4-1990	108 1-1988	83 29-2014	91 3-2001	82 9-2006	64 4-1998
Park Rapids	58 23-1942	61 24-1976	80 23-1910	96 21-1980	105 31-1934	101 22-1911	107 10-1936	103 4-1947	99 8-1931	90 5-1963	72 18-1953	60 6-1939
Pine River Dam	57 25-1973	59 9-1907	82 23-1910	96 21-1980	103 31-1934	100 30-1921	104 28-1917	101 18-1976	99 10-1931	88 2-1953	76 3-1922	66 24-1919
Pipestone	64 24-1981	67 24-1958	85 30-1943	93 29-1910	108 30-1934	106 6-1933	108 5-1936	108 3-1930	103 6-1922	92 3-1997	82 1-1950	62 6-1939
Pokegama Dam	53 25-1973	60 24-1976	81 28-1946	92 21-1980	101 31-1934	98 29-1931	103 13-1936	98 19-1976	99 8-1931	88 5-1963	72 3-1978	60 1-1962
Red Wing	55 11-2012	57 26-2000	81 19-2012	89 21-1902	97 15-2013	101 8-2011	106 24-1901	100 1-2006	95 10-2013	92 4-1997	78 9-1999	66 2-1998
Redwood Falls	68 24-1981	64 16-1981	83 18-2012	96 21-1980	107 31-1934	105 16-1933	110 13-1936	104 24-1936	103 14-1939	93 2-1953	82 8-1999	70 6-1939
Rochester	58 25-1944	63 17-1981	79 29-1986	92 24-2009	106 31-1934	105 27-1934	108 14-1936	100 15-1936	97 9-1955	93 3-1997	75 8-1999	63 16-1939
Roseau	50 20-1919	55 25-1958	81 23-1910	92 27-1952	101 30-1934	102 29-1931	107 12-1936	101 18-2003	98 11-1931	88 5-1970	72 5-1975	52 3-1941
St. Cloud	56 24-1981	58 27-1932	81 23-1910	96 21-1980	105 31-1934	102 24-1988	107 13-1936	105 15-1936	106 10-1931	90 2-1992	75 6-1999	63 6-1939
St. James	64 24-1981	63 17-1981	84 30-1968	93 21-1980	102 15-2013	104 21-1988	105 31-1988	104 1-1988	98 7-1976	92 5-1963	82 9-2006	68 2-1998
St. Peter	65 25-1944	68 26-1896	92 29-1913	94 21-1980	107 31-1934	104 21-1988	109 14-1936	104 4-1947	103 10-1931	92 6-1963	81 9-2006	69 2-1998
Sandy Lake Dam	54 6-2012	57 19-1981	78 27-2007	90 22-1980	101 31-1934	97 20-1910	102 28-1988	101 18-1976	95 7-1976	86 5-1963	74 3-1978	60 1-1962
Stillwater	57 9-2003	62 29-2000	83 29-1986	94 21-1980	96 22-1964	103 30-1963	106 3-1990	104 1-1988	98 6-1960	90 4-1997	78 8-1999	67 1-1998
Tower	52 27-1973	59 17-2011	77 20-2012	87 11-1977	95 31-1986	98 14-1897	101 14-1901	98 4-1900	95 8-1976	84 5-1963	75 9-1999	57 1-1962
Two Harbors	57 20-1900	59 8-1898	71 21-1910	88 29-1965	90 22-1964	98 28-1961	99 20-1901	99 14-1961	92 8-1906	86 3-1992	71 1-1990	62 22-1889
Wadena	58 23-1942	60 22-1961	79 28-1946	96 22-1980	104 31-1934	97 25-1936	112 10-1936	103 2-1938	98 14-1939	94 5-1963	74 3-1978	65 6-1939
Warroad	62 25-1944	63 24-1931	81 31-1968	92 22-1980	106 31-1934	105 27-1934	106 14-1936	103 1-1988	100 6-1922	93 4-1997	79 9-1999	68 2-1998
Waseca	45 24-1981	56 25-1958	73 31-1963	93 21-1980	94 21-1964	98 18-1995	100 6-1988	101 2-1989	94 5-1978	86 6-1961	80 9-2006	50 1-1962
Waskish	47 11-2013	58 24-1976	77 20-2012	92 22-1980	95 21-1964	95 18-1995	98 20-1977	95 20-1976	95 7-1976	85 5-1963	71 1-1999	52 2-1982
Wheaton	61 24-1981	66 2-1991	84 31-1963	98 21-1980	99 1-1959	106 25-1988	105 4-1949	104 19-1976	103 8-1959	96 5-1963	81 1-1950	59 1-1998
Willmar	62 24-1981	60 16-1981	79 23-1928	94 21-1980	105 31-1934	103 28-1931	107 21-1934	105 1-1988	105 10-1931	91 2-1953	80 9-1999	64 6-1939
Windom	65 24-1981	66 14-1954	84 30-1968	94 28-1910	98 5-1967	103 22-1988	103 9-1976	105 1-1988	100 6-1976	91 2-1953	80 14-1999	64 2-1998
Winnebago	63 26-1944	63 15-1954	85 23-1910	95 28-1910	106 31-1934	104 27-1934	107 14-1936	106 3-1930	100 7-1931	92 11-1910	82 9-2006	67 2-1998
Winnibigoshish Dam	52 22-1900	58 23-1961	80 28-1946	92 22-1980	100 31-1934	99 29-1931	103 14-1901	99 19-1976	100 8-1931	88 5-1963	70 8-1931	58 2-1962
Winona	64 25-1944	68 19-1981	88 27-2007	96 30-1952	107 31-1934	106 28-1931	108 12-1936	103 24-1948	102 6-1922	93 7-1963	84 1-1950	65 5-1998
Worthington	63 25-1981	64 18-1981	79 31-1978	91 22-1980	99 9-1998	103 22-1988	103 10-1976	104 1-1988	101 7-1976	92 3-1997	79 14-1999	64 4-1998
Wright	54 25-1973	62 24-1976	76 26-2007	63 21-1980	94 29-1986	95 24-1980	100 27-1988	100 18-1976	97 7-1976	86 5-1963	72 5-1975	60 1-1962
Zumbrota	58 25-1944	63 15-1921	82 29-1986	92 21-1980	106 31-1934	103 27-1934	109 14-1936	105 15-1936	102 6-1922	92 3-1997	78 1-1933	66 11-1913

MAP 6: Climate Stations with Four or More Statewide Daily Maximum Temperature Records

Ada (4)

Wheaton (11)

Browns Valley (14)

Beardsley (44)

Milan (12)

Madison (19)

Montevideo (16)

Canby (36)   Granite Falls (9)

Chaska (6)

Redwood Falls (9)

Marshall (5)

Lynd (5)

Springfield (5)   St. Peter (5)

Tracy (9)   Faribault (8)

Lamberton (4)   New Ulm (13)

Pipestone (10)   St. James (4)   Winona (10)

Amboy (6)

Winnebago (5)

Luverne (6)

Fairmont (16)

Q. On July 30, 1999, a new high dew-point record was set in the Twin Cities. What was it?

(answer on page 329)

Winter's warmest places are also typically in the southern counties, which, in comparison to northern Minnesota, are subject to slightly longer days, steeper sun angles, and more frequently bare landscape. The snow cover season in southern Minnesota is generally shorter, or even intermittent, thanks to cycles of freezing and thawing: the average number of days with 1 or more inches of snow cover in Brown, Freeborn, Mower, Redwood, and Winona counties is fewer than 90—compared to more than 120 across central Minnesota to the north. State record high temperatures in the winter months of December and February have exceeded 70°F, the highest being 73°F at Marshall, Beardsley, and Milan on December 6, 1939. Indeed, more than any other season, winter has seen temperatures trending upward in Minnesota: since 1980, 43 state record daily high temperatures have been tied or set during the winter months. Over the same period, only 17 new record highs have been measured in the summer, 29 in the spring, and 19 in the fall. Perhaps the state is becoming a "hot spot" after all.

Severe Convective Weather

Minnesota's Greatest Thunderstorm

The greatest thunderstorm in Minnesota's short documented history occurred in the west-central portions of the state in July 1867, when the landscape was dotted with just a few pioneer settlements and farms. Written records about this storm are scarce because firsthand witnesses were few, but St. Paul weather historian Tom St. Martin has compiled a narrative from the details available. Perhaps the best account is a paper titled "Notes of a Remarkable Storm" by George B. Wright, a land surveyor who witnessed the storm while doing fieldwork in Pope County.

Mr. Wright described the storm as he and his crew viewed it from the high ground along the southern border of Pope and Stevens counties near the present town of Hancock. Their camp was located on a hill about 110 to 150 feet above the surrounding countryside and was situated between the Chippewa River Valley to the east and the Pomme de Terre River Valley to the west, offering a viewing horizon of several miles over the prairie landscape in both directions.

The storm, probably a slow-moving mesoscale convective system associated with a stalled warm front, began around noon on July 17. At first there was little wind and some thunder and lightning. The rain started out at a modest rate, but its intensity increased during the afternoon and early evening. The storm continued all night, steady rain alternating with occasional violent outbursts accompanied by lightning. After 9:00 AM the next day, there were occasional lulls in the storm. Nevertheless, periods of showers continued all day, with another heavy, violent outbreak of rain just before sunset. Wright observed the storm to last about 30 hours, though settlers to the north and east reported that it rained for a duration of as many as 36 hours.

When the rain ceased and the low clouds lifted on the morning of July 19, Wright and his crew saw nothing but broad sheets of water covering the landscape. Foaming torrents streamed from upland areas down lines of drainage and poured into both the Chippewa and Pomme de Terre valleys. Previous to the storm the Chippewa River had been flowing in typical summer creek fashion, about 1 to 3 feet deep with a channel width of about 12 to 20 feet. The storm's impact on the river channel was almost unbelievable. The crew observed a river that occupied every inch of land between the prairie hills, a width ranging from 900 feet in the narrowest of channels between bluffs to three to four miles in the flats. Field notes even four weeks after the storm show that the Chippewa was still more than 600 feet wide over the flood plain, its main channel depth on the order of 15 to 20 feet.

Area lakes previously at very low levels—including those now known as Amelia, Emily, Long, Minnewaska, Page, and Reno—rose immediately and in some cases dramatically after the storm, some by as many as 10 to 12 feet, inundating all shoreline vegetation. Pioneer settlers encountered by Wright and his crew near the communities of Sauk Centre, Osakis, Westport, and St. Cloud estimated the storm's total rainfall to range from 30 to 36 inches, figures based on the catch in empty barrels left in open areas away from buildings and trees. This amount from a single storm is entirely unequaled in the Minnesota climate record and is somewhat analogous to the national 24-hour record storm total of 43 inches at Alvin, Texas (July 25–26, 1979, during tropical storm Claudette).

Q & A

What is the record annual number of confirmed tornadoes in the state of Minnesota?

a. 113

b. 74

c. 65

(answer on page 329)

The storm aftermath wreaked havoc on Wright's survey crew. Many river crossings were no longer fordable, and the mosquito populations were the worst Wright had seen in 20 years of camping and surveying. Western Minnesota's climate turned all but tropical for a time, conditions unprecedented for an inland area at 46 degrees north latitude.

The storm's geographic extent is not precisely known, but it was likely fairly limited. An observer to the west at Big Stone Lake reported light precipitation, while the records to the east at St. Paul and Fort Snelling show only about an inch of rainfall. Though somewhat confined in area, this remarkable storm greatly affected the Mississippi River. As many of the Mississippi's western tributaries reached an all-time flood stage, an unprecedented rise in the St. Cloud (Sherburne County) area was reported by the *St. Paul Pioneer*. On July 18 the observer at Fort Ripley (Crow Wing County) noted a one-day rainfall total of 7.50 inches and a river rise of more than 5 feet in 14 hours. On July 20 the river rose by 12 feet over just 24 hours. Even the American Indians living in central Minnesota had never seen the river rise so rapidly. A bridge at Sauk Rapids was carried off, ferries had to stop running between St. Cloud and St. Anthony, and by July 23 every boom on the river between St. Cloud and St. Paul had given way, releasing more than 35 million logs downstream. Levees failed as well, and West St. Paul was flooded beneath 2 feet of water. The *St. Paul Pioneer* reported that the Mississippi River had never been observed to be so high. Fortunately the flood crest was short lived, and by late July the river flow was receding.

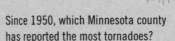

Since 1950, which Minnesota county has reported the most tornadoes?

a. Clay

b. Otter Tail

c. Polk

(answer on page 329)

Because the storm predated standardized weather observations and measurements, the credibility of these historical accounts has been challenged. However, multiple document sources and assorted citizens' written and verbal accounts attest to the real impacts of this storm on the Minnesota landscape. The storm's magnitude is beyond the realm of what has been documented in the modern climate record. The state's record 24-hour rainfall is 15.10 inches at Hokah (Houston County) on August 19, 2007, while the record storm total is close to 17 inches from the same system as it hovered over Houston and Winona counties on August 18–20.

TABLE 19: One-Day Maximum Precipitation Extremes (inches)

STATION	JAN	FEB	MAR	APR	MAY	JUN	JUL	AUG	SEP	OCT	NOV	DEC
Ada	1.05 1-1921	1.10 6-1946	1.24 3-1970	2.41 21-1964	4.51 22-1981	3.60 4-1902	3.80 20-1952	5.83 7-1941	2.98 24-1973	2.65 17-1971	1.58 1-1974	1.75 30-2006
Aitkin	1.00 5-1997	1.70 27-1971	1.80 1-2009	2.74 23-2001	3.09 31-1959	4.40 20-2012	5.29 18-1952	3.36 31-1951	4.07 3-1980	2.94 10-1973	1.84 1-1974	0.90 18-1977
Albert Lea	1.50 4-1949	1.50 21-1969	1.77 26-1950	3.82 24-1990	4.20 30-1980	7.50 15-1978	6.70 31-1961	5.06 15-1993	7.25 18-1926	2.37 4-2002	2.02 30-1991	1.70 10-1911
Alexandria	1.17 27-1948	0.84 27-1948	1.23 8-2000	2.07 7-2001	2.48 8-1986	4.22 23-2003	5.69 19-1945	3.45 25-2005	3.19 3-2005	2.64 14-1984	2.11 1-2000	1.34 3-1951
Argyle	0.72 3-2004	1.12 25-1930	1.45 27-1950	2.06 26-1924	4.26 12-2004	3.07 23-1948	4.80 30-1942	6.00 17-2006	3.05 23-1920	2.32 16-1934	1.60 14-1944	0.84 26-2009
Austin	2.53 22-1973	1.76 15-2005	2.50 6-2006	2.67 23-1990	3.50 30-1980	3.68 12-2008	4.32 2-1944	4.55 29-1947	4.06 15-2006	2.18 12-1966	2.37 13-2010	1.36 12-1965
Baudette	0.80 28-1956	0.80 11-1932	2.27 5-1966	2.04 20-1957	3.17 24-2010	4.18 10-2002	5.50 31-2001	5.05 1-2001	2.30 8-1991	3.00 1-1971	1.57 1-2000	1.30 15-1933
Bemidji	1.30 18-1996	1.40 13-1897	1.54 29-1933	2.59 13-1964	2.80 22-1962	3.90 22-1957	4.15 22-1914	3.35 7-1942	2.91 30-1995	2.49 7-1901	2.69 1-2004	2.10 3-1951
Big Falls	1.10 24-1950	0.80 5-1946	1.48 4-1966	1.84 29-1940	2.22 11-1985	2.75 21-1957	4.02 3-1966	3.90 30-1944	2.53 2-1999	2.72 31-1979	1.77 10-1977	0.95 5-1960
Brainerd	0.97 5-1997	1.86 12-1922	1.81 23-2009	3.04 28-1986	4.03 27-2012	3.92 14-2001	5.00 14-2009	3.46 1-1953	3.62 5-2004	3.11 26-2010	2.53 1-1971	0.88 4-1970
Browns Valley	1.87 5-1997	1.10 21-2011	2.12 20-1982	2.32 16-2012	1.60 12-1998	4.18 2-2007	3.40 7-1994	2.95 7-2009	3.30 2-2006	2.60 15-2013	1.46 1-2000	0.95 31-2006
Caledonia	1.56 29-1914	1.83 19-1971	3.57 7-1959	2.95 16-1973	3.96 18-2000	3.97 13-1950	6.60 21-1951	6.50 19-1907	4.53 1-2010	3.80 5-2013	2.67 18-1958	3.66 12-1899
Cambridge	1.51 4-1997	1.04 27-1948	1.54 15-1945	3.76 27-1975	3.20 20-1953	4.05 12-1984	5.47 22-1972	3.80 27-1977	4.84 22-1968	5.20 5-2005	2.34 26-1896	1.14 11-1949
Campbell	1.28 15-1952	1.56 7-1946	1.90 15-1989	2.26 20-1916	2.24 29-1926	4.01 29-2005	3.32 14-1937	3.56 21-1916	5.43 8-1981	2.33 13-1943	1.45 21-1947	1.14 11-1949
Canby	1.55 27-1944	2.10 26-1944	2.60 28-1924	2.16 23-2001	2.93 8-2000	3.36 13-1926	5.29 27-1963	5.35 19-1926	4.15 16-1921	2.95 31-1979	3.23 20-1975	1.60 28-1959
Caribou	1.00 9-1983	1.67 15-1998	1.07 4-1966	1.90 27-1966	2.20 18-1996	2.52 29-1991	4.93 2-1997	3.07 31-1966	2.87 9-1991	2.01 10-1949	2.00 18-1952	0.80 5-1960
Cloquet	1.29 11-1975	1.24 10-1965	1.86 12-1977	2.40 23-1960	3.36 10-1979	7.75 20-2012	3.58 26-1951	5.62 23-1978	8.44 6-1990	3.09 4-2005	3.08 9-1983	1.54 16-1984
Collegeville	1.40 4-1997	1.75 9-1909	2.58 23-2009	3.08 10-1893	5.84 22-1962	3.54 17-1957	4.73 2-1903	4.18 31-1977	3.42 10-1913	3.05 4-2005	2.68 9-1977	1.78 30-1936
Cook	1.33 11-1975	1.25 21-1971	1.20 4-1966	1.49 3-1982	2.18 24-2010	3.54 28-2011	4.30 5-1999	3.85 16-1972	5.00 30-1995	2.32 27-1971	1.75 9-1977	1.06 13-1968
Cotton	1.80 5-1997	1.77 25-2001	1.45 4-1985	2.49 23-2001	2.23 17-1987	4.40 23-2002	5.72 5-1999	5.84 13-1988	4.25 14-1998	2.16 16-1968	1.89 17-1996	1.34 16-1984
Crookston	1.00 30-1896	1.28 6-1946	1.42 16-1941	2.08 26-1924	4.95 11-1922	4.20 27-1960	5.40 2-1919	5.85 31-1908	3.12 2-1953	2.70 9-1997	1.90 1-2000	1.20 25-2009
Detroit Lakes	1.47 5-1949	2.05 11-2013	1.48 23-2011	2.25 30-1991	3.55 12-2004	4.25 21-2013	4.50 21-1909	5.40 11-1909	3.30 4-1913	3.49 10-1973	1.43 11-1919	1.00 4-1951
Duluth	1.18 23-1969	1.16 27-1998	2.25 15-1945	2.61 19-1948	3.93 23-2010	4.14 19-2012	5.20 21-1909	3.22 8-1939	3.91 17-1955	2.94 26-2010	2.33 10-1998	1.91 6-1950
Ely-Winton	0.81 10-1975	0.81 25-2001	1.00 25-1914	1.88 16-1994	4.83 20-1970	2.67 10-1970	3.00 23-1987	2.94 17-1988	4.78 10-1961	4.24 9-1973	1.49 7-1915	1.04 15-2008
Fairmont	1.40 13-1910	2.04 27-1948	1.87 23-1966	3.35 7-2006	3.73 18-2000	4.68 24-1993	4.79 19-1963	5.03 2-2006	6.20 15-2004	2.79 15-1998	2.30 26-1896	2.11 28-1982
Faribault	1.31 11-1975	2.23 29-2012	2.00 25-1996	1.90 30-1984	2.86 1-1973	5.34 30-1901	5.35 21-1951	4.13 25-1983	4.36 11-1986	2.65 24-1899	2.34 10-1975	1.61 28-1982
Farmington	1.35 18-1996	1.25 28-1902	1.70 1-1965	3.05 3-1934	3.50 29-1933	4.54 5-1994	5.10 8-1955	4.65 7-1984	6.05 16-1922	4.25 4-1900	1.80 9-1970	2.00 14-1891
Fergus Falls	3.00 6-1997	0.98 1-1922	2.25 29-1933	2.35 27-1942	4.65 18-1892	4.60 5-1896	3.48 2-1898	4.30 7-1992	3.28 10-1913	3.18 5-1946	1.68 9-1977	1.41 11-1949
Fosston	0.90 7-1989	1.20 25-1930	2.10 12-1928	1.97 30-1991	2.21 27-1963	6.00 25-1950	8.97 19-1909	3.27 12-1919	5.47 1-1973	4.71 13-1984	1.30 16-1930	1.25 11-1949
Grand Marais	1.05 15-1950	1.40 19-1924	2.20 29-1938	2.62 17-1967	2.78 5-1950	5.43 29-1996	3.73 20-2011	2.93 1-1987	3.49 10-1961	2.55 9-1968	2.57 23-2012	1.78 16-1984
Grand Meadow	1.78 25-1967	2.10 27-1893	2.00 23-1966	4.53 24-1990	4.70 18-2000	5.00 15-1892	5.22 19-1999	6.41 13-1911	7.20 15-2004	5.30 5-2013	3.14 1-1991	1.58 28-1952
Grand Rapids	1.25 4-1997	1.45 1-1922	2.26 23-2009	1.92 3-1940	3.14 1-1955	4.78 19-2012	4.16 2-1978	4.37 3-1983	3.94 2-1937	4.44 9-1973	2.10 2-1938	0.99 14-2008
Gunflint Lake	0.88 23-1982	1.40 22-1981	1.00 7-1973	2.45 17-1967	2.15 5-1950	2.42 17-1990	4.60 23-1987	4.05 1-1987	4.75 9-1977	2.58 1-1950	1.95 1-1991	2.57 16-1984
Hallock	0.92 9-1983	1.40 14-1915	1.90 28-2004	1.85 30-1937	2.16 17-1996	3.85 30-2010	4.73 28-2001	5.36 31-1956	5.50 4-1900	2.50 9-2007	2.00 14-1944	0.80 14-1933
Harmony	1.28 25-1967	1.48 8-1981	1.87 28-1956	3.14 10-2013	3.73 18-2000	5.71 1-2000	5.15 16-1950	4.26 31-1962	4.97 20-1983	3.10 15-1966	2.94 1-1991	1.75 6-1982
Hibbing	0.87 4-1997	0.67 22-1979	1.35 23-2009	3.57 13-2010	2.57 31-2014	3.66 22-2002	4.50 5-1999	5.75 13-1988	3.21 30-1995	3.68 9-1973	1.30 9-1977	0.93 16-1984
Hutchinson	0.92 18-1996	1.56 9-2012	3.42 4-1985	1.57 13-2010	2.81 5-1959	6.00 29-1997	3.48 23-1977	4.50 26-1967	3.70 8-1991	2.38 10-1973	2.57 1-1991	1.33 28-1959
Int'l Falls	0.95 10-1975	1.20 1-1911	1.29 4-1966	1.68 28-1895	2.67 23-1991	4.15 24-1898	4.20 2-1966	4.82 30-1942	3.34 1-1973	2.33 31-1979	2.62 3-1919	1.21 5-1960

STATION	JAN	FEB	MAR	APR	MAY	JUN	JUL	AUG	SEP	OCT	NOV	DEC
Itasca	0.88 24-1950	1.30 11-2013	2.65 30-1933	2.88 24-1993	2.98 31-1985	5.42 22-1957	4.61 3-1983	3.69 8-1984	2.50 21-2004	3.00 1-1924	1.96 9-1977	1.40 14-1927
Leech Lake Dam	0.96 9-1939	1.03 1-1893	2.50 25-1901	2.00 29-1940	2.65 23-1962	3.27 20-2012	7.02 10-1954	4.34 26-1942	3.22 10-1951	3.33 5-1969	2.02 12-1940	1.16 3-1891
Luverne	1.25 2-1999	1.27 28-1951	2.00 28-1907	3.17 2-1967	3.20 9-1951	6.51 6-1896	4.43 12-1908	3.60 29-1902	3.46 8-1960	4.12 31-1979	2.00 10-1998	1.35 19-1902
Marshall	1.19 30-1984	1.46 29-2012	2.45 15-2002	4.22 23-2001	3.68 8-1993	8.07 17-1957	3.42 26-1968	6.50 10-1994	3.51 2-2010	3.27 31-1979	2.16 12-1982	1.92 28-1959
Milan	1.24 4-1997	1.70 23-1977	2.39 12-1977	2.38 5-1997	3.16 3-1912	4.33 29-1971	9.78 4-1995	3.26 3-1953	3.53 20-2007	2.53 17-1996	2.56 20-1930	1.56 3-1951
Montevideo	1.86 17-1996	1.69 10-1965	3.60 23-1991	2.70 22-1985	4.80 7-1993	7.30 17-1957	3.95 3-1903	6.30 21-2002	3.27 22-1969	2.88 29-2004	8.50 27-2001	2.00 28-1982
Mora	1.20 4-1949	1.20 1-1922	1.91 12-1977	3.41 26-1954	2.82 24-1939	3.34 18-1954	3.00 20-1952	3.91 7-1995	2.51 12-1978	5.78 5-2005	2.05 28-1905	1.51 15-1984
Morris	1.21 31-1926	1.15 14-1926	1.93 28-1940	6.90 26-1954	3.64 31-1959	5.20 26-1914	4.84 8-1929	4.34 18-1935	3.30 12-1897	3.60 15-1984	1.92 20-1930	1.00 3-1891
MSP	1.21 24-1967	1.90 24-1930	1.62 1-1965	2.58 6-2006	3.16 21-1906	4.13 19-2014	9.15 23-1987	7.28 30-1977	4.36 12-1903	4.61 4-2005	2.52 11-1940	1.50 14-1891
New Ulm	1.50 22-1917	1.42 27-1971	4.91 12-1899	2.37 2-1967	3.32 21-1908	5.22 8-1953	7.37 15-1916	5.42 19-2007	4.35 12-1928	3.80 6-1911	2.55 26-1896	2.00 24-1893
Olivia	1.10 24-2010	1.63 29-2012	1.71 29-1981	3.25 23-2001	2.42 6-2012	4.53 17-1993	3.76 12-1991	2.70 7-2009	3.53 8-1991	1.95 4-2002	2.38 9-1977	1.10 10-2012
Park Rapids	1.20 4-1887	0.90 5-1908	3.39 30-1933	2.24 28-1970	4.54 23-1933	4.03 10-1994	6.08 18-1985	6.75 1-1906	2.36 4-1971	4.57 10-1973	1.59 1-2000	1.43 6-1987
Pine River Dam	1.75 11-1975	1.80 26-2001	2.20 4-1985	2.67 16-2012	4.45 20-1902	7.10 3-1898	5.01 28-1972	4.43 8-1899	3.38 14-1941	4.50 3-1903	2.75 27-1988	0.97 16-1902
Pipestone	1.50 21-1917	1.87 13-1915	2.16 29-1940	3.40 16-2003	4.23 25-1953	6.18 17-1957	4.65 11-1989	3.70 5-1955	4.17 8-1926	5.62 10-1973	2.10 12-1998	1.25 26-1916
Pokegama Dam	1.09 4-1897	0.80 24-1977	2.03 24-1901	2.12 24-1960	3.25 1-1955	4.51 8-1888	4.43 8-1926	5.08 25-1941	4.11 2-1937	5.62 10-1973	1.78 2-1938	0.86 13-1968
Red Wing	1.23 25-1967	1.78 29-2012	1.66 28-1998	3.14 28-1975	3.40 30-2004	5.46 14-2012	7.78 1-1978	6.60 11-1945	5.20 19-1907	3.30 6-1911	2.57 10-1975	1.45 28-1959
Redwood Falls	2.20 21-1917	1.53 7-1928	1.74 26-1921	3.30 22-2001	3.94 30-2004	4.31 16-1993	3.32 26-1990	3.83 10-1994	3.27 10-1933	2.92 6-1911	2.60 17-1931	1.41 26-1916
Rochester	1.42 24-1967	1.12 20-2014	2.20 22-2011	3.81 23-1990	4.02 17-2000	4.80 1-2000	7.47 11-1981	5.16 18-2007	5.38 12-1978	2.85 6-1911	2.30 1-1991	1.36 11-2010
Roseau	1.20 30-1916	0.94 27-1983	1.28 25-1945	2.60 5-1997	4.26 11-2004	4.04 10-2002	3.50 2-1919	4.75 4-1985	5.37 2-1957	2.15 15-1934	2.00 15-1909	1.00 4-1997
St. Cloud	1.90 10-1897	1.81 28-1951	2.49 23-2009	3.52 22-2001	5.00 16-1894	3.35 27-1920	4.80 1-1903	4.57 3-1956	3.32 8-1985	4.07 5-2005	2.02 9-1977	0.91 1-1945
St. James	1.20 18-1996	1.62 29-2012	1.96 9-1992	2.32 2-1967	4.32 25-1953	6.10 8-1953	3.50 8-1986	4.80 20-2007	7.50 23-2010	2.70 2-2009	2.97 20-1973	1.34 25-1982
Sandy Lake Dam	1.26 11-1975	2.14 27-1971	1.43 15-1945	2.96 15-1894	2.95 29-1978	5.06 20-2012	4.65 28-1972	4.23 31-1989	4.24 1-1986	4.33 2-1995	1.80 28-1905	1.09 15-1893
Spring Grove	1.42 5-1946	1.79 23-1959	2.05 4-1992	2.71 10-2013	3.72 18-2000	5.07 1-2000	4.58 6-1978	5.65 9-2007	4.20 1-2010	2.78 18-1998	2.32 17-1958	1.02 2-2007
Stillwater	1.77 18-1996	1.60 28-1948	1.70 6-1908	2.35 28-1986	2.77 28-1978	7.98 1-1967	5.14 24-1987	4.50 9-1993	3.56 16-1992	5.04 5-2005	2.68 15-1944	1.81 25-1982
Tower	1.40 4-1897	1.54 25-2001	1.91 4-1966	2.29 17-1961	2.45 29-1978	4.62 17-1931	3.50 6-1897	3.53 24-1932	8.70 7-2007	4.00 1-1995	3.21 17-1996	1.20 15-1902
Two Harbors	1.36 4-1982	2.02 21-2014	2.73 12-1977	2.76 23-2001	4.38 5-1950	10.45 20-2012	4.20 21-1909	5.25 9-1939	4.32 17-1955	3.45 10-1973	4.10 16-1909	2.12 1-1985
Wadena	1.86 5-1997	1.42 11-2013	1.95 15-1957	2.31 16-2012	3.40 27-2012	3.03 10-1968	4.33 8-1959	5.97 7-1995	3.23 12-1980	3.91 15-1984	1.73 9-1977	1.25 3-1951
Warroad	0.88 12-1936	1.20 13-1938	1.25 6-1916	2.49 21-1974	4.48 12-2004	5.16 11-2002	4.92 2-1919	2.84 16-2009	3.35 18-2003	2.50 16-1934	1.49 8-1915	1.01 7-1916
Waseca	1.40 20-1988	1.96 29-2012	3.00 11-1918	2.00 4-1945	2.95 19-1917	4.30 5-1930	4.11 18-2008	5.40 31-1962	5.33 23-2010	2.90 7-1931	2.64 30-1991	1.78 28-1982
Waskish	0.70 11-1975	0.84 24-1977	1.32 23-2009	1.59 11-2011	2.48 11-1985	3.46 12-2014	2.95 11-1972	4.65 1-2001	2.12 1-1925	2.33 19-1965	1.16 21-1977	0.95 6-2013
Wheaton	2.05 4-1997	1.50 19-1924	2.10 19-1982	2.20 26-1924	4.27 10-1920	7.02 2-2007	4.90 8-1950	3.95 31-1982	3.30 2-2006	2.90 5-1946	1.74 9-1977	1.25 28-1959
Willmar	1.50 18-1996	1.36 25-2007	2.05 4-1985	3.31 30-1936	2.35 22-1893	7.20 9-1895	3.96 11-1981	4.21 16-1926	5.22 25-1929	2.75 9-1898	2.00 12-1899	1.30 9-1899
Windom	1.84 29-1909	1.52 29-2012	2.62 4-1985	2.70 2-1967	3.89 29-1980	3.11 20-1983	3.84 13-1970	5.24 10-1994	8.34 23-2010	2.38 17-1968	2.39 9-1977	1.11 28-1987
Winnebago	1.87 4-1906	1.64 28-1948	1.65 14-1918	3.05 28-1981	3.56 18-2000	3.79 18-1956	4.55 19-1963	6.22 10-1948	8.34 25-2005	2.15 15-1998	1.85 10-1975	1.48 10-1911
Winnibigoshish Dam	1.55 11-1975	1.31 25-1930	2.10 10-1892	2.16 30-1942	2.77 23-1962	7.17 9-1888	3.85 22-1987	4.15 5-1916	3.70 19-1900	3.10 1-1995	1.63 9-1977	1.14 31-2010
Winona	1.74 4-1971	1.65 22-1922	2.00 25-1996	3.10 11-1995	3.83 1-1936	4.42 6-1914	4.30 1-1978	4.95 19-2007	3.70 16-1992	2.64 11-1986	3.45 8-1945	1.50 1-1985
Worthington	1.36 11-1975	1.60 29-2012	1.96 23-1987	3.55 23-2001	2.30 11-1992	3.73 15-2014	3.51 4-1978	4.80 10-1994	5.51 23-2010	3.33 1-1977	1.95 21-1975	1.39 28-1982
Wright	1.14 18-1996	1.12 21-2014	1.75 12-1977	2.18 23-2001	2.98 29-1977	6.11 20-2012	3.45 22-2012	3.80 31-1989	3.32 12-2003	2.79 9-1970	2.00 1-1974	1.30 16-1984
Young America	1.23 11-1975	1.38 10-1965	2.45 2-1965	2.32 28-1989	4.74 21-1960	4.65 22-1983	3.88 8-1955	4.27 31-1977	5.54 8-1991	2.61 8-1970	2.18 20-1975	1.48 28-1982
Zumbrota	1.36 7-1900	1.63 29-2012	1.75 29-1924	2.65 24-1900	3.44 29-1942	6.46 27-1998	4.85 16-1957	4.92 31-1962	5.16 23-2010	3.00 3-1954	2.61 10-1975	1.50 30-1923

MAP 7: Climate Stations with Four or More Statewide Daily Precipitation Records

Thief River Falls (4)

Beaver Bay (6)

Two Harbors (9)

Pine River Dam (4)

St. Cloud (4)

St. Francis (4)

Elk River (4)

Willmar (4)

Montevideo (4)

Minneapolis (4)

Red Wing (6)

Lake City (4)

Lynd (5)

Faribault (4)

Mankato (4)

Winona (7)

Grand Meadow (6)

Luverne (5)

Albert Lea (5)

Hokah (5)

Worthington (5)

Harmony (4)

Caledonia (5)

The adage in climatology is that what goes around comes around. It is conceivable that a rainstorm like the one of July 17–19, 1867, could recur in Minnesota. The powerful convective engine of a humid and unstable atmosphere drove this historical, perhaps one-of-a-kind thunderstorm. But such an engine drives many kinds of severe weather during the spring, summer, and fall seasons. Lightning, straight-line winds, tornadoes, and flash floods are all born of thunderstorms, and each can threaten human life and inflict serious damage on communities and landscapes. Of these threats, flash floods and tornadoes are the best documented historically, but in recent years straight-line windstorms, called derechos, have been well documented, too. All three of these storm types have had traumatic impacts on Minnesota's people and landscape; some of the most infamous are recorded here.

Significant Thunderstorms and Flash Floods

The Minnesota State Climatology Office defines a flash flood based on multiple criteria: a total rainfall of 6 or more inches within a 24-hour period is reported from at least one climate station, and the geographic area (in square miles) encompassing rainfall amounts of 4 or more inches is described. In much of Minnesota a 6- to 7-inch rainfall approximates the hundred-year return period for a 24-hour amount (in other words, there is a one percent probability of such rainfall), while 4 or more inches most often leads to soil erosion and economic damages from local flooding. Historically, about five major flash floods occur in the state each year, approximately 75 percent in the months of June, July, and August and more than 50 percent between 6:00 and 11:00 PM. A majority of flash floods are associated with stationary or slow-moving fronts oriented west-to-east or northwest-to-southeast across the state. In such situations the front parallels the wind flow in the upper atmosphere and individual thunderstorm cells can develop repeatedly along the frontal boundary and "train" across the same area of landscape, bringing multiple periods of heavy rain.

▓ In the nineteenth century flash floods were difficult to quantify because few observation sites existed in Minnesota's sparsely populated landscape. One of the first to be documented occurred on August 6, 1866, when a massive thunderstorm complex, likely associated with a stationary front, parked over south-central and southeastern counties. Sibley Indian Agency, located in the southwest corner of Sibley County, reported 10.30 inches of rainfall followed by more than 3 inches on the next two days. The Rush River rose abruptly and flooded the sur-

rounding landscape. St. Paul noted a rise of 4 feet in the Mississippi River's flow but only 1.72 inches of rainfall from this storm was reported by the local observer. However, to the southeast in Fillmore and Houston counties the storm turned tragic when a rapid and steep rise in the Root River swept away 30 settlers to their deaths. Among them were three members of the Wisel family from Preble Township, southeast of present-day Lanesboro in Fillmore County; the storm thus came to be known as the "Wisel Flood." Additionally, the town of Houston was completely submerged, and the Southern Minnesota and Root River Railroad could not run for days, the tracks inaccessible under 7 to 8 feet of water.

Another series of severe thunderstorms—this time very localized—on May 30–31, 1877, produced massive amounts of rain and runoff in the St. Paul area. Fort Snelling reported 4.30 inches on the thirtieth and another 5.12 inches on the thirty-first, the largest two-day total at the fort since it began its records in 1820. In fact, Fort Snelling's combined May and June total rainfall was 14.5 inches. However, these values are questioned by historian Tom St. Martin, who notes that other observers in St. Paul and Minneapolis did not record amounts comparable to those at Fort Snelling.

On July 2–3, 1879, a series of severe thunderstorms brought deadly tornadoes to Goodhue County but flooding rains to Ramsey County, especially St. Paul. Starting around 5:00 PM on July 2 and continuing throughout most of the night, torrents of rain produced more than 5 inches at the signal corps office in downtown St. Paul and 6.60 inches at Fort Snelling, a record that still stands today. Many St. Paul roads washed out; flood damages were noted along Rice, Wacouta, and Mackubin streets; and hundreds of people living on the flats along Como Road were forced to flee. A large boat filled with 150 cords of wood broke loose and was carried downstream.

July 26–27, 1892, brought a "remarkable display of electricity"—along with extremely heavy rainfall—to the Twin Cities area. St. Paul reported 5.69 inches, while Minneapolis saw 6.35 inches, a record for July 27 that still stands. Streets were flooded and streetcar service halted. Railroad beds were washed out, and summer cottages along Lake Phalen's shore were submerged by the rising waters.

June 11–13, 1899, brought a series of severe thunderstorms to southeastern Minnesota, resulting in flash flooding in Blooming Prairie (Steele County), Grand Meadow (Mower County), and Minnesota City and St. Charles (Winona County). Rainfall totals ranged from 6 to more than 10 inches at these locations. Roads were washed out and many crops had to be replanted—both situations presenting farmers with considerable challenges during the wettest June of the nineteenth century. Another event sparked by the severe thunderstorms overshadowed this flooding. In Wisconsin, New Richmond reported an F5 tornado (winds greater than 261 mph) which killed 117 people and destroyed more than 300 buildings. This remains the most lethal tornado to ever strike Wisconsin.

One of the heaviest early-twentieth-century flash floods occurred in northwestern Minnesota on July 19–20, 1909. Beaulieu in central Mahnomen County reported 11.72 inches of rainfall, Bagley in Clearwater County 11.28 inches, and Fosston in eastern Polk County 9.13 inches. Other rainfall reports included more than 7 inches at Walker (Cass County) and 5.88 inches at Park Rapids (Hubbard County). This flash flood was rather narrow in geographic extent, though it did wash out roads and railroads and severely damage small grain crops. Somewhat of a repeat scenario occurred in the Red River Valley on May 29, 1949, when severe thunderstorms passed across Pennington County and dumped 7.50 inches of rain on Thief River Falls in little more than six hours. The resulting damages and erosion losses amounted to just under $400,000.

One of the heaviest thunderstorms to ever pass over the Iron Range dropped enormous amounts of rain on September 9–10, 1947. Portions

Q & A

Out of more than 1,800 documented Minnesota tornadoes (from 1820–2005), how many have been classified as F5, with wind speeds greater than 261 miles per hour?

a. 7

b. 15

c. 32

(answer on page 329)

of St. Louis County, including the city of Virginia, reported more than 6 inches of rain, while an observer near Hibbing reported more than 8 inches. Streets and roads were flooded, and a number of businesses had to close. Strong winds accompanying the thunderstorms blew down trees as well.

■ Both Morris (Stevens County) and Milaca (Mille Lacs County) were hit by severe thunderstorms on April 26–27, 1954, the only known instance of flash flooding in the month of April, when spring snowmelt runoff is the usual culprit. Morris received 7.23 inches of rainfall, 6.90 in just 10.5 hours. Milaca received 5.05 inches. Lightning discharges were almost continuous during this storm, and golf ball–sized hail completely covered the ground at times. Basements, roads, and highways were flooded, as were freshly planted fields, which then needed to be resown. A number of small bridges washed out, and in Morris a woman was rescued from her flooded home by boat.

Flash flooding overtook a bridge in Jordan in 1960.

■ May 20–21, 1960, brought flash floods to east-central communities, including Jordan, Mapleton, Montgomery, New Prague, and New Richland, each of which reported 5 to 7 inches. Some unofficial observers with backyard rain gauges recorded as many as 10 inches from these storms. More than 60 percent of the properties in Jordan (Scott County) were flooded, and 115 families had to be evacuated. These rains not only produced street flooding and wet basements but also caused sharp rises in the Le Sueur and Blue Earth rivers. Some crops required replanting, and the Minnesota Department of Transportation estimated $150,000 in damages due to washed-out highways.

■ May 27, 1970, brought another bout of severe weather when within two to three hours 4- to 6-inch rains fell across Goodhue and Wabasha counties in southeastern Minnesota. Most rainfall occurred between 9:00 and 11:00 PM, with darkness making flooded roads difficult to see. Lake City and Wabasha, both Mississippi River towns, received more than 6 inches of rain. The Cannon and Zumbro rivers flooded out of their banks, sweeping away cars and drowning three people. At Zumbro Falls in Wabasha County the Zumbro River rose 11 feet in one hour. Damages to roads, bridges, and property as well as losses due to erosion totaled $6 million.

*   *   *

During the 1970s the Minnesota State Climatology Office expanded its weather observer network so that the overall number of volunteers grew from just under two hundred to well over a thousand. With this increase in climate station density—and with greater emphasis placed on recording precipitation—the geographic extent of flash floods could be more accurately noted. Indeed, analysis of most such storms is regularly published. As a result, many of the flash floods occurring in Minnesota since 1970 are more thoroughly described than those of earlier eras.

▦ Called the "granddaddy" of flash floods by former state climatologist Earl Kuehnast, the storm of July 21–22, 1972, encompassed much of central Minnesota, flooding portions of Crow Wing, Douglas, Isanti, Kanabec, Mille Lacs, Morrison, Otter Tail, and Todd counties. The area affected by 4 or more inches of rain was 6,800 square miles, and by 8 or more inches, 1,500 square miles—the largest flash flood in state history. Some parts of Morrison County reported 15 inches of rainfall; Fort Ripley (Crow Wing County) saw 10.84 inches in a 24-hour period. Every major highway from Alexandria east to the Wisconsin border was closed for three to 16 days because of flooding, and total damages exceeded $20 million.

▦ One of Duluth's heaviest flash floods occurred one month after the "granddaddy" storm. On August 20, 1972, intense rainfall crossed the area between 3:00 and 4:30 AM, delivering more than 4 inches in some places. The ground was already saturated from 3 to 4 inches of rain that had fallen on August 15 and 16, partially clogging some of the sewer drains with debris. Many streets flooded and washed out; damages were estimated at $12 million. This event was to remain Duluth's worst flash flood until June 2012.

▦ A rare case of October flash flooding occurred on the ninth and tenth in 1973, when severe weather crossed the state from southwest to northeast, delivering lightning, tornadoes, strong straight-line winds, hail, and intense rainfall. Unlike most other flash floods, the heaviest rainfalls were widely scattered. The largest amounts, ranging from 5 to 7 inches, fell in Aitkin (Aitkin County); Grand Rapids and Pokegama Dam (Itasca County); Hibbing (St. Louis County); Leech Lake, Remer, and Walker (Cass County); Park Rapids (Hubbard County); and Vesta (Redwood County). Basement flooding and some washed-out roads were noted in and around those communities.

The second-largest flash flood in total area occurred on June 28–29, 1975, when parts of Becker, Clay, Mahnomen, Norman, Otter Tail, and Wilkin counties received 4 to 13 inches of rain. Ulen in Clay County, northeast of Fargo-Moorhead, reported 13 inches of rainfall, which filled drainage ditches and closed Highway 32 for a time. The major impact, however, was on crop fields: the total area affected by rainfall of 4 or more inches was 6,000 square miles. Another flash flood, this one across Beltrami, Clearwater, Kittson, Marshall, and Pennington counties, delivered 4 to 8 inches of rain on July 1–2. Again, agricultural fields were flooded: heavy rains of 4 or more inches affected a total of 4,500 square miles. Property damage was reported in Thief River Falls (Pennington County) and Bemidji (Beltrami County), but because the storm targeted a sparsely populated area relative to the rest of the state it did not garner much media attention.

On average, how many hail loss claims are filed in Minnesota each year?

a. fewer than 25

b. fewer than 50

c. more than 100

(answer on page 329)

The famous state fair flash flood occurred on August 30–31, 1977. The Twin Cities airport recorded more than 7 inches of rainfall, while the fairgrounds saw nearly 5 inches, much of which arrived between 8:30 and 10:00 PM, washing out the grandstand show and leaving the fairgrounds ankle-deep in water. Other metropolitan area locations, including Afton, Bloomington, Lake Minnetonka, and Mendota Heights, also reported rainfall amounts greater than 7 inches. Hundreds of basements were flooded, but the area recording 4-inch amounts or greater was relatively small: 832 square miles.

Rochester saw not one but two significant flash floods in 1978. The first began about 6:00 PM on July 5 and ended about 2:00 AM on July 6, as areas southwest, south, and east of downtown received 6 to 7 inches of rainfall, topped by 7.30 inches in Quincy Township, just northeast of Eyota (Olmsted County). During the area's worst flooding in more than a century of record keeping, the Zumbro River's south branch rose at the remarkable rate of 1 foot per hour, resulting in a record crest on that watershed: 22.5 feet. One-fourth of the city was inundated under as much as 6 feet of water, and nearly 5,000 residents were evacuated. Roads were washed out, and five people drowned. The area encompassed by 4 or more inches of rain was 700 square miles. Then, a second flash flood came to Rochester on September 12, striking over six hours, from 10:00 AM to 4:00 PM. Downtown reported 7.07 inches, and the overall area receiving 4 or more inches was 1,200 square miles. Many of the same streets flooded again, and storm sewers were overwhelmed. The Zumbro River crested at 13.1 feet, once again above its 12-foot flood stage. These storms inspired one of the largest flood-control projects undertaken in Minnesota, its goal to mitigate future flood threats in the Rochester area.

• •

Freshet

Derived from Scottish and Middle English, a freshet is a sudden rise in a stream as it overflows its banks due to heavy rain or snowmelt runoff and causes local flooding, an occasional springtime event in Minnesota. In reference to the surplus flow of water across a landscape, freshet *has a gentler connotation than does* flood.

• •

▧ That same summer, another large flash flood struck northern Minnesota. Starting the afternoon of August 22, 1978, and lasting until the very early hours of the next day, heavy thunderstorms moved slowly across Aitkin, Cass, Hubbard, Itasca, and St. Louis counties. Rainfall amounts ranged from 4 to nearly 7 inches. Lightning knocked out the landing lights at Duluth International Airport, and many roads were inundated. Basements flooded, and residents were without phone service and power. The storm system delivered 4 or more inches of rainfall over 4,100 square miles.

▧ July 11, 1981, brought two flash floods to southeastern Minnesota, the storms moving through Olmsted and Fillmore counties between 2:00

Q & A

What does the term *bucketing* refer to?

a. an intense rainfall rate

b. a wind that blows buckets around the farm yard

c. a meteorologist trying to explain a blown forecast

(answer on page 329)

and 9:00 AM. Rochester International Airport reported 7.47 inches of rain—which closed the runway for a half hour—while just seven miles to the west an unofficial report of 11.04 inches was filed with the State Climatology Office. Roads were closed, and local watersheds south of the city reached flood stage. The area receiving 4 or more inches of rainfall approached 540 square miles. Further to the southeast, Fillmore County reported 7 to 8 inches, 7.70 in Preston, to be exact. The Root River flooded, and a dozen homes were evacuated. The area encompassed by 4 or more inches of rainfall was 1,080 square miles.

◼ An intense thunderstorm and consequent flash flood occurred on June 21, 1983, across Kandiyohi, Meeker, Sherburne, Stearns, and Wright counties. The system produced a double pulse of very heavy rainfall, one coming between 2:00 and 6:00 AM and another between 11:00 AM and 2:00 PM. Most rainfall reports ranged from 4 to 12 inches: a Meeker County Soil and Water Conservation District observer near Watkins reported 9.68 inches. The familiar results followed: flooded basements, washed-out roads, drowned crops, the latter damages calculated in the millions of dollars. The area that experienced 4 or more inches of rainfall was 1,330 square miles.

◼ July 23–24, 1987, brought the Twin Cities' worst-ever flash flood. Thunderstorms "trained" along a front stalled over the metropolitan area, dumping up to 10 inches of rain over six hours, from 7:00 PM to 1:00 AM, sometimes as much as 2 to 3 inches per hour. Another severe thunderstorm 72 hours earlier had dumped 4 to 9 inches over nearly the same area, leaving the soils saturated. Roads became rivers; storm sewers spouted like geysers. All Twin Cities watersheds flooded, including Nine Mile and Minnehaha creeks. Two people were killed, one by a collapsed basement wall, another swept away while trying to cross Nine Mile Creek. Portions of the interstate highway system through the Twin Cities were closed for days. More than $21 million in damages was done to 8,643 residential and 264 commercial properties, with another $6 million in damages to public properties. The area encompassed by 4 or more inches of rainfall was 1,460 square miles, mostly over Hennepin and Ramsey counties.

▨ Portions of Blue Earth, Cottonwood, Faribault, Martin, Murray, and Nobles counties received from 4 to 9 inches of rain over 11 hours on June 3–4, 1991. Amboy in Blue Earth County reported 9.85 inches. Many watersheds that feed the Minnesota River went out of their banks, flooding roads as well as numerous basements. Croplands were under water or severely eroded, prompting some replanting. The storm contributed to one of the wettest growing seasons seen in southwestern and south-central Minnesota.

▨ Minnesota's wettest-ever May through August occurred in 1993: the statewide average rainfall was well over 21 inches, brought in part by several flash floods. The first was on May 7–8, when a series of thunderstorms dropped 3 to 7 inches of rainfall in just three hours on an already wet landscape in Chippewa, Cottonwood, Lincoln, Lyon, Murray, Nobles, Pipestone, Redwood, and Rock counties. Thousands of acres of cropland were flooded, as were many basements in Lamberton and Redwood Falls (Redwood County), Marshall (Lyon County), and Montevideo (Chippewa County). Several hundred homes and businesses reported flood damage; near Marshall a small mobile home was dislodged from its foundation and washed down the Redwood River. An even larger complex of thunderstorms pelted southern Minnesota with heavy rain and hail on June 16–17, as 4 to 6 inches of rain fell across a swath covering the Minnesota River Valley from Lincoln County east to Rice County. As a result, the Minnesota River remained above flood stage for much of the rest of June. Additional, smaller-scale flash floods occurred in July over Becker, Big Stone, Clay, Lyon, and Mahnomen counties. The historically wet summer was capped with yet another intense storm on August 14–15, when 4 to 10 inches of rain fell over Faribault, Fillmore, Freeborn, and Mower counties. Austin in Mower County was particularly hard hit: a thousand homes suffered water damage, and the Cedar River reached its second-highest flood crest, 21.3 feet.

▨ Two unusual and intense flash floods occurred in northwestern Minnesota in June 2002. The first struck on June 9 and 10, dumping 6 to 14 inches of rain across portions of Beltrami, Clay, Koochiching, Lake of the Woods, Marshall, Norman, and Roseau counties—an area rivaling that of the "granddaddy" storm of 1972. All watersheds flooded out of their banks. Nearly every property in Roseau was damaged, despite desperate efforts to protect the city with sandbags holding back the Roseau River. The storm produced the largest flood crest ever measured on that river—23.3 feet—exceeding even the worst examples of spring snowmelt flooding. Many buildings in Ada (Norman County) also suffered damage. Two weeks later another flash flood struck on June 22–23, partially overlapping the area affected by the earlier storm system. Rainfall amounts of 4 to 8 inches fell over parts of Becker, Beltrami, Clearwater, Itasca, Koochiching, Mahnomen, Norman, and St. Louis counties. Basements flooded and roads washed out, but damages were less severe compared to the earlier storm. Nevertheless, a second flood crest washed through most northern watersheds.

■ One of the largest and most intense September flash floods on record occurred on September 14–15, 2004, when rainfall of 6 or more inches fell over portions of Blue Earth, Dodge, Faribault, Fillmore, Freeborn, Houston, Jackson, Martin, Mower, Olmsted, Steele, Waseca, and Winona counties—an area of more than 4,000 square miles. Up to 13 inches of rain was reported southwest of Alden in Freeborn County. The storm inflicted more than $6 million in property damage and $22 million in crop damage across the area. The communities of Albert Lea and Hollandale in Freeborn County, Austin in Mower County, and Ellendale and Blooming Prairie in Steele County suffered the most: many homes were flooded and roads impassable.

■ A rare October flash flood occurred in the east-central counties of Benton, Chisago, Dakota, Isanti, Kanabec, Mille Lacs, and Pine, where between 6 and 9 inches of rain fell over October 3 and 4, 2005. In Chisago County an automated gauge recorded 9.59 inches in Rush City; 6.61 inches was reported from Wild River State Park. Many roads were closed and basements flooded by this rare storm. In terms of geographic scale and rainfall intensity, it is one of only two such October storms—along with one in 1973, mentioned above—to visit Minnesota.

Late in the drought-stricken summer of 2007, a slow, northward-moving warm front over south-central and southeastern Minnesota brought a series of heavy thunderstorms over August 18–20, with the most intense occurring on August 19. Many weather observers reported new daily record rainfall amounts, including an all-time state record of 15.10 inches at Hokah (Houston County). The three-day storm's total rainfall exceeded 8 inches at a number of locations, and Fillmore, Houston, Olmsted, Wabasha, and Winona counties were declared federal disaster areas by FEMA due to flash flooding. The community of Rushford (Fillmore County) was hardest hit, as 490 of the 766 homes were flooded or had damage, along with 58 of the 70 businesses in town. Nearly 300 people had to be rescued by boat, and some roads and bridges were closed for days. Officials reported seven fatalities due to the storm, five in Winona County and two in Houston County. By the end of the month Hokah had reported a new statewide monthly rainfall total of 23.86 inches.

A rare September flash flood occurred in portions of Lake and St. Louis counties in far northeastern Minnesota on September 6, 2007, as a narrow band of intense thunderstorms developed late in the day. Babbitt, Embarrass, and Ely reported more than 6 inches of rain while Tower reported more than 8 inches. Many township roads were washed out, along with a section of Minnesota Highway 1 near Ely. Some of the flooding was greatly tempered by the drought situation that had dominated the summer season; the dry landscape was able to absorb a good deal of the excessive rainfall.

June 7–9, 2008, brought flash flooding again to southeastern Minnesota, this time concentrated in Fillmore and Houston counties. Strong and persistent thunderstorms yielded rainfall totals of greater than 5 inches at Hokah and La Crescent, more than 6 inches at Brownsville, more than 7 inches at Harmony, and more than 10 inches at Reno. For a time nearly all roads in Houston County were flooded, and the Root River crested well above flood stage. This storm was the first in a series that collectively produced one of the worst floods ever in northeastern Iowa, bringing widespread damages to Cedar Rapids and Iowa City.

Another sluggish warm front draped across southern Minnesota over September 22–23, 2010, producing a series of intense thunderstorms that brought record-setting amounts of rain and flash flooding to many counties. Observers reported 6 to 8 inches of rain, topped by Amboy (Blue Earth County), which reported 10.68 inches. The area receiving 6 or more inches of rainfall encompassed more than 5,000 square miles, marking one of the largest flash floods in state history. The Hormel corporate offices in Austin (Mower County) had to close due to flooding, and there were hundreds of reports of flooded basements throughout the southern communities. Many roads were closed, and most area stream gauges exceeded flood stage. Even the Mississippi River at St. Paul rose above flood stage (14 feet), a rare event for the autumn season. The year 2010 concluded as the wettest overall in state history, with more than 30 communities receiving 40 or more inches of precipitation.

The summer of 2012 brought drought to many parts of the state. Yet in the middle of the dry spell, a flash flood occurred across portions of Dakota, Goodhue, and Rice counties on June 14–15. A narrow band of intense thunderstorms crossed the area on the afternoon of June 14, producing some record-setting rainfall amounts. Dundas, Vasa, and Red Wing reported more than 6 inches of rainfall, while Cannon Falls reported nearly 9 inches. The Cannon River at Welch reached its third-highest flood stage of all time, and the Little Cannon River near Cannon Falls reached an all-time flood crest. Many basements were flooded, and Minnesota Highway 52 was closed for a time. Some flooded homes in Cannon Falls were evacuated as well.

The very next week, over June 19–20, 2012, portions of northeastern Minnesota experienced the worst flash flooding in history. Intense thunderstorms developed along a slow-moving warm front, especially over Aitkin, Carlton, Lake, and St. Louis counties. Rainfall amounts of 6 to 8 inches were common, and a few spots around Duluth and Two Harbors reported more than 10 inches, with the bulk of the rain coming over a 12-hour period from late on the nineteenth to the early morning of the twentieth. The governor's office declared a state of emergency in these counties, and National Guard troops were deployed for rescue and clean-up operations. Raging Kingsbury Creek flooded the Duluth Zoo and drowned several animals. Some seals were washed out onto Duluth streets but later discovered and returned. An eight-year-old boy was swept into a culvert in Duluth and traveled for eight blocks before being recovered relatively unharmed. Minnesota Highway 210 was completely washed out near Jay Cooke State Park, as was the historic swinging pedestrian bridge over the St. Louis River, which reached an all-time record flood crest. The Knife River flooded, closing Highway 61 between Duluth and Two Harbors, and several residents of Knife River were evacuated. Interestingly enough, this area of the Minnesota landscape went back into drought for the remainder of the year 2012.

Q & A

What type of cloud forms a "thunderhead," producing lightning, heavy rain, and sometimes hail?

a. cirrus

b. cumulonimbus

c. stratus

(answer on page 329)

Around the summer solstice in 2013, a series of severe thunderstorms brought flash flooding to different parts of the state. Over June 20–21, heavy rains fell across portions of western and central Minnesota. Artichoke Lake (Big Stone County), Glenwood (Pope County), and Morris (Stevens County) received more than 5 inches of rain, while Lake Park (Becker County) reported 6.68 inches and Breezy Point (Crow Wing County), 7.75 inches. Many basements were flooded, and many streets had to be closed. These thunderstorms also brought damaging winds, with gusts up to 71 miles per hour at Morris and 85 miles per hour at Benson (Swift County). Then over June 21–22, another series of strong thunderstorms brought hail, damaging winds, and intense rains to south-central counties. Wells (Faribault County) and Spring Grove (Houston County) saw more than 5 inches of rain, with reports of flooded basements, streets, and surrounding farm fields. The real disruption from these storms occurred as a result of widespread power outages, said by an Xcel Energy spokesperson to be one of the largest such episodes in company history.

One of the most remarkable intense thunderstorms occurred on May 31, 2014. Widespread rainfall was reported that day across most of the state, but in Stearns County near Sauk Centre at a Discovery Farm operated by the Minnesota Department of Agriculture the automated rain gauge recorded 4.60 inches of rainfall in one hour. This corresponds to about a once-in-a-thousand-years rainfall intensity. Overall the storms delivered 3- to 5-inch rainfall amounts around the state over May 31 to June 2, 2014, causing some isolated flooding.

■ June 2014 was the wettest month in recorded state history, with a statewide average rainfall of more than 8 inches. Many individual observers reported their wettest month in history with more than a foot of rain. Not surprisingly, there were several flash floods during the month. Perhaps the most notable of these occurred over June 14–15 in southwestern counties. Total rainfall from the storm typically ranged from 4 to 6 inches, with Woodstock (Pipestone County) reporting 6.37. The Rock River near Luverne (Rock County) reached an all-time high flood crest of 14.79 feet, and many roads had to be closed.

Minnesota's Significant, Early Tornadoes

Minnesota's first documented tornado occurred on April 19, 1820, when a small funnel descended upon cantonment New Hope, precursor to Fort Snelling, at the juncture of the Minnesota and Mississippi rivers. It damaged the barracks roof but caused no injuries. One of the first documented tornadoes following statehood occurred in Dodge County on July 3, 1860, when a settler's home was unroofed by a waterspout. On June 17, 1865, a tornado near Red Wing in Goodhue County crossed the Mississippi River into Pierce County, Wisconsin, and was said to have safely carried a horse an eighth of a mile. In Fillmore County on June 28 another tornado destroyed four farmhouses and left behind many dead and featherless chickens and turkeys. The first documented deaths from a tornado came on July 8, 1868, as a storm passed across Pope and Stearns counties in central Minnesota and destroyed the Richardson farm in Raymond Township (Stearns County), killing two.

As Minnesota's population density increased during the second half of the nineteenth century, more settlers meant more tornado sightings. In

Q & A

What is a tail-end Charlie?

a. the last winter storm of the season

b. the southernmost thunderstorm cell in a squall line

c. the instrument package attached to a radiosonde, or balloon

(answer on page 329)

the 1870s the area that would become the Twin Cities recorded two destructive tornadoes, one in Mendota on June 24, 1875, which damaged 20 buildings and carried off several sheep, and another that struck on June 29, 1877, killing a woman and destroying three homes just south of Minneapolis. The latter storm was part of a tornado outbreak that produced additional destructive storms in Douglas, Sherburne, and Wright counties, killing three and leaving evidence of winds in the F3 range (158–206 mph). Another outbreak occurred on July 2, 1879, as three tornadoes set down across Goodhue County near the towns of Frontenac, Vasa, and Wanamingo. Two F4 tornadoes (207–260 mph) caused 14 deaths and 43 injuries and destroyed an orphans' home in Vasa. One tornado crossed the Mississippi River into the Wisconsin town of Diamond Bluff, depositing debris from the Minnesota side there. The storm produced 5.11 inches of rainfall in St. Paul and 6.60 at Fort Snelling, leading to flash flooding.

With the birth of the National Weather Service in 1870 and organization of the Minnesota Weather Service under Dr. William Payne during the 1880s, the number of weather observers grew significantly and, likewise, eyewitness accounts as well as detailed documentation increased. Weather Service personnel assessed the storms' character by collecting photos, usually of storm aftermaths, and taking statements from eyewitnesses or survivors. Martin Hovde, during his tenure as head of the Weather Service Office in Minnesota (1934–55), initiated an effort to more fully document the character of storm damage and distinguish tornadic winds from downburst or straight-line winds by examining damage evidence and the footprints storms left on the landscape—a precursor to today's routine assessments utilizing the Fujita Scale (now called the Enhanced Fujita Scale, a modification made by the NWS in 2007) to classify tornado intensity. Despite the modern technology applied to detecting tornadoes, a storm's intensity is not forecast nor really known until NWS meteorologists can conduct a post-storm survey.

Numerous studies of Minnesota tornadoes show that their peak occurrence is from 4:00 to 6:00 PM, when just over a third strike. Since a strong thunderstorm's convective energy—its vertical transport of heat and water vapor—is typically needed to produce a tornado, the vast majority occur during daytime hours and, with the longer days of summer, during the late afternoon and early evening. But tornadoes have also occurred at midnight in Minnesota, most recently at 11:55 PM on July 12,

2004, near Dent in Otter Tail County and right on the stroke of twelve in Albertville, Wright County, on September 10, 2002.

An assessment of tornado damage paths published in *Significant Tornadoes: 1680–1991* and *Significant Tornadoes: Update 1992–1995*, both by Thomas P. Grazulis, shows the vast majority of trajectories oriented southwest to northeast over most Minnesota counties. The average path length for a tornado—its time on the ground—is fewer than five miles, but storm touchdowns range from very brief—less than 0.1 miles—to extremely long—as many as 60 to 80 miles. The damage swath cut by high winds may be quite small, fewer than 50 yards wide, but a few surveys have revealed damage along a corridor of up to one mile in width.

· ·

Recurrence interval and exceedance interval

In climatology and hydrology the terms recurrence interval *and* exceedance interval *are used to describe frequencies of weather factors and events. The recurrence interval, often called the return period, refers to the average period of time between the incidence of a given quantity and that of an equal or greater quantity. For example, in southern Minnesota the recurrence interval for a rainfall rate of 2 inches per hour is about 10 years; in southern Minnesota the recurrence interval for a 6-inch rainstorm in a 24-hour period is about 50 years. The exceedance interval is similar to the recurrence interval but refers to the average number of years between a given quantity or event and one of a greater magnitude. Both parameters require analysis of long-term measurements.*

· ·

Studies of tornado statistics in Minnesota also show that, while they have occurred as early as March 18 and as late as November 16, more than 80 percent appear during the months of May through August. In terms of tornado intensity, Minnesota's weather history is heavily skewed toward weaker storms. On the Fujita Scale, F0 tornadoes (winds less than 72 mph) make up two-thirds of Minnesota's storms; F1s (73–112 mph) comprise about 24 percent of storms historically; and F2s (113–157 mph) account for only about eight percent. Thus the most intense tornadoes—rated F3, F4, or F5—account for less than one percent of all historical occurrences. Though quite rare, the most intense tornadoes, categories F4 and F5, have on occasion left scars—emotional and physical—on the people and the landscape.

The frequency of tornadoes in Minnesota appears to be increasing, though this impression likely stems from growing numbers of storm spotters and the deployment of more sensitive Doppler radar systems. At one time the annual average number of tornadoes was about 18 state-wide, but since the late 1980s, with more Doppler radar systems in place, the statewide average has risen to 30 to 40 tornadoes per year. As for geographical frequency, statistics compiled by Thomas P. Grazulis and Todd Krause show extremely large variability across the state's 87 counties. Otter Tail and Polk counties each documented more than 60 tornadoes during the 1950–2013 period, while Cook County reported only two. Both county size and population density—not to mention location—affect these figures. From 1950 to 2013 Minnesota's annual tornado count was generally 27, with a decrease in the average annual death toll and injury tally. The peak annual frequency occurred in 2010, when 113 tornadoes were reported statewide, the first and only time Minnesota has reported more tornadoes than any other state. As a tribute to NWS forecasting skill and warning capability, only 46 injuries and three deaths occurred during that record-setting year.

Q & A

What important plant nutrient is released by lightning?

a. phosphorus

b. sulfur

c. nitrogen

(answer on page 329)

Doppler radar

Doppler radar is a type of weather surveillance that takes advantage of the Doppler effect. Based on the frequency change between outgoing and reflected radar signals, it determines the velocity of atmospheric targets moving directly toward or away from the unit. Doppler radar allows meteorologists to interpret wind speeds accompanying thunderstorms and to view rotating winds associated with funnel clouds.

It is worth mentioning that starting in the 1970s the NOAA–National Weather Service began to use the Fujita Scale developed and published by Dr. Theodore Fujita (1971) to assess and rate tornadoes based on estimates of wind-induced damages. His system was given the name of F-Scale. Much of the published literature and historical reconstruction of significant tornadoes (including Grazulis 1993) is based on the F-Scale. However, starting in 2007 NOAA-NWS began to use an Enhanced F-Scale based on a modification of Fujita's original work but fine-tuned for damages inflicted by three-second wind gusts. The Enhanced F-Scale (Table 6, page 31) is used today to assess and evaluate tornado damages, and its standards are reflected in the narrative descriptions and research studies.

Tornadoes leave lasting impressions on people and communities. Out of more than 1,800 tornadoes reported in Minnesota since the formation of the NWS, some of the most memorable and destructive ones are described here.

Other Significant Tornadoes

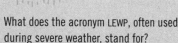

What does the acronym LEWP, often used during severe weather, stand for?

a. Low Easterly Wave Pattern

b. Line Echo Wave Pattern

c. Lost Energy Wave Pattern

(answer on page 329)

■ Between 2:00 and 6:00 PM on July 15, 1881, an outbreak of tornadoes numbering at least six sped across southern Minnesota, killing 24 people and injuring 123. The family of tornadoes moved to the southeast, an unusual trajectory, and passed through Blue Earth, Brown, Faribault, Renville, Sibley, and Watonwan counties. Two of the tornadoes were F4s (207–260 mph) and nearly a half mile in diameter. New Ulm, where outlying settlers may have gathered in fear of Indian raids, was especially hard hit: 47 buildings were destroyed and 200 damaged, and six people were killed and 53 injured.

■ On July 21, 1883, at about noon a family of tornadoes crossed Dodge, Olmsted, and Wabasha counties. At least one was of F4 intensity (207–260 mph), while the damage swath through Dodge County suggested that two tornadoes had traveled in parallel paths about eight miles apart. Four people were killed and 300 injured.

■ Later that summer, around 5:30 PM on August 21, 1883, Rochester was hit by a large F5 (greater than 260 mph) tornado that killed 37 people and injured 200 in both Olmsted and Dodge counties. The massive vortex appeared to be nearly a mile in diameter as it approached Rochester. Its audible roar warned even unsuspecting citizens of impending danger as a very low black cloud descended from the upland southwest of the

TABLE 20: Documented Tornadoes by County, 1950–2014					
Aitkin	16	Isanti	11	Pipestone	12
Anoka	23	Itasca	10	Polk	63
Becker	30	Jackson	21	Pope	19
Beltrami	29	Kanabec	10	Ramsey	8
Benton	7	Kandiyohi	43	Red Lake	18
Big Stone	7	Kittson	32	Redwood	27
Blue Earth	37	Koochiching	8	Renville	24
Brown	25	Lac qui Parle	11	Rice	21
Carlton	6	Lake	6	Rock	14
Carver	19	Lake of the Woods	14	Roseau	33
Cass	23	Le Sueur	21	St. Louis	32
Chippewa	15	Lincoln	17	Scott	14
Chisago	15	Lyon	30	Sherburne	12
Clay	33	McLeod	17	Sibley	28
Clearwater	22	Mahnomen	19	Stearns	46
Cook	2	Marshall	32	Steele	21
Cottonwood	17	Martin	18	Stevens	15
Crow Wing	22	Meeker	19	Swift	33
Dakota	28	Mille Lacs	13	Todd	18
Dodge	17	Morrison	21	Traverse	10
Douglas	20	Mower	31	Wabasha	6
Faribault	34	Murray	24	Wadena	16
Fillmore	14	Nicollet	19	Waseca	19
Freeborn	48	Nobles	35	Washington	22
Goodhue	20	Norman	16	Watonwan	20
Grant	19	Olmsted	31	Wilkin	27
Hennepin	31	Otter Tail	69	Winona	14
Houston	9	Pennington	7	Wright	25
Hubbard	20	Pine	11	Yellow Medicine	22

city. After destroying many structures, the tornado lifted a train off the tracks and tipped it over. Attending to the injured and rebuilding the town required a team effort: out of the wreckage came the idea to build a local medical center, today's Mayo Clinic.

▦ Between 3:00 and 5:00 PM on April 14, 1886, was another devastating tornado outbreak, the most lethal in Minnesota history. At least four funnels touched down across Benton, Douglas, Hubbard, Morrison, Pope, and Stearns counties. An F4 (207–260 mph) on the ground for 14 miles struck the St. Cloud (Sherburne County) and Sauk Rapids (Benton County) area. Seventy-two people were killed, including several members of a wedding party conducting a ceremony near the Mississippi River. The tornado, nearly 800 yards wide at its maximum intensity, momentarily sucked all the water out of the Mississippi as it crossed over near Sauk Rapids. It destroyed a school, two churches, a bridge, and a post office, inflicting injuries on 213 people.

▦ At 5:30 PM on July 13, 1890, clusters of severe thunderstorms passed over east-central and southeastern Minnesota. An F3 (158–206 mph) born over Anoka County crossed into Ramsey County and stayed on the ground for ten miles. It destroyed a number of summer cottages on Lake Gervais and killed six people, some of whom drowned after being blown into the lake. Among the victims were the Reverend W. Pfaffle of Texas, who had served as a guest preacher that morning at the First German Methodist Church in St. Paul; and Emmanuel Good and Caroline Schurmeier, also

A large F3 tornado passing over Lake Gervais near St. Paul in 1890 was traveling northwest to southeast, an unusual path orientation for such storms. This tornado killed six people.

of that congregation and members of prominent St. Paul families. Later that decade, stained glass windows commissioned in their memory were installed in the church, today known as the Fairmount Avenue United Methodist Church. That 1890 evening also turned deadly in southeastern Minnesota as strong downburst winds struck and overturned the excursion vessel *Sea Wing* on Lake Pepin, drowning more than half of the 200 passengers aboard, the worst boating disaster in state history.

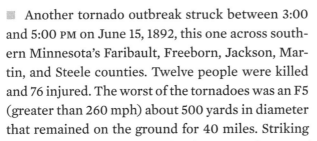

When the National Weather Service issues a tornado warning in Minnesota, who is responsible for sounding the sirens placed in the various counties?

a. city police chiefs

b. the county sheriff

c. the county emergency manager

(answer on page 329)

Another tornado outbreak struck between 3:00 and 5:00 PM on June 15, 1892, this one across southern Minnesota's Faribault, Freeborn, Jackson, Martin, and Steele counties. Twelve people were killed and 76 injured. The worst of the tornadoes was an F5 (greater than 260 mph) about 500 yards in diameter that remained on the ground for 40 miles. Striking near Easton in northern Faribault County, the tornado obliterated many farms, depositing barn timbers up to three miles away from their foundations. A schoolhouse near Sherburn in Martin County was destroyed, but the teacher and 16 students escaped with only injuries.

On June 27, 1894, from 5:00 to 9:00 PM, during one of state's largest outbreaks, at least 12 destructive tornadoes—all either F2 or F3 intensity—were reported from 13 southern and central counties. The shortest path was just one mile in Hennepin County, while the longest was 30 miles across Meeker and Stearns counties. Four deaths and 86 injuries were reported. Then, nearly three months later, on September 21 a tornado outbreak swept across northern Iowa and southeastern Minnesota, including Faribault, Fillmore, Mower, and Winona counties. The storms arrived shrouded in darkness between 7:00 and 11:00 PM. Worst hit were the towns of Spring Valley in Fillmore County, where five people were killed and many homes demolished, and Le Roy in Mower County, where four people died and 80 buildings were damaged or destroyed.

The fall of 1900 brought two memorable tornadoes to Minnesota, one in the south and one in the north. Shortly after 5:00 PM on September 24, an F2 (winds 113–157 mph) formed southwest of Morristown in Rice

County, moved northeast at 60 miles per hour, and was on the ground for eight miles. Many people sought shelter in Gattske's Saloon of Morristown, but the brick building collapsed, killing seven. One additional death was reported from a farm north of town, where several hogs were also killed. In one of the first NWS efforts to document the character of such storms, Thomas Outram, section director of the Minneapolis office, published eyewitness accounts in the "Monthly Climate and Crop Service Report." Later that fall, at 3:50 PM on October 6, a very unusual late-season F3 (158–206 mph) took an odd trajectory, traveling northwest through Biwabik in St. Louis County and destroying homes and mining camp buildings. Carrying with it a deafening roar, the storm stayed on the ground for ten miles, ultimately killing ten and injuring eight.

• •

Fulgurite

Derived from the Latin fulgur, "lightning," fulgurite is a glassy, rootlike tube formed when lightning strikes sandy soil. The intense heat vaporizes soil moisture, and the remaining molten material fuses into a tubelike structure an inch or two in diameter and inches to several feet in length. Because the material holding the tubes together is fragile, they typically crumble when dug up. University of Florida lightning researchers discovered the world-record fulgurite, which had three branches, one extending 16 feet into the soil.

• •

At 1:50 PM on October 3, 1903, an F4 (207–260 mph) began its journey across Olmsted and Winona counties and traveled a northeastern path into western Wisconsin. The storm, its funnel up to 300 yards wide, stayed on the ground for 55 miles, inflicting the most damage on St. Charles, where seven people were killed, 30 injured, and 50 homes and businesses damaged or destroyed.

August 20, 1904, brought numerous evening tornadoes and treacherous downburst winds to sections of eastern Minnesota. Tornadoes reportedly blew through Minneapolis's and St. Paul's downtowns, though funnels were not observed. Before being torn from its mountings, the Weather Bureau anemometer, located on the roof of the Pioneer Press building at the corner of Fourth and Robert Street in St. Paul, recorded a one-minute sustained wind of 110 miles per hour and a gust of 180 miles per hour—the highest wind speed ever measured in Minnesota. An F4 (207–260 mph) cut a path through Waconia in Carver County, and damage was also reported in St. Louis Park in Hennepin County. A section of the High Bridge, which spans the Mississippi River, was blown down, and hundreds of buildings and houses near and within St. Paul were damaged or destroyed. Downtown, three people were killed and more than 50 injured, mostly members of the Tivoli Concert Hall audience. Other tornadoes cut a swath from Glencoe in McLeod County to Stillwater in Washington County.

On August 21, 1918, at 9:00 PM an F4 (207–260 mph) struck Tyler in Lincoln County, destroying or damaging 600 homes and businesses. The tornado vortex, its diameter measuring 400 yards, was on the ground for 20 miles and dissipated in Lyon County, three miles northeast of Tyler. In all 225 people were injured and 36 died, including three patients in the town's only hospital and eight more in the pool hall.

About 4:00 PM on June 22, 1919, an F5 (greater than 260 mph) swept across Fergus Falls in Otter Tail County, leaving a three-block-wide swath described as "a vast acreage of kindling." The funnel was 400 yards wide at times and traveled for 20 miles, ending ten miles east-northeast of town. Approximately 400 buildings were damaged or destroyed, including the Lake Alice Grand Hotel. Deaths numbered 57 and injuries 200, while damages were estimated at $3.5 million, at the time a single-storm record.

On June 22, 1924, an F4 (207–260 mph) passed across Cottonwood, Lincoln, Lyon, and Redwood counties, destroying farmsteads but missing most of the towns over a 55-mile course, one of the longest storm paths in history. The highly visible funnel, at times 800 yards wide, hit 54 farms and killed 300 head of livestock, but just two people were killed and 20 injured.

Fergus Falls sustained significant damage after a 1919 tornado, an F5 that killed 57 people.

▧ On August 20, 1928, an F4 (207–260 mph) traveled from northeastern Iowa into Freeborn and Mower counties, remaining on the ground for 40 miles. It struck between 4:00 and 5:00 PM, moving from near Emmons and Twin Lakes in Freeborn County toward Austin in Mower County and causing approximately $1 million in property damage, mostly to farms and homes south of Austin. Six people were killed and 60 injured.

▧ April 5, 1929, brought a long-lived tornado to east-central Minnesota and to western Wisconsin. The F4 (207–260 mph) descended around 5:30 PM just northeast of Lake Minnetonka in Hennepin County and remained on the ground for approximately 80 miles, traveling across the communities of Columbia Heights and Fridley in Anoka County and Forest Lake in Washington County and passing just north of Shafer and near Chisago City, both in Chisago County. It crossed the St. Croix River and inflicted further damage near Milltown, Wisconsin. All told, the tornado killed two people, injured 40, and caused property damages of nearly $1 million.

▧ Around 4:20 PM on May 27, 1931, an F3 (158–206 mph) moved across Clay and Norman counties, traveling on the ground for 40 miles and damaging or destroying several farms. Perhaps the most amazing story surrounding this tornado has to do with the *Empire Builder* train, which was moving at 55 miles per hour near Sabin in Clay County when it was hit broadside by the funnel. Five coach cars were lifted off the tracks and thrown 80 feet, but of the 175 passengers, only one person was killed and 57 injured. Later that year, on November 16 at about 9:35 PM, a small F2 (113–157 mph) passed near Maple Plain in Hennepin County, destroying a farm. This late-season tornado, the latest ever observed in Minnesota, came as quite a surprise.

Across Minnesota, severe weather sirens are sounded as a warning for which of the following weather features?

a. tornadoes

b. blizzards

c. severe thunderstorms

(answer on page 329)

▧ April 30, 1936, brought at least four tornadoes to the southern counties of Faribault, Fillmore, and Martin. Following a track from Ceylon in Martin County toward Blue Earth in Faribault County, the most destructive of these tornadoes—an F4 (207–260 mph)—was on the ground for about 40 miles between 5:00 and 6:00 PM. The funnel, at times 800 yards wide, cut a substantial swath of damage, affecting 200 farms, destroying about 100 buildings, killing two people, and injuring 61, though some of the destruction and injuries were later found to be the result of downburst winds.

• •

Heat lightning

The term heat lightning *is derived from a mistaken belief that lightning is produced by an excessively heated atmosphere, based on observations of lightning under otherwise clear summer skies. What is viewed as heat lightning is actually a reflection of distant lightning flashes off the horizon. All lightning technically produces heat: a single stroke can warm the surrounding air to more than 50,000°F. The air's rapid expansion causes sound waves, which are later heard as thunder. Sound travels approximately a mile every five seconds: to gauge the distance of the lightning flashes, count the number of seconds that pass between the flash and the resulting thunder, assuming about one-fifth mile for every second. Thus a 15-second interval between observed lightning and the sound of thunder indicates that the flash occurred about three miles away. Lightning strokes from more than ten miles away are rarely heard as thunder.*

• •

▓ Between 3:00 and 4:00 PM on Sunday, June 18, 1939, a destructive F4 (207–260 mph) began its path across Hennepin and Anoka counties. With a funnel diameter of 250 yards, it was on the ground for 25 miles, traveling at 30 miles per hour. Nine deaths and 222 injuries were reported, mostly in the towns of Anoka and Champlin (Hennepin County). Forty homes were completely destroyed and 300 damaged; clothing, paper, and small articles of furniture were found 50 to 70 miles away. Many witnessed the Mississippi River temporarily sucked dry as the tornado crossed it between Champlin and Anoka. Martin Hovde, NWS section director in Minneapolis, conducted one of the first comprehensive surveys of storm damage following this tornado, seeking to better quantify the wind's destructive power. He also began tabulating a history of the state's tornadoes, setting a precedent continued by other NWS administrators in Minnesota.

▓ On September 4, 1941, an F2 (113–157 mph) moved 30 miles across Hennepin, Ramsey, and Washington counties in east-central Minnesota, striking relatively early in the day, just after noon, and killing four people and injuring 50. Two of the deaths were at Lake Gervais: similar to the 1890 tornado, cottages along the shore were damaged and one man drowned after being blown into the lake. At least 20 cottages on White Bear Lake were also damaged, and the tornado overturned 200 Soo Line freight cars, some containing more than 40 tons of coal.

▨ A huge F5 (greater than 260 mph) came out of South Dakota and struck near the town of Beardsley in Big Stone County between 5:00 and 6:00 PM on June 17, 1944. The funnel—1,500 yards wide at times and on the ground for 30 miles—completely wiped out dozens of farms. Downburst wind damages were also inflicted on 72 farmsteads.

▨ The famous Mankato tornado struck on August 17, 1946, between 6:00 and 7:00 PM. The F4 (207–260 mph), its funnel 400 yards wide, traveled nine miles from just southwest of Mankato in Blue Earth County to North Mankato in Nicollet County, completely destroying 26 cabins at the Green Gables Tourist Camp, three miles southwest of Mankato. It damaged homes and farms in the area, killing more than a thousand turkeys, and hurled a 27-ton road grader 100 feet. Injuries numbered 100, and 11 people were killed. About one hour later, a second tornado of equal intensity struck the town of Wells in Faribault County. Thirty people were injured when the roof of the town theater collapsed with 400 people inside.

▨ Between 5:00 and 6:00 PM on June 19, 1951, a powerful F4 (207–260 mph) with a funnel up to 300 yards wide traveled 20 miles across Hennepin and Anoka counties, destroying or damaging 50 homes, many in Brooklyn Center (Hennepin) and Fridley (Anoka). Twenty people were injured and one killed. One month later, on July 20 at around 9:00 PM, a series of strong thunderstorms brought tornadoes and downburst winds to Hennepin County, this time killing three and injuring 40. One tornado, an F3 (158–206 mph), traveled 18 miles. The most serious damage was done at Fort Snelling and the Wold-Chamberlain Airport (now Minneapolis–St. Paul International): 63 planes were destroyed and 37 planes plus a number of military buildings damaged. All told, the storm inflicted a then record-setting $6 million in damages.

Another outbreak occurred on May 10, 1953, between 4:00 and 5:00 PM. An F2 (113–157 mph) damaged several farms and killed livestock in Pope County between Cyrus and Starbuck, staying on the ground for about 20 miles. Later, another F2 traveled eight miles across Free-born County southeast of Hollandale, damaging or destroying many rural buildings, including a home where six migrant workers were killed. Still later, two more tornadoes, these F3s (158–206 mph), struck in Fillmore, Olmsted, and Winona counties. One snapped hundreds of trees in Olmsted's White-water State Park; the other leveled a rural school building near Chatfield. Many farms suffered damage; these storms caused 19 injuries and eight deaths.

According to the native cultures of Australia and Canada, what is a three-dog night?

a. it's so cold you have to sleep with three dogs to stay warm

b. dogs take turns howling at the moon

c. it's raining so loudly it drowns out howling dogs

(answer on page 329)

May 20, 1957, brought five separate funnels across Clay County's Fargo-Moorhead area, the worst-ever outbreak to affect this Red River Valley population center. Between 6:30 and 8:00 PM, F3, F4, and F5 tornadoes cut a wide swath, damaging or destroying more than 1,300 homes in a 20-block section of Fargo-Moorhead. Many farms were damaged; debris was carried all the way to Detroit Lakes; and ten deaths and 103 injuries were reported. Dr. Theodore Fujita, the world-famous tornado researcher, found evidence for five different tornado paths varying in length from seven to ten miles.

On September 16, 1962, an extremely short-lived F4 (207–260 mph) devastated a neighborhood southeast of Rochester in Olmsted County. On the ground for only a few minutes, the 75-yard-wide funnel traveled a single mile. However, it completely destroyed 11 homes and injured 34 people, most of whom had been sleeping, for this powerful tornado struck at a very unusual time: from 4:10 to 4:15 AM.

May 6, 1965, brought the worst tornado outbreak ever to hit the Twin Cities metropolitan area, with multiple funnels appearing in Anoka, Carver, Hennepin, and Ramsey counties between 6:30 and 9:00 PM. Because tornadoes had struck across Becker, Fillmore, Houston, Mower, Otter Tail, and Winona counties the preceding day, many were alert to

NWS personnel survey the damage in Fridley after the 1965 tornado outbreak, the worst in Twin Cities history.

possible severe weather on May 6 as well. The NWS gave a 20-minute warning for the first tornado. Six separate tornadoes, their paths varying from seven to 40 miles, crossed the area; the four classified as F4s (207–260 mph) were responsible for most of the fatalities and damages. In all 14 people were killed and more than 500 injured. Thousands were left homeless, hundreds of buildings were destroyed or damaged, and 25 planes at Anoka airport were demolished. The hardest hit communities—some targeted by more than one tornado—were Chanhassen, Deephaven, Fridley, Mound, Mounds View, Navarre, and Spring Lake Park. Total damages exceeded $60 million, Minnesota's most costly tornado outbreak to that point.

▨ April 30, 1967, brought another cluster of eight tornadoes to southern Minnesota. Freeborn, Mower, Olmsted, Steele, and Waseca counties all reported damaging storms between 6:00 and 8:30 PM. The path length of these tornadoes varied from six to 40 miles and their strength from F2 (113–157 mph) to F4 (207–260 mph). In all 13 deaths and 81 injuries were attributed to these storms. Towns hardest hit included Waseca in Waseca County and Albert Lea in Freeborn County.

▨ March 18, 1968, marks the earliest annual calendar date for a tornado sighting in Minnesota. An F2 (113–157 mph) near Truman in Watonwan County was on the ground for only four miles, destroying three farms shortly after 5:30 PM. A piece of barn wood was found embedded in the side of a farmhouse a quarter mile away.

▨ At 6:50 PM on June 13, 1968, a rare F5 (greater than 260 mph) began to pass across portions of Lyon, Murray, and Redwood counties in southwestern Minnesota. On the ground for 13 miles, its funnel as wide as 600 yards, this intense tornado threw two railroad boxcars 300 yards and carried a steel I-beam two miles. The town of Tracy in southeastern

Lyon County took the brunt of the storm: nine people were killed and 111 homes destroyed. Total damages exceeded $4 million.

An outbreak between 3:00 and 6:00 PM on August 6, 1969, produced eight separate tornadoes, five of which passed across portions of St. Louis County in northeastern Minnesota. The paths of these tornadoes varied in length from five to 33 miles, and six were classified as F2 (113–157 mph), one as F3 (158–206 mph), and one as F4 (207–260 mph). In all 15 people were killed and 106 injured, with damages totaling approximately $5.5 million. Much of the destruction was in the Hill City area of northern Aitkin County and in Outing of southern Cass County. Many lake cabins were destroyed and trees debarked, and in St. Louis and Cook counties thousands of trees were downed or snapped off.

In the early afternoon on July 5, 1978, the Red River Valley experienced tornadic winds from a storm that traveled an estimated 61 miles across Clearwater, Mahnomen, Norman, and Polk counties. This F3 (158–206 mph), at times 400 yards wide, was highly visible on the prairie landscape. Four people were killed and 38 injured. The storm caused extensive damage in the towns of Gary (Norman County), Fosston (Polk),

An F5 twister hit Tracy in June 1968, leaving nine dead.

and Clearbrook (Clearwater), tallying an estimated $12 million for the loss of 250 farm buildings, 200 homes, and 32 mobile homes.

▓ On September 3, 1980, an F3 (158–206 mph) on the ground for five miles in Stearns County near St. Cloud caused one death and 15 injuries. The tornado was particularly devastating to Waite Park, where 6 homes, 14 trailers, and 16 apartment buildings were lost, part of an estimated $9 million in damages.

▓ Around 4:50 PM on June 14, 1981, an F3 (158–206 mph) touched down near Edina in Hennepin County and traveled for 15 miles through portions of Minneapolis and Roseville until it dissipated near Lake Owasso in Ramsey County. It damaged residences in the St. Anthony Park neighborhood near the historic Gibbs Farm as well as the Har-Mar Shopping Center in Roseville, where most of the 83 injuries occurred. Total damages were estimated at $47 million.

▓ On April 26, 1984, an F3 (158–206 mph) traveled for 5.5 miles across northeast Minneapolis and through St. Anthony in Hennepin County and New Brighton and Mounds View in Ramsey County, killing one and injuring 52. This short-lived tornado, on the ground from 8:30 to 8:41 PM, damaged many homes and businesses, including the Apache Plaza Shopping Center.

▓ Minnesota's first "celebrity tornado" occurred on July 18, 1986, when a cameraman aboard a helicopter filmed the storm for nearly its 30-minute life span and the footage was broadcast live. The F2 (113–157 mph) moved slowly across Anoka and Hennepin counties, primarily affecting the cities of Brooklyn Park (Hennepin) and Coon Rapids and Fridley (Anoka) and particularly damaging the Springbrook Nature Center. Damages to property were estimated at $650,000.

▓ July 5, 1991, brought another "celebrity tornado" to the state, an F3 (158–206 mph) that tracked through Cass and Crow Wing counties between 6:30 and 7:00 PM. An amateur filmed the storm—later broadcast on national television—as it moved across Gull Lake, destroying boats and a marina in the Brainerd area of Crow Wing County. The storm struck a classic car show at the Brainerd International Raceway, flipping trailers and autos. Estimated losses: $2 million.

▪ June 16, 1992, brought one of the state's largest tornado outbreaks: between 4:00 and 9:00 PM, 22 tornadoes were reported in 11 southwestern counties, causing well over $60 million in damages. Among these storms was the most recent F5 (greater than 260 mph) to strike Minnesota. Starting around 4:00 PM, the well-forecast tornado touched down near Leota in Nobles County and traveled northeast through Chandler in Murray County, staying on the ground for 16 miles, at times up to a width of 300 yards. In Chandler 40 homes were destroyed and 47 damaged; one person was killed and more than 40 injured. A check from Chandler was found in Willmar, Kandiyohi County—about 95 miles away.

▪ At 1:35 AM on August 9, 1993, a weak F0 tornado (45–78 mph) struck near Littlefork, Koochiching County, dislodging and carrying a mobile home a quarter of a mile before destroying it and killing its two occupants. This storm was unusual for far northern Minnesota and especially for striking so late at night. It did nearly $50,000 in damages to other structures.

▪ The worst-ever March outbreak of tornadoes occurred on March 29, 1998, when 14 separate tornadoes were reported around southern Minnesota—in Blue Earth, Brown, Cottonwood, Le Sueur, Murray, Nicollet, Nobles, and Watonwan counties—between 3:30 and 7:00 PM. Among the worst hit areas was the town of Comfrey: 75 percent of its structures were damaged or destroyed. This F4 (207–260 mph) was on the ground for 67 miles, appearing as a large dust cloud or rolling fog bank, at times up to 1.25 miles wide. A second destructive tornado—similarly shaped and sized, but an F3 (158–206 mph)—ripped through St. Peter in Nicollet County, damaging or destroying hundreds of homes and businesses and uprooting more than a thousand trees, almost completely denuding the beautiful campus of Gustavus Adolphus College. These tornadoes caused just two deaths and 21 injuries—thanks to detailed forecasting by the NWS—but total estimated damages exceeded $200 million.

On July 25, 2000, an F4 (207–260 mph) passed through Granite Falls in Chippewa/Yellow Medicine counties. On the ground for nine miles between 6:10 and 6:25 PM, it carved a 500-foot-wide swath through a residential area. Scores of homes were destroyed, blown right off their foundations; damage estimates ranged as high as $20 million. One person was killed and 15 injured. Also that year, on November 1, a rare late-season F1 tornado (73–112 mph) struck near Prinsburg in Kandiyohi County. On the ground for only a half mile, it nevertheless seriously damaged two farms.

Q & A

Scientists monitoring and forecasting air quality sometimes use the terms *socks* and *knocks* to refer to what?

a. wind socks and coastal wave action

b. amounts of sulfur and nitrogen oxides in the air

c. odors and noises carried by the wind

(answer on page 329)

The year 2001 brought 74 tornadoes to the Minnesota landscape, a record annual number at the time. More than 90 percent of these tornadoes were rated F0 (winds less than 73 mph) or F1 (73–112 mph), and more than half of them occurred in the month of June. The strongest tornado, an F3 (158–206 mph), was on the ground for 12 miles across Otter Tail County near Parkers Prairie and into Todd County. It destroyed a number of farm buildings but caused no injuries.

A most rare tornado struck near Albertville (Wright County) at midnight on September 9, 2002. This tornado was rated F1 (73–112 mph). Less than 0.1 percent of all tornadoes occur at this hour. The tornado was on the ground for 1.1 miles, starting near the Cedar Creek Golf Course. It damaged about 20 homes along its path.

The year 2006 brought some damaging tornadoes to various parts of the state. An F3 tornado (158–206 mph) near Butterfield (Watonwan County) destroyed a farmstead and several outbuildings, including a silo, between 6:00 and 6:30 PM on August 1. Thankfully it was on the ground for only 1.5 miles. Another F3 tornado traveled on the ground for a little over 4 miles near Warroad (Roseau County) between 6:00 and 6:30 PM on August 5. It seriously damaged the Marvin Windows factory, overturning a number of semi-tractor trailers. At the city park and campground, 30 to 40 camper units were destroyed along with a number of boats. Several homes in the area sustained significant damage, and a gift shop was completely demolished. Yet another F3 tornado raced across Le Sueur County between Kasota and Waterville between 5:00 and 5:30 PM on August 24. It was on the ground for more than 17 miles, destroying many farm buildings and damaging a number of homes. Insurance claims were filed by more than 4,000 homeowners, and there were thousands of automobile claims as well for damages due to the wind or flying projectiles. The last tornado of the year, an F2 (113–157 mph), occurred after dark on the evening of September 16 as it moved across parts of Hennepin County from Rogers to Dayton. A ten-year-old child was killed, and 200 to 300 homes were damaged by the storm.

May 25, 2008, brought a strong tornado to portions of Anoka and Washington counties near Centerville and Hugo. The Enhanced F-Scale had been deployed by the NWS the previous year, so this tornado was rated EF-3 (136–165 mph) based on damage surveys. This storm occurred around 4:00 PM and brought damages to 794 homes, 27 of which were completely destroyed. A two-year-old boy was killed after being blown out of the first floor of a house, and a 62-year-old woman died of a heart attack while trying to clean up after the storm. In the countryside, a number of farm buildings were damaged as well.

The year 2010 brought 113 tornadoes to Minnesota, more than any other state in the nation (for the first time) and the most in state history. The worst outbreak occurred on June 17 when 48 tornadoes were reported, three being rated EF-4 (166–200 mph). Three fatalities occurred that day: one in Polk County, one in Otter Tail County, and one in Freeborn County. By far the most damaging tornado swept through Wadena (Wadena County) at about 5:00 PM. This tornado, described as a multiple vortex wedge, traveled about 10 miles, destroying a number of farm buildings and homes to the west of town. Many public and industrial buildings were wrecked, including the high school. The town was full of visitors for the Wadena all-school reunion, and 20 people were injured. School buses and other vehicles were propelled through the air for hundreds of yards. Fortunately, there were no fatalities.

Between 1:00 and 1:30 PM on May 22, 2011, an EF-1 tornado (86–110 mph) touched down near the intersection of I-394 and Highway 100 in

St. Louis Park (Hennepin County). It traveled northeast for more than 6 miles, causing widespread damages to businesses and homes in and around Wirth Park and North Minneapolis. More than 600 buildings were damaged, and 35 homes could not be immediately reoccupied. This storm uncharacteristically tracked right through the Minneapolis area.

On November 10, 2012, a very unusual weather system brought high temperatures and dew points to southern Minnesota, producing thunderstorms and four rare late-season tornadoes that briefly touched down near the communities of Burnsville, Eagan, and Mendota Heights (Dakota County) and Mahtomedi (Washington County) shortly after sundown. All four tornadoes were rated EF-0 (65–85 mph), and most caused tree damage and some downed power lines. These storms were noteworthy because less than 0.1 percent of all tornadoes occur in the month of November.

Derechos

Minnesota's convective storm season can also bring a rare but large and destructive windstorm known as a *derecho,* named by the University of Iowa's Dr. Gustavus Hinrichs in 1888. The Spanish term means "direct" or "straight ahead," referring to the wind's impact, which differs from that of a rotating tornado. Now used to describe any downburst wind, a derecho usually arises out of a mesoscale convective complex composed of an organized cluster of interacting thunderstorms. These systems typically show up as a bow echo on Doppler radar, representing high velocity downburst winds flowing outward along a squall line. With velocities of 57 or more miles per hour, derechos cut wide and long damage paths, sometimes tens of miles wide and hundreds of miles long. They occur most often in June, July, and August.

In comparison to severe thunderstorms and tornadoes, each numbering from 30 to 60 annually, derechos are relatively rare in Minnesota, inflicting their widespread damage only every few years. With modern technology, especially Doppler radar systems and satellite imagery, it is possible to detect the birth of a derecho and track

Q & A

On June 5, 2014, what very unusual observation was reported along the shores of Lake Superior?

a. chunks of winter's leftover lake ice washing up on shore

b. a triple rainbow

c. a dramatic waterspout

(answer on page 329)

it as it moves over the landscape. However, before the 1970s these storms were often mislabeled as tornadoes or severe thunderstorms, making historical documentation difficult to find. Nevertheless, weather observations over the decades offer evidence of some significant derecho windstorms in Minnesota.

. . . .

Mesoscale Convective System (MCS)

A Mesoscale Convective System is a cluster of thunderstorms larger than any individual cumulonimbus cloud but smaller than a frontal system. MCSs appear on satellite imagery as circular or linear cloud forms with very bright tops, indicating cold air. Severe weather including hail, damaging winds, heavy rainfall, and tornadoes often accompanies an MCS.

. . . .

Significant Derechos

Perhaps one of the earliest derechos documented in Minnesota arrived on July 15–16, 1902. The storm began in the eastern Dakotas on a very hot, humid afternoon as massive lines of thunderstorms organized along a frontal boundary and traveled northwest to southeast across Minnesota. Though the Twin Cities Weather Service reported wind gusts of only 46 miles per hour, evidence in rural areas dramatically suggested a much greater magnitude. In a narrow swath of damage ranging from 10 to 25 miles wide, many observers across the state noted blown down buildings, barns, and windmills. The Minnesota Climate and Crop Service reported flattened fields of wheat and oats. Observers at Ada in Norman County called it a "terrible wind" and at Crookston in Polk County, a "heavy wind." At Grand Meadow in Mower County the observer noted that crops were a complete loss and that he had never seen such a destructive wind in 31 years of weather work. In Faribault, Rice County, an anemometer measured a wind speed of 80 miles per hour and several large trees were uprooted by the straight winds.

Considerable non-tornadic wind damage was noted in reports filed by the NWS during the 1930s and 1940s, prompting the conclusion that some major dust bowl–era storms may have been derechos. One report, from September 11, 1942, leaves little doubt. That afternoon, a powerful squall line developed over Yellow Medicine County and produced damaging winds at Granite Falls and Hanley Falls. The storm system moved across the state to North Branch in Chisago County before dissipating. Willmar in Kandiyohi County reported winds of 66 miles per hour, while Monticello in Wright County recorded a wind gust of 70 miles per hour. Farm buildings and groves of trees were flattened, and homes and office buildings lost portions of their roofs. More than 650 barns were destroyed and more than 1,700 houses damaged. Estimated losses exceeded $2.3 million across portions of Anoka, Chippewa, Chisago, Kandiyohi, Meeker, Wright, and Yellow Medicine counties.

A similar report of straight-line wind damage was filed in June 1947. On June 27 at about 9:00 PM a severe thunderstorm squall line entered Minnesota from South Dakota, causing damage in Yellow Medicine County. The storm raced due east, finally exiting the state at Hastings in Dakota County around 4:30 AM on June 28. No tornadoes or funnel clouds were sighted, but straight-line wind damage was abundantly apparent along a path nearly 200 miles long and from 10 to 20 miles wide. Farm buildings and silos were blown down, and planes and hangars at Southport Airfield near Rosemount in Dakota County were damaged. Winds may have exceeded 70 miles per hour in Dakota County, although the NWS recorded gusts to only 50 miles per hour in the Twin Cities. Property damages were estimated at nearly $700,000 across a 16-county area.

The first well-documented derecho to strike Minnesota in the age of satellite meteorology occurred on July 4, 1977. Forming in western Minnesota, the storm intensified over Cass County shortly after 10:00 AM, producing winds of 70 to 100 miles per hour. These strong winds caused forest blowdowns in the Brainerd (Crow Wing County) and Mille Lacs (Mille Lacs County) areas; additionally, Gull Lake Lodge's roof was blown off near Brainerd, and several boats attempting to ride out the storm on Mille Lacs were destroyed by 8- to 10-foot waves. Eight counties from Crow Wing to Chisago suffered significant damage to trees and property; the storm also wreaked havoc in Wisconsin and Michigan, traveling a total of 800 miles before dissipating. Dr. Theodore Fujita of the University of Chicago determined that the storm was a strong derecho and described it in detail using radar data and ground surveys of the damages.

▓ Another well-documented derecho entered Wilkin and Traverse counties about noon on July 19, 1983. Initially boasting 58-mile-per-hour wind gusts, this storm expanded and intensified as it crossed the state, following the Interstate 94 corridor. In Douglas County, at Alexandria's airport the anemometer registered a wind gust of 117 miles per hour. Planes and hangars there and at Princeton's airport (Mille Lacs and Sherburne counties) were seriously damaged by this storm. Scores of houses and farm buildings sustained damage as the storm moved at 50 miles per hour across the central counties, through the Twin Cities, and into Wisconsin over a period of four hours. Northern States Power (today's Xcel Energy) reported the largest power outage in its 74-year history, costing $2.5 million in repairs. The storm later struck Madison, Wisconsin, and Chicago, Illinois, before dissipating.

▓ July 7, 1991, saw a derecho form in South Dakota and race along the Iowa-Minnesota border to Wisconsin with a forward speed of about 50 miles per hour. Winds up to 70 miles per hour were measured at Ostrander (Fillmore County), and elsewhere a few brief tornadoes touched down. Thousands of acres of corn and soybeans were flattened in Fillmore County, and a number of power poles were blown down. More serious damage was inflicted in Iowa and Wisconsin.

> **Q & A**
>
> **Q.** On July 20, 1999, a new weather record was set in the Twin Cities. The high temperature was 72°F and the low was 68°F. What was the record?
>
> *(answer on page 329)*

▓ A destructive and long-lived derecho, with winds estimated at 80 miles per hour, passed across Becker, Clearwater, Hubbard, and Mahnomen counties on July 13, 1995. More than a thousand old-growth trees were blown down in Itasca State Park; the Minnesota Department of Natural Resources estimated that 129,000 acres of timber—or about 5.2 million trees—were destroyed. Trucks equipped with plows were called on to clear debris from roads and highways. Scores of homes and farm buildings were damaged, contributing to a total exceeding $30 million in damages to property, crops, and timber. This derecho traveled from eastern Montana to Pennsylvania, a distance of 1,400 miles, in 27 hours.

▓ On the evening of May 30, 1998, a derecho traveled across central Minnesota, inflicting damages in Carver, Dakota, Goodhue, Hennepin,

McLeod, Ramsey, Scott, Sibley, and Washington counties and particularly the St. Paul neighborhood of Highland Park. The windstorm moved with great speed, averaging 50 to 60 miles per hour, while its 80- to 100-mile-per-hour winds damaged more than 2,000 homes, removed roofs from apartment buildings, overturned semi-trailer trucks, and uprooted or snapped thousands of trees. Damages were estimated at more than $48 million.

Portaging through rough terrain in the Boundary Waters after the July 4, 1999, derecho

▨ Perhaps Minnesota's most devastating derecho swept through the Boundary Waters on July 4, 1999. Winds of 80 to 100 miles per hour passed over a 600-square-mile swath of the Superior National Forest, leveling a quarter million acres of timber valued at $12 to 18 million. Sixty people vacationing in the Boundary Waters Canoe Area Wilderness (BWCAW) were injured and 20 rescued by floatplane. Significant wind damage also occurred at Ada (Norman County); Chisholm, Ely, and Hibbing (St. Louis County); Moose Lake (Carlton County); and Walker (Cass County). The exceptionally large and long-lived windstorm began in eastern North Dakota and ended in New England, traveling a distance of 1,300 miles in 22 hours. Notably, it was the most northerly derecho documented in the United States. The tremendous amount of downed timber in the BWCAW elevated the fire danger, which has been partially managed through controlled burns.

▨ On July 31, 2008, a derecho moved out of South Dakota into Minnesota during the morning hours, bringing 60 to 70 mile-per-hour winds. As the day heated up and the convective cloud system grew stronger, winds blew harder, peaking at 86 miles per hour near Madison Airport in Lac qui Parle County. There were many reports of tree damage and trucks being blown off the roads. Strong winds blew down the Minnesota Vikings football training camp tents in Mankato (Blue Earth County). This derecho traveled all the way to Chicago by the end of the day.

The Growing Season

Minnesota's inherent climate variability has always posed a challenge to those who farm the land. But thanks to the University of Minnesota's research and extension work coupled with experiential lessons passed from farmer to farmer, agriculture has thrived in the state since the grain milling industry was established in the nineteenth century and the barge transportation system developed on the Mississippi River. Once, more than 200,000 farms dotted the Minnesota landscape, their owners engaged in a mixture of crop and livestock production, often on parcels of just 100 to 200 acres. Today the number of farms is about 75,000, with an average size closer to 350 acres. Most producers specialize in crops only, or dairy, or poultry, or livestock. Though fewer people are involved, the overall value of agriculture to the Minnesota economy is enormous: in 2013, more than $75 billion in economic activity associated with agriculture was reported by the Minnesota Department of Agriculture and more than 340,000 jobs were tied to the industry. Seven of Minnesota's top 20 public companies have roots in the farm and food sector.

Many of Minnesota's soils are well suited for crop production, being deep and rich in organic matter and nutrients and having high water-holding capacities. The vast majority of the state's 27 million acres of cropland is rain-fed and lies in the southern, central, and western counties. Only about five percent of this acreage is irrigated, mostly that located on coarse textured soils with lower water-holding capacity. The annual value of crop production in Minnesota is estimated at a little over $13 billion. This total in 2012 earned a ranking of fourth nationally among states. Major crops include corn, soybeans, spring wheat, oats, barley, sunflowers, edible beans, potatoes, sugar beets, flaxseed, canola, hay, wild rice, and sweet corn and green peas for processing.

Livestock, dairy, and poultry production comprise the other half of Minnesota agriculture, yielding more than $7.5 billion annually and ranking seventh nationally among states in 2012. Minnesota's animal agriculture is heavily weighted to dairy operations, hogs, cattle, and turkeys, which account for a majority of the cash receipts. Generally dairy is more concentrated in central and southeastern counties, while poultry and hogs are more dominant in southern and southwestern sections.

Land application of animal manures in these areas is common; most farmers soil test to determine crop fertilizer needs and factor in nutrient credits for manure applied to their land.

Crop and animal agriculture are closely linked to Minnesota's food industry, also a dominant force in the state's economy, with such companies as Cargill, Land O'Lakes, Hormel, Jennie-O, General Mills, Pillsbury, International Multifoods, and American Crystal Sugar. Perhaps more than any other economic sector, agriculture is dependent on the vagaries of the weather during each growing season. When Mother Nature deals a detrimental blow, the downstream effect on the food industry tightens margins and raises prices. Conversely, favorable weather conditions for agricultural production generally prompt lowered commodity prices, squeezing individual farmers. These and other factors make for significant variability in annual farm income, but the best farmers have found ways to cope.

DID YOU KNOW?

In evapotranspiration, water vapor enters the atmosphere through two avenues: evaporation from wet surfaces and water bodies, and plant transpiration, or release of moisture via leaf spores.

The technology and management skills deployed in Minnesota's agriculture help buffer the effects of weather and climate. Thanks to biotechnology and plant breeding, farmers can select crops from varieties that are heat tolerant, frost tolerant, and disease resistant; that respond well to fertilization; that tolerate the use of herbicides and pesticides; that mature and dry down well before harvest; and that can be planted in dense populations. Modern equipment technologies enable farmers to till, plant, and fertilize in one pass across a field, even varying plant population and fertilizer on the go using digitized soils maps. Similarly, harvesters can monitor yields to help determine variations in production on different soil types. Larger and more efficient equipment allows more acres to be planted or harvested in a single day than ever before, maximizing the farmer's use of favorable weather opportunities to do field work. Animals are inoculated to protect them from diseases and parasites; their food rations and water are carefully regulated to match metabolic needs influenced by weather conditions and age. A whole range of other technologies and knowledge is applied to other agricultural endeavors such as crop spraying, crop drying, animal breeding, odor control, soil management, marketing, and storage. Much of this husbandry is directly related to coping with weather conditions.

Minnesota represents the northernmost reaches of the American corn belt, a blessing in some respects: most winters are cold enough to kill off many crop disease spores and insects. Corn has been Minnesota's dominant crop for at least three generations, but it increasingly shares the spotlight with soybeans: both are planted on about 7 million acres. Both corn and soybeans are still essential elements of the North American food chain, but they also have been harvested for other uses: corn for ethanol production and soybeans for biodiesel fuel production. Planting time for these crops, along with all the others, has progressively moved to earlier dates.

Keck's Swamp Angel

In the late nineteenth century, when Minnesota farmers were plagued by wet and ponded soils that postponed crop planting, pioneer settler Harry Keck devised a deep plow that came to be known as Keck's Swamp Angel. With this plow, settlers cut drainage ditches to allow excess water to run off and also drained a number of the state's wetland areas, especially in southern and central counties. A team of up to 20 oxen pulled the heavy, awkward plow through the soil. Its name may have been derived from its enormous weight, perhaps similar to that of the eight-ton Swamp Angel cannons used in the Civil War.

Soils typically thaw by April in most areas of the state. Moisture is subsequently redistributed within the root zone and added to the topsoil by snowmelt and spring rains. With significant tile drainage established across many agricultural areas, soils shed excess water quickly, making them tillable earlier in the spring and also allowing their temperatures to warm more quickly to levels sufficient for seed germination. Studies of soil temperatures at the 4-inch depth from University of Minnesota Research and Outreach Centers at Crookston, Lamberton, Morris, and Waseca for the 20 years from 1984 to 2003 show that average daily soil temperatures reach 50°F or higher between April 15 and 20. In most years, stored soil moisture in the root zone—typically the 5-foot depth—at this time ranges from 7 to 9 inches of water. Much of this storage comes from the previous fall's precipitation, before the soil freezes for the winter. Studies have shown that rainfall storage is most efficient in the fall after crops have been harvested, when moisture loss from surface evaporation is reduced and rainfall intensity more closely matches soil infiltration rates, leading to less runoff. Measurements have shown that as much as 60 to 80 percent of fall precipitation gets stored in the soil.

The typical level of soil water storage—7 to 9 inches—that occurs before spring planting provides a form of moisture insurance for crop production. Corn and soybeans typically require 18 to 22 inches of moisture to produce good yields. Median precipitation values during the growing season—May through September—in the state's primary agricultural counties range from 15 to 19 inches, but variation can be great. Climate records for west-central Minnesota show that in one year out of five, May-September precipitation may total only 13 or fewer inches. Moisture stored in the root zone helps ensure that the total water needs of primary crops are met under these circumstances, for plant root systems will extract all the available soil water if necessary.

A combine brings in a bumper soybean crop, an important ingredient in the region's biodiesel fuels industry.

Many farmers use a minimum amount of tillage, and some combine tillage with planting and fertilizing in a single pass across the field. These practices allow for rapid planting when favorable spring weather and reasonable field conditions arrive. Recent years have shown that close to half the acreage of corn planted in the state can be seeded in as little as one week of favorable weather. Over the most recent three decades, the average planting date for spring wheat is the last week of April; for corn, the first week of May; and for soybeans, mid-May. Sometimes very cold and wet spring conditions have delayed these planting dates by as many as three weeks, as was the case in 1979. Minnesota's other crops are planted at various times, mostly during April and May. Rarely does any planting occur as late as June unless fields have been flooded out (as was the case in some counties in 2013 and 2014) or a late spring freeze has severely damaged emerging plants (as was the case in 1992).

Damage from spring frosts is unusual. When late frosts occur in May, the growing point of many crops is still below the soil surface, protected from serious damage even though parts of emerged leaves or stems may be killed. Under these circumstances the crop continues to develop and will produce new vegetative growth, often without affecting the final yield. June frosts are exceptionally rare but more damaging, killing off not only plant tissue, visible as "burned" or yellowed leaves and stems, but sometimes the entire plant, including the crown or growing point. The most recent example of such late-season frost occurred on the summer solstice, June 20, 1992, across southern and central counties. Though frost damage was significant in low-lying fields, many crops recovered and produced respectable yields, though below projected trends.

· ·

Leaching

Leaching is the movement of soluble soil elements such as chloride, bromide, sulfate, or nitrate by percolation, as moisture gravitates through successive layers of soil. In many agricultural soils moisture is stored until plant roots remove it during the growing season; however, in some soils deeper percolation can deposit soluble materials into aquifers that may be drinking water or irrigation sources. Potential leaching losses are governed by soil moisture, soil texture, and rainfall frequency and intensity.

· ·

It is unusual for first fall frosts to come before crops have fully matured, as early as the beginning or middle of September. On occasion when crops have been planted very late in the spring and their growth has been slowed by a cool summer, even late September frost may catch them in an immature state and consequently reduce the yield. However, losses from fall frost damage tend to be minimal in Minnesota and are sometimes compensated for by reduced moisture content; since crops dry faster after freezing, the cost of drying prior to storage is somewhat reduced, saving the farmer on that particular task.

Frost damage is dependent on both the severity of the temperature—how cold it gets—and the duration below freezing—how many hours at 32°F or colder. Oftentimes a ground-level frost is produced by official temperature readings that never reach 32°F. Calm overnight air settled near the surface develops a strong temperature gradient such that a thermometer placed 5 feet above ground—a typical height for official weather observations—will read as many as four to six degrees warmer than the temperature at the surface. Thus a climate station's official overnight minimum temperature reading of 34 to 38°F may coincide with a ground frost in the area, circumstances that inspire the National Weather Service to issue a frost advisory. Air temperature readings of 32°F or colder are considered official freezes, and the NWS will issue a freeze warning when such conditions are expected. Oftentimes these values are observed for only minutes or perhaps up to one hour before sunrise; their consequent damages may be visible, but plants will recover. A killing freeze is said to occur with an official minimum temperature reading of 28°F or colder, sufficient to kill off most vegetation, though it may take days to show it. A hard freeze, when even the hardiest vegetation dies, is associated with a minimum temperature reading of 24°F or colder.

Frost heaving

Usually the result of freeze-thaw cycles, frost heaving is the lifting of a surface by the internal action of frost. Frost heaving can also occur, to a lesser extent, by sublimation, when ice forms from water vapor at high humidity. During a thaw period, water droplets fill the pore spaces, especially the larger cracks or channels, in soils or road pavements. Subsequent freezing produces "ice lenses," which expand to force the overlying material upward. Because moisture is not evenly distributed, this expansion is irregular and may produce considerable bumpiness—good for agricultural soils, helping with drainage and aeration, but destructive to pavements and roadbeds.

The median spring and fall dates for these various temperature thresholds at several Minnesota climate stations can be found in Table 21, along with the median seasonal duration—that is, the number of days—between freezes. Note that the median frost-free season—the period between lows of 32°F—is highly variable: from fewer than 100 days in the frost pockets of northeastern Minnesota to more than 170 days in the southeast. Almost all of the state's crop production takes place in areas with a median frost-free season of more than 110 days. The counties with longer seasons, from 130 to 170 days, are generally the most favorable for corn and soybean crops as well as for alfalfa hay, which can be harvested three to four times per season in such areas. But frost-free growing season length is not the only criteria used to select which crops or varieties to plant: seasonal patterns in soil and air temperature are also important factors.

Both air and soil temperature tend to govern how rapidly most crops grow. Certain soil temperature thresholds are necessary for the rapid germination, development, and emergence of each crop. For example, most farmers wait until soil temperatures are 40°F or higher to plant small grains, 50°F or higher to plant corn, and 55°F or higher to plant soybeans. These guidelines need not be strictly adhered to since more than anything it is important to plant seed in a timely manner.

DID YOU KNOW?

Heliotropic plants, or "sun trackers," change orientation with the position of the sun. Crops like sunflowers and some species of cotton are heliotropic—the Latin *helio* means "sun" and *tropos* "to turn": they face east to greet the sun in the morning and west to say good-bye in the evening. By tracking the sun, a sunflower may receive up to 40 percent more sunlight on its leaves than it would if in a fixed orientation. Some desert plants exhibit heliotropic behavior during the winter, when the day length is shorter and the sun's elevation angle lower.

Once crops have emerged from the soil, air temperature begins to regulate their pace of growth and development. Research has shown that the accumulation of mean daily temperatures above various thresholds is closely related to the number of calendar days required to reach critical stages of plant development. For some crops, even maturation is closely tied to the accumulation of daily mean temperature during the growing season. Because of this relationship, the Growing Degree Day (GDD) concept has evolved to help characterize the match between climate and crop varieties. A common example is hybrid corn, whose seeds are marketed with a Relative Maturity (RM) rating (used throughout Minnesota), a scale that typically ranges from 80 to 115 days. The scale, as its name indicates, is relative: an 80-day RM hybrid will not necessarily mature 80 days from planting but will do so earlier than a 90-day RM hybrid, and so forth. Minnesota is divided into Relative Maturity Zones based on the frost-free season's length and the accumulation of temperature, or GDD, for corn.

The GDD method for corn is based on temperature thresholds of 50°F and 86°F, called the 50/86 method in the corn industry. These two values are used as reference points to compute the daily GDD values from

TABLE 21: Median Dates of Critical Low Temperature Thresholds in Minnesota (1981–2010)

DISTRICT AND STATION	MEDIAN DATE OF LAST MINIMUM TEMPERATURE IN SPRING			MEDIAN DATE OF FIRST MINIMUM TEMPERATURE IN FALL			DAYS BETWEEN MEDIAN DATES		
	24°F	28°F	32°F	24°F	28°F	32°F	24°F	28°F	32°F
NORTHWEST									
Ada	4/23	5/1	5/10	9/26	10/4	10/14	137	154	173
Agassiz Refuge	4/20	4/29	5/7	9/26	10/4	10/13	139	157	175
Argyle	4/25	5/4	5/15	9/23	10/2	10/12	130	150	170
Crookston	4/20	4/30	5/9	9/25	10/4	10/15	137	155	175
Detroit Lakes	4/21	4/29	5/8	9/25	10/3	10/14	137	156	175
Fosston	4/24	5/3	5/17	9/21	10/1	10/12	126	149	169
Hallock	4/27	5/5	5/15	9/22	10/1	10/8	129	147	164
Mahnomen	4/21	4/30	5/11	9/25	10/4	10/15	135	155	176
Roseau	4/27	5/5	5/17	9/24	10/3	10/12	127	147	166
Warroad	4/27	5/5	5/16	9/26	10/5	10/17	131	151	171
NORTH-CENTRAL									
Baudette	4/25	5/4	5/17	9/23	10/3	10/14	126	149	170
Bemidji	4/26	5/3	5/14	9/25	10/4	10/15	132	151	170
Big Falls	4/29	5/11	5/26	9/20	10/1	10/10	116	141	163
Cass Lake	4/30	5/10	5/23	9/21	10/1	10/11	119	142	162
Grand Rapids	4/24	5/4	5/17	9/25	10/3	10/15	129	151	174
Gull Lake	4/18	4/27	5/5	10/4	10/15	10/26	150	170	191
International Falls	5/1	5/13	5/28	9/16	9/27	10/6	111	135	157
Leech Lake	4/21	4/30	5/12	9/29	10/8	10/20	138	158	181
Park Rapids	4/28	5/7	5/19	9/22	9/30	10/9	124	144	163
Thorhult	4/29	5/12	5/27	9/17	9/26	10/6	112	136	159
Walker AGC	4/17	4/27	5/5	10/3	10/14	10/26	149	169	190
NORTHEAST									
Cook	4/28	5/8	5/22	9/22	10/2	10/14	122	145	168
Duluth Airport	4/21	5/2	5/17	9/27	10/6	10/20	131	156	181
Ely-Winton	4/28	5/5	5/16	9/26	10/6	10/18	132	152	173
Grand Marais	4/18	5/1	5/22	10/6	10/20	11/1	137	170	196
Hibbing*	NC	NC	NC	NC	NC	NC	97	122	146
Two Harbors	4/15	4/27	5/9	10/6	10/19	10/31	148	175	199

*Hibbing data not continuous

DISTRICT AND STATION	MEDIAN DATE OF LAST MINIMUM TEMPERATURE IN SPRING			MEDIAN DATE OF FIRST MINIMUM TEMPERATURE IN FALL			DAYS BETWEEN MEDIAN DATES		
	24°F	28°F	32°F	24°F	28°F	32°F	24°F	28°F	32°F
WEST-CENTRAL									
Alexandria	4/13	4/23	5/2	10/5	10/14	10/25	155	173	194
Artichoke Lake	4/9	4/19	4/29	10/5	10/15	10/26	158	178	197
Benson	4/16	4/27	5/6	9/29	10/7	10/18	143	162	183
Campbell	4/17	4/26	5/5	9/28	10/6	10/16	144	162	181
Canby	4/14	4/23	5/2	10/3	10/13	10/24	152	171	192
Fergus Falls	4/14	4/24	5/3	10/3	10/12	10/23	151	170	190
Glenwood	4/19	4/29	5/8	9/28	10/5	10/16	140	159	180
Madison	4/14	4/24	5/2	9/30	10/8	10/19	149	166	186
Milan	4/13	4/24	5/2	9/28	10/7	10/19	146	160	182
Montevideo	4/16	4/26	5/3	9/28	10/5	10/16	145	162	182
Morris	4/12	4/24	5/3	9/29	10/7	10/20	146	166	188
Otter Tail	4/20	4/29	5/8	10/1	10/9	10/20	144	162	182
Rothsay	4/19	4/28	5/6	9/29	10/6	10/17	144	160	179
Wheaton	4/16	4/26	5/3	9/30	10/9	10/19	148	165	184
CENTRAL									
Buffalo	4/12	4/22	5/1	10/3	10/12	10/24	152	172	193
Chaska	4/9	4/18	4/28	10/5	10/16	10/28	160	180	201
Collegeville	4/10	4/20	4/30	10/7	10/18	10/29	159	180	200
Gaylord	4/10	4/19	4/29	10/4	10/14	10/24	156	176	196
Hutchinson	4/11	4/22	5/2	10/2	10/11	10/23	151	171	194
Jordan	4/14	4/25	5/5	9/29	10/7	10/19	145	164	186
Litchfield	4/12	4/24	5/3	10/1	10/8	10/18	148	166	188
Long Prairie	4/24	5/3	5/14	9/24	10/3	10/14	130	150	170
Melrose	4/19	4/29	5/8	9/27	10/5	10/16	139	158	178
New London	4/14	4/23	5/1	10/5	10/14	10/25	154	173	192
St. Cloud	4/18	4/29	5/10	9/26	10/4	10/15	137	158	180
Santiago	4/17	4/28	5/9	9/26	10/3	10/14	137	157	179
Stewart	4/11	4/21	5/1	10/1	10/11	10/23	152	172	194
Wadena	4/21	4/30	5/11	9/27	10/6	10/17	138	157	178
Willmar	4/9	4/20	4/30	10/4	10/14	10/25	155	176	198

TABLE 21: Median Dates of Critical Low Temperature Thresholds in Minnesota (1981–2010) *cont'd*

TABLE 21: Median Dates of Critical Low Temperature Thresholds in Minnesota (1981–2010) *cont'd*								
DISTRICT AND STATION	MEDIAN DATE OF LAST MINIMUM TEMPERATURE IN SPRING			MEDIAN DATE OF FIRST MINIMUM TEMPERATURE IN FALL			DAYS BETWEEN MEDIAN DATES		
	24°F	28°F	32°F	24°F	28°F	32°F	24°F	28°F	32°F
EAST-CENTRAL									
Aitkin	4/23	5/4	5/19	9/24	10/3	10/15	126	149	173
Brainerd	4/22	5/1	5/11	9/23	10/1	10/12	133	152	172
Cloquet	4/29	5/13	5/28	9/22	10/3	10/15	116	141	167
Forest Lake	4/16	4/25	5/5	10/1	10/10	10/23	147	167	189
Hinckley	4/26	5/6	5/19	9/23	9/30	10/10	125	144	165
Milaca	4/19	4/30	5/11	9/26	10/4	10/15	135	156	178
Moose Lake	4/19	4/30	5/13	9/30	10/9	10/23	138	161	185
Mora	4/19	4/30	5/12	9/25	10/4	10/15	135	155	178
MSP-Airport	4/6	4/15	4/25	10/10	10/22	11/1	167	190	209
Pine River Dam	4/22	5/1	5/10	9/30	10/8	10/20	140	159	180
Stillwater	4/6	4/15	4/25	10/12	10/25	11/3	170	192	211
SOUTHWEST									
Lamberton	4/15	4/25	5/3	9/29	10/7	10/18	146	165	186
Luverne	4/14	4/23	5/1	9/30	10/9	10/20	149	168	188
Marshall	4/14	4/24	5/2	10/1	10/9	10/19	149	167	187
Pipestone	4/18	4/27	5/7	9/26	10/4	10/14	140	158	178
Redwood Falls	4/8	4/19	4/30	10/2	10/12	10/23	154	175	196
Tracy	4/11	4/21	4/30	10/3	10/12	10/23	156	175	194
Windom	4/12	4/23	5/1	10/3	10/14	10/25	153	174	195
SOUTH-CENTRAL									
Albert Lea	4/7	4/17	4/28	10/5	10/15	10/26	160	181	201
Fairmont	4/7	4/15	4/24	10/9	10/19	10/30	167	187	205
Faribault	4/16	4/27	5/6	9/29	10/8	10/21	144	163	184
New Ulm	4/10	4/19	4/29	10/1	10/11	10/22	155	174	195
Owatonna	4/14	4/24	5/3	10/2	10/12	10/25	150	169	191
St. James	4/10	4/21	4/30	10/2	10/12	10/22	154	173	193
Springfield	4/13	4/23	5/2	9/28	10/7	10/20	148	167	189
Waseca	4/11	4/20	4/30	10/1	10/12	10/23	153	173	195
Winnebago	4/8	4/17	4/28	10/5	10/16	10/27	159	180	201

TABLE 21: **Median Dates of Critical Low Temperature Thresholds in Minnesota (1981–2010)**								
DISTRICT AND STATION	MEDIAN DATE OF LAST MINIMUM TEMPERATURE IN SPRING			MEDIAN DATE OF FIRST MINIMUM TEMPERATURE IN FALL			DAYS BETWEEN MEDIAN DATES		
	24°F	28°F	32°F	24°F	28°F	32°F	24°F	28°F	32°F
SOUTHEAST									
Austin	4/9	4/20	4/30	9/30	10/9	10/22	151	172	194
Farmington	4/7	4/17	4/27	10/6	10/17	10/29	160	182	203
Grand Meadow	4/15	4/26	5/5	9/29	10/8	10/20	145	164	186
Preston	4/18	4/29	5/10	9/28	10/7	10/18	139	159	182
Rochester	4/8	4/17	4/28	10/5	10/16	10/27	159	181	201
Rosemount	4/11	4/22	5/1	10/3	10/14	10/25	154	174	195
Theilman	4/18	4/29	5/9	9/27	10/5	10/16	138	157	180
Winona	4/3	4/11	4/21	10/14	10/26	11/5	174	196	215
Zumbrota	4/20	5/1	5/12	9/24	10/2	10/13	133	152	174

maximum and minimum air temperatures. For example, on a day when the maximum temperature is 86°F and the minimum is 50°F, the daily mean (maximum plus minimum divided by two) equals 68°F. This value is 18 degrees above the lower base of 50°F for corn, translating to 18 GDD for the day. On a day when the actual maximum temperature exceeds 86°F, the upper limit for corn is set as the maximum and substituted in the calculation. Similarly, if the day's actual minimum temperature falls below the lower threshold of 50°F, it is reset to the lower limit for corn. Thus, a day with a maximum temperature of 88°F (reset to 86°F) and a minimum temperature of 46°F (reset to 50°F) computes to a daily mean value of 68°F and a daily GDD value of 18 as well. These temperatures are used to calculate daily and seasonal GDD totals.

Across Minnesota the total seasonal GDD values for field corn from May through September typically range from 1,600 to 2,600. Values below 2,000 GDD generally indicate a climate unsuitable for cultivating corn for grain, though it may be grown for silage that can be harvested when immature and chopped for cattle feed. The RM rating for corn hybrids is closely linked to seasonal GDD values. Climate regions in

Minnesota that show a seasonal GDD value of 2,000 to 2,200 are more suited to a corn hybrid with an 85 to 95 RM rating, while areas with 2,400 or more GDD tend to accommodate the 105 to 115 RM ratings. Year-to-year climate variability may cause the seasonal GDD totals to vary by 10 to 12 percent. Generally, in years when GDD values are higher than normal, crops mature early, while in years when GDD values are lower than normal, crops mature late.

The GDD concept has been applied to other crops and even to the life cycle and development of diseases and insects. Temperature appears to regulate the pace of activity in many biological organisms; aside from soil moisture, it is the single most important climatic element to affect the Minnesota growing season. It influences the chemical and biological activity in the soil, the volatility of pesticides and herbicides used on crops, the metabolism and weight gain of livestock, and, when in the extreme, even the day-to-day uncertainty of commodity prices on the Chicago Board of Trade. On occasion extreme periods of temperature, whether hot or cold, can negatively affect the growing season. The heat waves of July 1995 and 1999 caused serious losses to the state's turkey producers due to high mortality rates; the early frost in the first few days of September 1974 translated into reduced yields for corn growers. For the most part, however, temperature extremes and stresses during the growing season are short lived in Minnesota.

Other environmental stresses during the growing season can come from severe storms that produce hail or wind damage and flash floods that drown out crops in low-lying areas. Most farmers carry all-hazard insurance for protection from such events. Sometimes persistent rains, wet soils, or high winds prevent cultivation and spraying for weeds, which can lead to significant yield loss as well. The worst possible weather and climate hazard, though, is drought. Fortunately, drought in Minnesota has been rare historically; the last serious, widespread, and damaging episode played out in 1988, though as recently as 2012 extreme drought plagued some parts of the state. Other droughts have been widely spaced except for the 1930s: impacts were severe in 1894, 1910, 1917, 1922, 1932, 1933, 1934, 1936, 1958, and 1976. Drought is the one environmental threat for which there is little cure or opportunity for mitigation; Minnesota farmers count themselves fortunate that it does not come visiting very often.

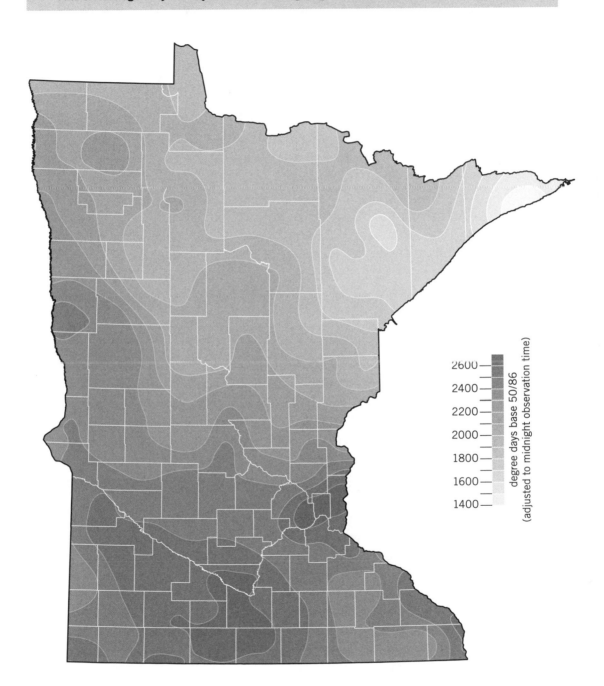

MAP 8: Average May-to-September Growing Degree Days (base 50/86°F), 1981–2010

Farmers assess a near-complete crop loss from hail damage that arrived during corn silking time in southern Minnesota, 1978.

Because their livelihood is so dependent on the weather during the growing season, Minnesota farmers have a history of being weather observers. Many belong to Soil and Water Conservation District observing networks, and some are even NWS Cooperative Weather Observers. For the most part they show great respect for the weather and for the soil resource they manage. Little wonder, then, that weather remains the most common theme of everyday conversation in the rural landscape, a characteristic that has not changed in more than a hundred years.

Mother Nature Blessed the Grand Excursion

A highlight of summer 2004 was the 150th anniversary and re-creation of the 1854 Grand Excursion from Rock Island, Illinois, to St. Paul from June 25 to July 5, 2004. The original trip—June 5–10, 1854—celebrated completion of the railroad from Chicago west to the upper Mississippi River and touted the virtues and potential of the western territories, acquired by the Louisiana Purchase nearly a half century earlier. According to the *Galena Jeffersonian,* the intent was "to make a thousand, more or less, men of capital and influence acquainted with the enchanting beauty, the boundless resources . . . of the Great West." Former president

Millard Fillmore (his term had ended in 1853) and more than 1,200 citizens—politicians, artists, writers, reporters, and business leaders—took the railroad from Chicago to Rock Island, then boarded a small fleet of steamboats bound upriver to St. Paul.

Though it is difficult to prove cause and effect, the region west of the Mississippi River did grow enormously shortly after the Grand Excursion, attracting millions of dollars in investments and thousands of new immigrants. The majestic beauty of the river valley landscape had a great deal to do with this success. In addition, an unusual and serendipitous episode of near-perfect June weather probably amplified the adventurous spirit, revelry, and enthusiasm of the excursionists, attitudes which were, in turn, reported in the press. Though the excursion occurred before any standard weather-observing system was in place in Minnesota Territory, documentation of that June's weather can be found in the records of Fort Snelling and in the daily readings of Dr. Charles L. Anderson, an early citizen of St. Anthony and a weather observer for the Smithsonian Institution. These data, when combined with comments in the *Minnesota Pioneer,* yield a very interesting picture indeed.

What Twin Cities record was set on August 12, 2000?

a. most rainfall

b. highest dew point

c. lowest temperature

(answer on page 329)

By 1854 Minnesota Territory had already acquired a reputation as the American Siberia. Much of this reputation was staked upon weather records and diaries from Fort Snelling, established in October 1819. The first winter was so harsh that 40 soldiers perished, primarily from pneumonia and scurvy. Below-zero temperature readings were the rule rather than the exception. And 1819–20 was only a precursor to a number of severe winters with temperatures as cold as -41°F and abundant snowfall, followed by late springs and frequent flooding on the Minnesota and Mississippi rivers (in 1822, 1826, 1831, 1838, 1843, and 1849, for example). To excursionists preparing to visit Minnesota Territory for the first time, the stories and events from the previous year alone may have been disconcerting. Snowy conditions and cold spring temperatures dominated 1853. Snow and frost came on May 10 and May 18, followed by a very wet June, fully half the days being rainy. Travelers familiar with Minnesota's reputation through stories of weather extremes and hardships—and perhaps a few tall tales—must have felt considerable apprehension concerning the weather they might face. Surely, most brought warm clothing.

Following a lack of late-winter snowfall (3 inches for March and April compared to the modern average of nearly 14 inches), the spring of 1854 came in haste. Warmer-than-normal spells of weather were frequent in March and April, and ice on the Mississippi broke up early. The river was running very low and the landscape looking dry as a result of minimal precipitation the previous fall. Prairie fires were observed throughout most of March and again in early May. Snow flurries were reported on May 1. Then came a blessing in disguise: heavy thunderstorms with frequent winds in the 30- to 50-mile-per-hour range brought the combined April and May rainfall to 7 to 9 inches, well above the normal of 5 to 6 inches, and produced a good deal of runoff. These conditions not only alleviated the threat of prairie fire but also caused the Mississippi River to rise significantly, probably eliminating concerns about snags, sandbars, or shallow waters. The needed rain also made the river valley abundantly green and lush by the time the Grand Excursion departed on June 6. Area strawberries were ripe and ready to pick by June 2, and weather observers noted that "all the common prairie flowers were in bloom."

In the week before the riverboats left Illinois, the weather had been unsettled and sultry, with thunderstorms and three consecutive days of measurable rainfall, totaling nearly an inch at Fort Snelling. Weather records from both the fort and St. Anthony show overnight lows in the mid- to high sixties on June 5, indicating an uncomfortably high dew point for early June. Despite the weather, "The boats, decorated with prairie flowers and evergreens left Davenport at ten o'clock and sailed with music on their decks . . . saluted by the gay fireworks from the Old Fort," according to historian William J. Petersen. He further noted, "Everyone was delighted with the bright moonlight and the refreshing river breeze" that provided a respite from the oppressive dew point. Shortly after midnight on the overnight trip toward Galena, the passengers witnessed a violent thunderstorm, its frequent lightning strikes illuminating the boats, bluffs, and riverbank vegetation. A short, rainy stop in Dubuque on the afternoon of June 6 would mark the end of their discomfort.

Following the storms, a strong Canadian high-pressure system settled in overnight, bringing clear skies, brisk northerly winds, low dew points, and cool temperatures—some 15 to 25 degrees below those of the previous day. The crisp air was probably invigorating. Though the overnight low of 40°F on the morning of June 7 was the month's coldest reading at Fort Snelling, air temperatures modified by the valley landscape were likely several degrees warmer for the excursionists. Days were bright and sunny, with highs in the mid- to upper sixties, about six to 12 degrees cooler than normal for that time of June.

Wednesday night found the Grand Excursion on Lake Pepin, where four of the riverboats were lashed together to travel across the lake as barge tows do now. The high-pressure system provided a cool, calm, and beautiful setting as the lights from the boats "danced and streamed on the waters." The low dew points and cool temperatures likely prohibited much mosquito or pesky insect activity, while a nearly full June moon (waxing, 86 percent full until setting about 3:00 AM) lit up the surrounding bluffs. The night was spent in "dancing, music, and flirtations."

The boats reached St. Paul earlier than expected on Thursday, June 8. The weather was still magnificent, with morning lows in the forties and highs in the mid- to upper sixties, moving the *Minnesota Pioneer* to comment on the pleasant contrast with the stormy, rainy period earlier in the week. Passengers were treated to visits to St. Anthony Falls, Minnehaha Falls, Fort Snelling, and Fountain Cave, then to a gala dinner and dance. The excursionists departed shortly after midnight on Friday, June 9, to return to Rock Island under the month's full moon. Again, the weather was near perfect, still dominated by Canadian high pressure. Temperatures were slightly warmer, with highs from 70 to 80°F and lows in the fifties under mostly clear skies.

Journalists and guests wrote wonderful stories of the beauty and majesty of the landscape, probably never realizing that they had traveled the river valley under perhaps the best weather conditions possible at that time of year. Thanks to the three-day dominance of a vast, cool, and dry Canadian air mass, temperatures were 10 to 12 degrees below normal, with comfortably low dew points, invigorating northerly winds, and beautifully clear skies. A check of June 6–8 climate records from 1891 to 2003 shows that comparable conditions have occurred only three times since: in 1897, 1937, and 1998. The river was deep enough but not too fast flowing for easy navigation, and the landscape was blanketed with lush vegetation and flowering plants.

Had they stayed a bit longer in St. Paul, excursionists might have seen another side of the weather. A thunderstorm was noted again on June 11, and a Fort Snelling observer reported on June 15, "The weather is awful hot these days, only good for growing crops and mad dogs." (Dew points were in the high sixties to low seventies, with temperatures in the mid- to upper eighties.) A severe thunderstorm on July 4 damaged trees and buildings in St. Paul.

The Grand Excursion produced many accounts promoting the western territories, and money and people streamed into the region. One settler was Ira O. Seeley, who explored the Zumbro River Valley and settled in Wabasha County with his family in fall 1854, just months after the historic river trip. He founded the town of Mazeppa and was the first District 12 representative elected to the territorial and state legislature in 1857–58. He was my great-great-grandfather.

Significant Days

Independence Day

Minnesotans generally celebrate July Fourth amid warm and humid conditions. With average high temperatures in the eighties and lows in the fifties to sixties, residents may feel uncomfortable even during evening fireworks. The statewide high temperature record for this national holiday dates to 1936: 107°F at Pipestone in Pipestone County and at Worthington in Nobles County. Those extreme temperatures were confined to the southwest, however, and many other communities enjoyed more comfortable afternoon readings in the seventies and eighties. Very hot conditions were also posted for Independence Day 1966—105°F at Redwood Falls in Redwood County—and 1988—106°F at Browns Valley in Traverse County. But by far the most widespread heat and high humidity occurred on July 4, 1949, when 18 communities reported daytime highs of 100°F or better, topped by 105°F at Wheaton in Traverse County. The humidity was excessive that day, as well, with

Choose your viewing location wisely: fireworks over the Stillwater bridge, 1934

tropical-like dew points in the low seventies that produced Heat Index values ranging from 110 to 115°F. The humidity remained high well into the evening, maintaining sultry overnight lows of 83°F at Jordan in Scott County and 80°F at Winona in Winona County. The most recent oppressively hot July 4 occurred in 2012, when a number of climate stations set new record high temperature values. Eleven climate stations reached 100°F, and scores of others reported mid- to upper nineties. Combining these temperatures with high dew points, many observers reported Heat Index values ranging from 102 to 118°F. The afternoon dew point in the Twin Cities reached a new record value of 77°F, producing a Heat Index of 108°F. The low temperature in the Twin Cities also set a record for warmth by dropping only to 81°F.

Conversely, 1972 brought Minnesota's coldest-ever July Fourth. Many northern communities actually saw a midsummer frost, and Tower in St. Louis County reported a low of 27°F—the state record for the date. Other St. Louis County locations recording abnormally low temperatures were Cook and Hoyt Lakes, 30°F, Cotton, 31°F, and Virginia, 32°F—there the afternoon high barely touched 70°F. In Itasca County, Bigfork recorded 30°F, while Roseau County's Wannaska saw

31°F and Koochiching County's Big Falls 32°F. More geographically isolated frosts occurred in 1932 and 1950, but perhaps the most remarkable celebration of this holiday was in 1876, when Duluth residents made ice cream with ice harvested from the harbor.

Although usually brief in nature, showers and thunderstorms frequent Independence Days around the state. In the Twin Cities, July Fourth brought rain each year from 1900 to 1905, the wettest being 1900, with 2.27 inches. In fact, since 1891 more than 40 percent of these holidays have produced rain in the metropolitan area. Over the same period, even Morris, in comparatively arid Stevens County, has reported July Fourth rainfall more than 40 percent of the time, including a 1993 thunderstorm that delivered 3.90 inches. By far the wettest Independence Day occurred in 1995, when a thunderstorm dumped 9.78 inches of rain on Milan in Chippewa County and the Chippewa River rose 9 feet, reaching its second-highest flood crest.

One of the worst storms to hit Minnesota on July Fourth was the famous 1999 derecho, or straight-line windstorm, that cut a swath through the Chippewa and Superior national forests and severely damaged trees in the Boundary Waters Canoe Area Wilderness (BWCAW). As the storm grew early in the day, its winds caused an estimated $85 million in damages to the Fargo-Moorhead area. By afternoon, winds of 80 to 100 miles per hour wreaked havoc on 600 square miles—the Hibbing Airport anemometer recorded 81 miles per hour just before nearby hangars were damaged. Tens of millions of trees were blown down, and 60 people were injured in the BWCAW, some requiring rescue by floatplane. The storm left a tremendous fuel load of downed trees; fortunately, wet weather combined with good fire management by the U.S. Forest Service and the Minnesota Department of Natural Resources have diminished the threat of serious fire.

For the ultimate in Independence Day festivities—fireworks viewing—one important climate element to watch is the prevailing wind direction. The dominant direction on this day is usually south to southeast; therefore, one's best bet is an upwind perspective to the south or east of where the fireworks are being shown. On rare occasion, summer dry spells can present serious fire danger, as they must have in 1894, 1910, and 1976, years with persistent midsummer drought that later produced abundant fall fires in the state.

Labor Day

The first Monday in September has been recognized as Labor Day in many American cities since the 1880s and as a national holiday since 1894. The nation's biggest Labor Day weather story occurred on September 2, 1935, when a category 5 hurricane struck Florida with winds over 150 miles per hour, taking hundreds of lives and destroying the Flagler Railroad system, which ran the length of the Keys. This hurricane—which predated the tradition of naming tropical storms—still ranks as one of the strongest to ever make landfall in the United States.

That same year, the northern hemisphere weather pattern brought a late-summer polar air mass that descended from high latitude over northern portions of Minnesota to produce one of the state's coldest-ever Labor Days. The statewide record low temperature of just 22°F on September 2, 1935, was set at the Sawbill Camp in Cook County. Record lows by location were also measured at Alborn in St. Louis County, with 29°F; Baudette in Lake of the Woods County, 32°F; Grand Rapids in Itasca County, 31°F; and Pine River Dam in Crow Wing County, 32°F.

While it has never snowed on Labor Day in Minnesota, rainfall occurs about 30 to 40 percent of the time. The holiday has seen heavy rains and thunderstorms: both 1957 and 1964 brought record-setting precipitation to many areas. On September 2, 1957, northwestern Minnesota communities were hit with heavy thunderstorms that produced rainfall of more than 2 inches in much of the Red River Valley: Hallock in Kittson County reported 4.39 inches, Roseau in Roseau County 5.07 inches, and Halstad in Norman County 5.91 inches—a Labor Day statewide record. These rains interrupted the hay harvest and caused long delays in bringing in grain crops. On another wet Labor Day, September 7, 1964, heavy thunderstorms once again dumped upwards of 2 inches of rainfall across the state. The most flooding occurred in the Duluth area, which received a double dose of thunderstorms—both early morning and late

DID YOU KNOW?

State Fair Records

- Warmest temperature measured during the fair: 104°F on September 10, 1931
- Coldest temperature measured during the fair: 33°F on September 13, 1890
- Driest fair: 1968; rainfall total: 0.08 inches
- Wettest fair: 1977; rainfall total: 9.48 inches
- Highest dew point measured during the fair: 77°F on August 28, 1955
- The most common prevailing wind direction during the fair: south to southwest
- Average number of rainy days during the 12-day fair: 3 to 4
- Highest Heat Index—combined effects of temperature and humidity—during the fair: 105–110°F, recorded on Monday, September 1, 1913; Friday, August 27, 1926; Sunday, August 28, 1955; and Thursday, August 22, 1968

afternoon showers. Duluth harbor reported 4.35 inches of rain, while Two Harbors in Lake County measured 3.81 inches. In Duluth the heavy rains flooded more than 60 basements and a local elementary school and backed up the city's storm sewer runoff system, causing damages estimated at more than $5 million.

Labor Day has traditionally marked the last day of the Minnesota State Fair. In a pleasant end to summer, daytime high temperatures typically range in the seventies and nighttime lows in the fifties, but hot, muggy weather can make an appearance, too. In 1913 the state fair's temperature reached 97°F on Labor Day (September 1), but it was even hotter in southeastern communities: Grand Meadow in Mower County and Red Wing in Goodhue County reported 100°F, and Winona in Winona County, 101°F. September 4, 1922, and September 2, 1929, produced state record high temperatures for the dates with 103°F, both reported at Beardsley (Big Stone County). But in 1931, seven communities reported temperatures of 100°F or higher on Labor Day (September 7), with a new state record high of 104°F set at Wadena (Wadena County) during what was arguably Minnesota's warmest year on record.

In a shift from extreme heat to unusual cold, the frosts of Labor Day 1974 are still remembered for their devastating impact on Minnesota crops. Similar to Labor Day 1935, September 2, 1974, saw a polar high-pressure system descend from high latitude and bring very cool and dry conditions to the state. Although few morning temperatures dipped below the freezing mark, many communities reported lows ranging from 33 to 36°F, with widespread ground frosts and damage to immature crops. A low temperature of 34°F was reported as far south as Austin (Mower County). These temperatures and frosts stand out as a climate singularity, arriving as they did two or more weeks earlier than any previous year for most communities, especially across the south.

Minnesota's Fall

Of the four seasons, fall may have the greatest impact on the Minnesota psyche. Residents react to the typical fall weather pattern in a classic love-hate manner. Beautiful crisp, clear mornings perfect for viewing early fall colors give way to foggy mornings, abundant clouds, and biting winds. Fall signals an end to the growing season, but it concludes the pesky insect season as well. The vacation season draws to a close; parents are relieved to see the school year begin. The boating and golfing seasons wind down, but the allergy season ends, too. Despite these conflicting effects, fall is most Minnesotans' favorite season, and with good reason. It offers a glorious, invigorating environment; then, like a finger-wagging parent it becomes dull, darkened, windy, and frigid, reminding us to prepare for winter.

Summer's uncomfortable dew points and risk of severe weather abate sharply in September and October. Daytime temperatures are generally more suitable for outdoor activity; nighttime temperatures are quite nice for sleeping. More often than not climate conditions match optimum comfort zones and it feels good to be outside. Comfortable temperatures are just a start: many more climatic details often escape attention or fade from memory.

September can bring summer-like weather, with highs in the nineties and the occasional thundershowers. Less than five percent of the state's recorded tornadoes have occurred during this month. The most recent September tornado was an EF-0 (65–85 mph) near Williams (Lake of the Woods County) along Minnesota Highway 11 on September 11, 2013, at about 3:30 PM. It was fairy inconsequential, only felling a few trees. Multiple tornadoes were reported during September in both 2005 (7) and 2006 (4), the most storied inflicting significant damage to Rogers (Hennepin County) on September 16, 2006.

Since most Minnesota school buildings are not air-conditioned, summer-like conditions that persist into September are quite noticeable. Daytime temperatures have been as high as 111°F (September 11, 1931, at Beardsley in Big Stone County) and at least 100°F as late as September 22 (at Ada in Norman County, in 1936). September usually brings just one or two days with temperatures in the nineties. In 2013, high temperatures in the mid-nineties and dew points in the seventies in southern Minnesota, including the Twin Cities metropolitan area, caused many school cancellations because of high Heat Index values and lack of air conditioning.

Like in the summer months, September rain showers are generally short lived: the Twin Cities' monthly average number of thunderstorm days is 4.2, while the average number of days with measurable precipitation (.01 or more inches) is 9.4. In September 1894 rain came on only three days, contributing to the year's terrible fire season. Both 2009 and 2012 also brought only three rainy days to the Twin Cities area, amplifying the drought that encompassed much of the Minnesota landscape in those years. The heaviest 24-hour rainfall recorded in the Twin Cities during the month is 4.96 inches (September 12, 1903), while on a statewide basis it is 9.48 inches at Amboy in Blue Earth County (September 23, 2010).

On the other hand, September can also be winter-like, with frosts and even snow. September 1974 brought the earliest-ever frost to many southern Minnesota communities; on the third, fully two weeks before the earliest frost of previous record, even Twin Cities residents witnessed damage to garden plants. Many northern Minnesota locations routinely see frost in September, a feature often linked to the start of fall leaf color. Though no single-digit low temperatures have been recorded in the state during this month, places like Mahnomen (Mahnomen County),

Thorhult (Beltrami County), and Big Falls (Koochiching County) have seen values as cold as 10°F. Significant September snowfalls have been observed as well. Early measurable snow occurred at International Falls in Koochiching County on September 14, 1964 (0.3 inches) and at Warroad in Roseau County on September 15, 1916 (0.2 inches). Appreciable snows have fallen at many locations during the last week of the month, including Fosston in Polk County, 6.5 inches on September 25, 1912, and Bird Island in Renville County, 8 inches on September 26, 1942. In fact, September 1942 brought measurable snowfall to most Minnesota communities, including 1.7 inches in the Twin Cities on the twenty-sixth, the earliest winter wake-up call of record.

October, the favored fall month, hosts the harvest moon, Octoberfest celebrations, Halloween, peak leaf color, homecoming football games, harvest dinners, the Twin Cities Marathon, and a number of other treasured events—a final appeal to enjoy the outdoors. In October 1895 the *Minneapolis Journal* waxed eloquent about the season:

> October is generally a kingly month in Minnesota. It opens with the usual affluence of sunshine and quickening, bracing air, which stimulates like the ichor of the Olympian gods. Day after day, the transformation of summer greenery into the royal and gorgeous tones of autumn will go on and summer's silent fingering will be overwoven with pageantry of color which no human art can call into being. The recessional of the year is grander than the processional.

Many Minnesotans cherish October for its pageantry of landscape color and its frosty mornings with clear, blue skies. The bounty of harvest fills the pantry in the form of homemade pickles, sauces, and jams. Outdoor activities abound: harvest festivals and dinners, visits to the apple house and glasses of fresh cider, song-filled evening hayrides with friends and family, a bonfire rally, the last boat trip on the lake or river, football on a Friday night or comfortable Saturday afternoon, selecting pumpkins for Halloween, or watching the formations of migrating birds. Compared to the leftover heat and thunderstorms sometimes prevalent in September and the first winter storms that appear in November, October's weather rarely disappoints.

Q & A

What is the lowest relative humidity ever measured during the month of October in Minnesota?

a. 6 percent

b. 26 percent

c. 46 percent

(answer on page 329)

Fall leaf color usually peaks in October, as the period between sunrise and sunset decreases by two to three minutes each day and chlorophyll breaks down faster than it is produced, making yellow, red, and purple pigments more prominent. The total reduction in day length amounts to about 1.5 hours, but the changeover from daylight saving time, setting clocks back one hour on the last weekend of the month, amplifies this difference. While the change in day length initiates leaf color change, the pace at which it occurs is modified by temperature. Former state climatologist Earl Kuehnast established a relationship between overnight minimum temperature and the rate of leaf color change, explaining that three nights with temperatures in the thirties trigger color change and seven to ten nights yield peak color. Based on the geographic distribution of the state's temperatures, peak leaf color usually occurs in the far northern counties during the third or fourth week of September and is far more evident in central and southern Minnesota during October. The Minnesota Department of Natural Resources recommends five travel routes for viewing fall colors: Highway 34 between Walker and Detroit Lakes; the Pillsbury State Forest near Brainerd; Highway 61 in southeast Minnesota; Highway 95 between Stillwater and Taylors Falls; and County Roads 23 and 16 near Orr and Ely.

October is typically drier and quieter than September. In the Twin Cities area, measurable precipitation occurs on around 8.4 days, about one day less than an average September. Thunderstorms may be seen on one or two days, while an October tornado is reported only every other year. Mean daily temperature declines by 15 to 17° from the first day of October to the last, a sharp drop equal to March's climb in temperature; climatologists refer to these changes as seasonal symmetry. October temperatures have been as high as 98°F in western communities (Beardsley in Big Stone County on October 5, 1963) and have even reached 90°F at the end of the month (Canby in Yellow Medicine County on October 30, 1950). Some of the lowest-ever relative humidity measurements—afternoon readings in the single digits—have been made in October, when vegetation has died and the landscape dries out. Such conditions increase fire danger. Indeed, during prolonged dry fall seasons the threat can be as significant as in late winter and early spring, while vegetation is still dormant.

Even pleasant October can bring cold and snowy weather, generally during the second half of the month. All-time lowest temperatures, often below zero degrees, include -16°F at Roseau in Roseau County on October 26, 1936. Snowfall in the northern counties is not unusual and can be quite significant: Baudette (Lake of the Woods County) reported 16 inches on October 18, 1916 (perhaps lake-effect snow from cold air passing over a relatively warmer Lake of the Woods); Sandy Lake Dam (Aitkin County) reported 15 inches on October 25, 1942; and Badger (Roseau County) reported 14 inches on October 4, 2012. For the southern half of the state, October snowfalls are few and far between, the most recent occurring over October 12–13, 2009, when many observers reported 2 to 4 inches. On rare occasions blizzards may strike the state, as they did in 1820 at Fort Snelling; in 1880 in southwestern Minnesota—isolating Canby for weeks; in 1916 in western and northern Minnesota—shutting down railroads; and on Halloween 1991 across eastern Minnesota.

October brings pumpkins, apples, and falling leaves. Cynthia Shields posed with her carved pumpkin in 1935.

Affection for October may be inspired by its proximity to what is arguably one of the worst months of the year. November certainly has its share of negative connotations: freezing rain, frozen ground, and icing lakes; howling winds and the first wind chills; low, gray decks of clouds; pile after pile of leaves to rake; short days getting shorter; the transition to layered clothing; the onset of flu season and Seasonal Affective Disorder (SAD); haze and morning fog; all-day rains; the first winter storm watches and warnings. Despite these and other depressing qualities, November is redeemed somewhat by the bountiful holiday of Thanksgiving.

In November, with its shorter days and lower sun angle, the daily course of the sun has little effect on air temperature. Cloudiness increases significantly: by the end of the month the average number of cloudy and partly cloudy days has doubled from the beginning. The freezing level of the atmosphere as well as cloud ceilings, or heights, become lower, and advection begins to control the temperature. Winds from the northwest bring polar air masses whose time at high latitudes has made them cool and dry. Wind shifts to the southeast or southwest transport warm, moist air into the region, often with clouds that stabilize overnight minimum temperatures, keeping them from dipping too low. The give-and-take of these air masses often has a two- to three-day life cycle, creating the so-called roller coaster of fall temperatures, alternating periods of warm Indian summer–like days and cold, blustery, winter-like conditions.

Though November will occasionally tease with a 70 or 80°F reading, more often than not the weather reminds Minnesotans to prepare for winter. The last of the fall chores must be done: plowing, soil testing, composting, mulching, bringing in the dock, winterizing the car, bagging fallen leaves, putting up storm windows, finishing outdoor painting, putting away patio furniture, unpacking the winter wardrobe, and reviewing the snowplowing contract. The Novembers of 1999, 2001, and 2009—the warmest in recorded state history—were so mild that most of the month could be utilized to wrap up these fall chores.

Q & A

What category of weather records was set around the state on October 19, 2000?

a. earliest-ever below zero temperature

b. wettest-ever October 19

c. warmest-ever October 19

(answer on page 329)

Minimum temperatures can drop well below zero during the month, as cold as -45°F, measured at Pokegama Dam in Itasca County on November 30, 1896. The first winter storm watches and warnings usually occur in November, wind-chill warnings may be broadcast, and blizzards are not uncommon. Some counties of western Minnesota show a historical blizzard frequency during the month of November of once in every four to five years. Numerous November blizzards have delivered more than a foot of snow in one day, while some have produced snowfalls of more than 20 inches. Further, a number of climate stations have recorded more than 40 inches of snowfall during the month, and a handful, including Duluth and Two Harbors, have reported more than 50 inches.

Many area lakes start to freeze in November; soils begin to freeze as well. The rate at which soils freeze depends on many factors, including air temperature, snow cover, soil moisture, soil texture, and vegetative cover. Generally the first to freeze are drier soils that lack vegetative cover or crop residue and therefore more readily release heat to the atmosphere. Limited historical records for initial soil freezing dates (Table 22) show that the earliest in Morris (Stevens County) is November 7, while the average there is November 24. Another example is Rochester (Olmsted County): earliest, November 9; average, December 6.

TABLE 22: Initial Soil-Freeze Dates

LOCATION	AVERAGE DATE	EARLIEST DATE	LATEST DATE
Bemidji	November 24	November 6	December 8
Brainerd	December 1	November 18	December 14
Crookston	November 28	November 10	December 26
Duluth	December 2	October 31	January 18
Fergus Falls	November 26	November 9	December 30
International Falls	December 8	November 23	December 26
Lamberton	December 4	November 10	December 30
Mankato	December 21	December 7	January 12
Minneapolis	December 8	November 13	January 8
Morris	November 24	November 7	January 7
Rochester	December 6	November 9	January 10
Roseau	December 6	October 28	December 30
St. Cloud	November 27	November 11	December 11
St. Paul	December 6	November 7	January 2
Wheaton	November 24	November 10	December 3
Winnebago	December 2	November 11	December 25
Winona	December 20	November 18	January 10

Finally, November brings the dark or SAD days. As shorter days become more evident, some people begin to suffer from the deprivation of light, a malady called Seasonal Affective Disorder, the effects of which are both physical and mental, in some cases resulting in severe depression. The somewhat rapid reduction in day length is magnified by two other factors: a lowering sun angle and increased cloudiness. This change in sun angle, called declination, creates very long and lasting shadows, especially on northerly slopes, shading some parts of the landscape for much of the day. The degree of cloudiness also increases, peaking during November, when two-thirds of the days are mostly cloudy (.8 or greater sky cover) and most of the remaining days are partly cloudy (.4 to .7 sky cover). The combination of sun angle and relative cloudiness creates highly diffuse light. The average possible sunshine is less than 40 percent during November in most Minnesota communities, and perfectly clear days are extremely rare. Thus, both the quantity of light—in terms of day length—and the quality of light—in terms of direct sunlight—are diminished.

· ·

Percent possible sunshine

Percent possible sunshine, a standard measurement made at National Weather Service offices, is the ratio of the actual duration of bright sunshine—that is, unobstructed sunlight measured by a sunshine recorder—in hours and minutes compared to the astronomically possible duration from sunrise to sunset for a particular location. In the winter, when only 8 hours of sunshine are possible for the Twin Cities, 7 hours of actual sunshine would equal 87.5 percent possible sunshine, while in summer, when 15 hours of sunshine are possible, 7 hours of actual sunshine would equal only about 47 percent possible sunshine. Long-term averages show that late September and early October bring the maximum percent possible sunshine to Minnesota.

· ·

Though often a month frantic with winter preparations and depressing for lack of sunlight, November closes with Thanksgiving, an invitation to pause and express gratitude for all that we have. Climatically it is analogous to a recessional process, as the *Minneapolis Journal* noted, for as winter begins Minnesotans turn the last page of the last chapter of another outdoor season. In rural Minnesota, Thanksgiving is often a benediction upon the bounty provided by the year's climate conditions, expressing a sense of gratitude that spans multiple farm family generations.

Indian Summer

The term *Indian summer* dates back to at least 1778 and probably originated in reference to the American Indian practice of using the last good spells of autumn weather to increase winter food stores. In New England this definition applies even today, except that Indian summer is said to occur after the fall's first killing frost. Residents in most northern climates define it as a period from mid- to late fall when skies are generally clear, winds are calm, days are sunny and hazy with comfortably warm temperatures, and nights are cool. In Minnesota, Indian summer may occur before or after the first fall frost. There are occasional years when the first frost does not arrive until late October or early November, after residents have already enjoyed a lovely Indian summer. In the Twin Cities, for example, there is historically a ten percent chance that the first fall frost will not occur until after October 26; sometimes it has arrived as late as November 7.

Minnesota state climatologist Jim Zandlo coined the term *retro summer* in the fall of 2003 to refer to remarkably warm spells that produced summer-like conditions and summer-like ambitions in state residents in October. That year, frigid temperatures with frequent frosts during the first part of October were followed by unusually sunny and warm days, with temperatures nearly 20 degrees above normal, hitting the high seventies and even the mid-eighties. A near-record number of rounds of golf were played on area courses. Roofing jobs and painting tasks were finished, gardens were mulched, and leaves were bagged in some of the most pleasant weather in recent memory. Indian summer is regarded as a last chance to relive some summer activities and prepare for winter. Excepting the rapidly declining day length, the collaboration of wind, temperature, and humidity can produce a summer déjà vu that entices people to play hooky from work and finish projects that might have been put off until next year.

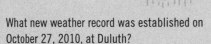

What new weather record was established on October 27, 2010, at Duluth?

a. high temperature of 71°F

b. snowfall of 4.3 inches

c. low temperature of 5°F

(answer on page 329)

Though the derivation and meaning of *Indian summer* is unquestionably North American, other regions of the northern hemisphere have similarly distinctive phrases for unseasonable late fall weather. According to the *Glossary of Meteorology,* the corresponding period in Europe is sometimes referred to as "Old Wives Summer" or "halcyon days." The former credits women who prepared for the onset of winter by stocking up on food, clothing, firewood, and other supplies; the latter points to a nautical period of tranquility associated in ancient times with the nesting of the halcyon, or kingfisher, on or about the winter solstice. In England, this period is also given other names depending on when it occurs: "St. Austin's Summer" if in September, "St. Luke's Summer" if in October, "Allhallow Summer" if in early November, and "St. Martin's

Summer" if in mid-November. Presumably these names derive from holidays or festivities observed in bygone times.

In Minnesota, Indian summer usually arrives after the first fall frost, but on rare occasions it may come after a period of cold, cloudy weather that does not produce a frost. Though more frequent in October, Indian summer spells have developed in November, when their unexpected appearance makes them all the more welcome. As an example of why these fine, sunny conditions are so cherished, in southern Minnesota average daytime temperatures in October reach 70°F or higher with a 25 percent frequency (one day in four), but by November the historical frequency for such temperatures is less than three percent (about one day in 40).

Of course, there is a price to pay for such fine weather. The fire danger usually increases substantially, as it did in the fall of 2003, and prairie or forest fires may be ignited more frequently by lightning strikes or careless campers. High temperatures are not the only factor, however: a significant change in the interaction between the atmosphere and the Minnesota landscape occurs in the fall. Agricultural crops are harvested, denuding the landscape of vegetation that emits water vapor into the lower atmosphere. At the same time, perennial vegetation and forests are going dormant, diminishing their transpiration stream as well. The absence of surface moisture leads to a drier atmosphere and can produce desert-like conditions. For example, in October 1999 some places in Minnesota exhibited conditions that matched those of the desert around Tucson, Arizona. The fire danger at this time was arguably greater in Minnesota because of the abundance of dried-out vegetation—in other words, fuel—when compared to the customarily dry Tucson region.

Another feature of Indian summer is the sometimes dramatic and much appreciated character of the day. Fresh, clear, and clean air and bright sunshine prompt many a noontime walk or outdoor coffee break to enjoy the gift from Mother Nature. Though perfectly clear, cloudless days are somewhat rare across the Midwest—an annual average of less than one day in ten—in Minnesota the peak occurrence of clear days falls between September 27 and October 3, when the mean frequency

Q & A

March 31, 2014, brought highly unusual weather to the residents of Yellow Medicine and Lac qui Parle counties in western Minnesota. What was the nature of this unusual weather?

a. a large hail storm

b. a heat wave

c. tornado and blizzard warnings on the same day

(answer on page 329)

TABLE 23: October 25, 1999, Climate Report, 2–7 PM CDT				
LOCATION	AIR TEMP (°F)	DEW POINT (°F)	RELATIVE HUMIDITY (%)	PEAK WIND GUST (MPH)
Twin Cities	65	11	12	9
Albert Lea	63	6	10	30
Faribault	64	0	7	29
Litchfield	62	0	8	31
Rosemount	63	4	9	28
Tucson, AZ	88	16	7	8

exceeds 20 percent, or two days in ten. These dates correspond closely with Indian summer and tend to show other characteristic features such as lower humidity, higher air temperature, and the dominance of high-pressure systems. This period also roughly coincides with the Twin Cities Marathon, which draws thousands of competitors during the first weekend of October each year. Perhaps as much or more than any other highly visible public event in Minnesota, the Twin Cities Marathon is usually blessed with nice weather. Those event planners must have wisely taken the state's climatology into account when they set up this race.

Fire Weather

Minnesota's native prairie and forested landscapes have a long history of fires: evidence of their passing exists in American Indian oral histories as well as in paleo-climatological records such as tree rings and charcoal sediments. Recent data from the Minnesota Department of Natural Resources (DNR) Forestry Division show that the number of reported wildfires, including grassland, brush, and forest fires, exceeds 1,600 annually; this estimate is likely conservative, since many fires go unreported to the DNR.

A 50-year data set covering the period 1927–76 and compiled by former state climatologist Earl Kuehnast showed that of nearly 62,000 fires reported in Minnesota the vast majority occurred in the spring. In fact, the months with the highest percentage were May with 32 percent; April, 28 percent; October, 11 percent; and August, 10 percent. All other months showed relatively few fires, with a negligible number in December, January, and February, when snow typically covers the landscape.

On May 11, 2007, very poor visibility was reported by a number of eastern Minnesota weather observers, due to

a. dense fog

b. an I-35 traffic jam

c. smoke from the Ham Lake Fire along the Gunflint Trail

(answer on page 329)

Climatic elements tend to relate closely to wildfire frequency. In the case of spring fires, lush fall vegetation followed by an open winter or early loss of continuous snow cover are contributing factors. When combined with a dry spring, above-normal temperatures, and persistent winds, the risk of wildfire is higher, especially in April and May, when daytime humidity levels can fall below 30 percent. Spring wildfires are mostly prairie grassland or brush fires. The likelihood of fall fires—which are equally tied to high temperatures, low humidity, and persistent winds—is greatly enhanced by summer drought. The worst fall fires are most often in forested areas, with their abundant dry fuel loads in the tree canopies, the undergrowth, and the litter and duff of the forest floor.

Duff

In the fire weather program of the U.S. Forest Service, duff *describes partially decayed organic matter on the forest floor that can become highly combustible during a drought, contributing to the longevity and spread of forest fires. In fact, the Keetch-Byram Drought Index is sometimes referred to as the soil/duff drought index because it measures how dry the soil and duff layers are.*

During the nineteenth century and particularly during certain times of year, daily weather observers at Fort Snelling and others in the present Twin Cities area remarked with some frequency about a smoky atmosphere or about seeing distant prairie fires light the nighttime skies. These fires were especially visible to the south and west, where the Big Woods dominated the Minnesota River Valley landscape prior to the establishment of agriculture. The Big Woods was often subject to drought, and fires were fed by savannah-type vegetation, including, among other species, big and little bluestem, Indian grass, compass plant, aster, bur and red oak, aspen, elm, sugar maple, and basswood. The Fort Snelling records summarized by Charles Fisk and Tom St. Martin are chock-full of observed fires, particularly from the 1830s to the 1870s and also most notably in the spring and fall seasons.

Sometimes the nighttime observer would note that the sky was bright in all directions from prairie fires, but more often the glow came from the south and west, toward the Big Woods. The cause behind these fires is speculative. Certainly some could have been started by soldiers, settlers, or American Indians. Undoubtedly some were ignited by lightning. There are several cases where drought likely turned the landscape vegetation into a tinderbox, as with the spring and fall fires of 1844 and 1853, years the Fort Snelling observers characterized as very dry. The only example of prairie fire in January came in 1849, following an extremely dry November and December. In October 1856, the Big Woods communities of Henderson and Le Sueur were affected by both prairie and forest fires, some of which proved quite destructive.

Relatively more forest fires were reported during the 1860s. On October 29, 1861, observers noted both prairie and forest fires burning in the Dayton's Bluff area above and to the northeast of St. Paul. In July 1863 the *St. Paul Pioneer* reported on nearby forest fires whose byproducts were such that residents of the city "ate smoke, drank smoke, and inhaled smoke." October 15, 1864, was also smoky thanks to fires in the Big

TABLE 24: **Fires Observed Near Fort Snelling by Month and Year**	
January	1849
February	(none)
March	1844, 1845, 1847, 1848, 1851, 1854, 1858
April	1846, 1847, 1848, 1850, 1851, 1853, 1855, 1856, 1857, 1861, 1872, 1874
May	1845, 1850, 1854, 1857, 1869
June	(none)
July	1833, 1863
August	1854
September	1854, 1871
October	1834, 1835, 1842, 1844, 1845, 1846, 1847, 1848, 1850, 1852, 1853, 1854, 1856, 1861, 1864, 1871, 1874
November	1835, 1844, 1845, 1846, 1847, 1850, 1853, 1854
December	1851

Woods, while timber fires to the southeast, in the Hastings area (Dakota County), could be seen at night during May 1869.

Widespread throughout the region during the terrible fall of 1871 were the anxiety, helplessness, and trauma that accompany fire. Across the Upper Midwest the summer had brought drought: Fort Snelling reported only 5.87 inches of rainfall during the months of July, August, and September, less than half the usual amount. The moisture deficiency was even worse in other parts of Minnesota and in Wisconsin and Illinois, the situation further aggravated by summer temperatures six to eight degrees above normal in July, August, and September. In the first few days of October a prairie fire started near Breckenridge in Wilkin County and spread rapidly east and south across the state, traveling an estimated 150 miles. By October 7 fires were reported in Cokato and Howard Lake in Wright County and Dassel in Meeker County. Raging fires swept through Lynd and Marshall in Lyon County, New Ulm in Brown County, and Windom in Cottonwood County. The observer at St. Paul noted, "smoke hangs like fog over the city . . . the air is full of cinders . . . and burnt spears of grass and twig fill everywhere."

A summary of death and damage in Minnesota was never completed, but the impacts in Wisconsin, Michigan, and Illinois were more traumatic and better documented. These October fires became known as the Peshtigo Fire, named for the east-central Wisconsin town that experienced the greatest loss of life. Over the period from Sunday, October 8, to Monday, October 9, fire raced across Wisconsin, destroying 1.25 million acres of timber and killing as many as 2,000 people. Western Michigan caught fire as well, the flames consuming up to 2.5 million acres of timber and killing 200 people. Coincidentally, the great Chicago fire reportedly started in Mrs. O'Leary's barn on October 8, destroying more than 17,000 buildings and taking 200 lives. These fires were enhanced by very dry air and driven by high winds, 28 to 34 miles per hour, the effect of a midlatitude cyclone.

The great fires of 1871 exposed the vulnerability of settlers on the midwestern prairies and in the northern forests, but it was just a first chapter. Another summer drought brought a terrible fire to Hinckley in

Q&A

In a daily forecast, what does the National Weather Service's Red Flag Warning mean?

a. conditions are dangerous for burning

b. there is a threat of flooding

c. weather will make driving more difficult

(answer on page 329)

Pine County on September 1, 1894. Many parts of Minnesota's central and eastern forested areas had reported fewer than 3 inches of rainfall during June, July, and August. In fact, in Minneapolis and St. Paul the total rainfall for June through August was just 1.73 inches, still the all-time driest such period. The St. Paul city engineer remarked on the abnormally low water levels: the city was "struggling along" with about two-thirds of the water needed to meet demands. It was also a hot summer, with many communities reporting 25 to 30 daytime highs of 90°F or greater. Similar to 1871, by the end of summer the landscape vegetation was tinderbox-dry. Saturday, September 1, brought another hot day as an inversion layer capped the smoke plumes emitted by small forest fires near Hinckley. A low-pressure system passing to the north brought southerly winds of 20 to 25 miles per hour, which helped whip the flames from earlier fires into one large firestorm that burned more than 400 square miles in fewer than four hours. More than 400 people perished, and several towns, including Hinckley, burned to the ground.

Once again in 1910, summer drought set up a dangerous fall fire season. During the driest overall year in Minnesota history, many north-central areas received fewer than 4 inches of rainfall during June, July, and August. Baudette in Lake of the Woods County tallied only 3.76 inches for the period, an all-time low that remains unchallenged.

September, too, remained drier than normal. Then on October 7 a strong low-pressure system tracked across southern Canada, producing southerly winds of 25 to 34 miles per hour. These winds whipped up smaller fires burning west of Baudette, and by evening a wall of fire sped toward town. People evacuated by train and by boat, but most of Baudette was burned, as was its cross-river neighbor, Spooner. Miraculously, no deaths were reported in Baudette, although more than 40 bodies were found around the nearby countryside.

The forest fires of 1910 incinerated all but a few trees and some lucky buildings near Baudette.

Fire returned to the North Woods in 1918. Though records for 1918 show a modest summer drought over most of Minnesota—less extensive than those of 1871, 1894, and 1910—the north-central and northeastern counties saw extreme summer drought. In Brainerd (Crow Wing County), Duluth (St. Louis County), Fort Ripley (Crow Wing County), Moose Lake (Carlton County), and Two Harbors (Lake County), rainfall during June-August was less than half of normal. Just 5.60 inches of rainfall at Cloquet in Carlton County made for the third-driest summer on record. The moisture deficiency continued throughout September and into the first half of October as many northeastern communities received only 1 to 2 inches of rain. The prolonged dryness caused water tables in some of northern Minnesota's peat land to decline, leaving on the surface a good deal of decayed and highly flammable vegetation. Many small fires started during the fall and smoldered for weeks. On October 12 a strong low-pressure system tracked across southern Canada, bringing with it southerly winds of 25 to 40 miles per hour. The smoldering fires were fanned into raging blazes, which engulfed nearby forests and spread rapidly east. Major forest fires were reported in parts of Beltrami, Carlton, Itasca, Koochiching, Lake, and St. Louis counties, encompassing an area of more than 2,000 square miles. Between 500 and 600 people lost their lives, mostly in the Cloquet area, and property losses were estimated at $50 million. An additional 3,000 people suffered bad burns, and 5,000 were left homeless. Fortunately, the second half of October proved to be quite wet, and rainfall extinguished all the fires that remained.

Devastating fire swept through Cloquet in 1918, leaving only ruins of the bustling lumber town.

Another notable fall fire season occurred in 1976, a year dominated by widespread summer drought. In some counties the rainfall deficit was worse than any of the 1930s: Ortonville in Big Stone County reported a total annual precipitation of 6.37 inches, a state record low, and a summer total of just 2.76 inches for June through August. Many areas of central Minnesota reported summer rainfall totals that were only 30 to 40 percent of normal. The summer was also hot: several communities saw numerous days with temperatures of 100°F or higher. During the first week of September, a deep low-pressure system passing across southern Canada brought strong southerly winds to Minnesota. About noon on September 7, winds were blowing 20 to 30 miles per hour across Cass, Hubbard, and Wadena counties and temperatures hovered in the nineties with relative humidity in the low twenties. Fires developed in the Huntersville State Forest and spread rapidly through the jack pine crowns, sending fireballs as high as 200 feet. More than 22,000 acres burned in these counties: the value of lost timber exceeded $700,000, and wildlife losses were heavy. Ten homes were destroyed or damaged but no people were killed. Around 9:30 PM rain began to fall, helping bring the fire under control. The dry pattern continued throughout September; many areas received less than 1 inch of rainfall for the month. In fact, on a statewide basis it was the sixth-driest September on record. Other traumatic though less damaging fires occurred near Blaine in Anoka County and Palisade in Aitkin County during October 1976. Minnesota's final tally for the year was more than 3,000.

What does the term *mush-raker* refer to?

a. the lead sled dog

b. dew drops on mushrooms

c. a person who repairs umbrellas

(answer on page 329)

Wildfire frequency in Minnesota has varied over time. In the long-term records, 1919 saw the fewest reported: 151. However, that figure is likely a drastic underestimate, for reporting at that time was not as widespread and comprehensive as it is today. In the most recent four decades of wildfire statistics, the fewest reports show up in the very wet year of 1979: 713. Conversely, there have been years when the number of wildfires exceeded 2,000. Short- or long-term severe to extreme droughts were generally the rule for those years: 1929, 1930, 1931, 1932, 1933, 1934, 1936, 1976, 1980, 1987, 1988, 1990, 2003, 2004, 2006, and 2012. The overall wettest year in Minnesota history, 2010, also shows a wildfire frequency of more than 2,000; however, the northeastern forested region of the state was in severe drought for part of that year.

Three recent large wildfires in Minnesota were the Cavity Lake Fire of July 2006, the Ham Lake Fire in May 2007, and the Pagami Creek Fire of August and September 2011. All of these fires affected portions of the Boundary Waters Canoe Area and at various times were amplified by weather conditions of low humidity and high wind speeds. The Cavity Lake Fire started with a lightning strike and burned more than 31,000 acres. At the time that region of Minnesota was in severe drought. The Ham Lake Fire was started by an escaped campfire and consumed more than 75,000 acres. Occurring early in May and prior to full canopy green-up, this fire was difficult to control and produced so much smoke that for a time Interstate 35 had to be closed to traffic. The northeastern portion of the state was in extreme drought at the time. The Pagami Creek Fire was started by a lightning strike in August, but it smoldered in a boggy landscape for a prolonged period before unusually low relative humidity in late August and early September, coupled with persistent high winds, caused the fire to spread throughout the understory growth and then jump through the forests as a crown fire, covering several miles per day. Difficult to control, it ended up burning more than 92,000 acres, mostly during the month of September, when that region of the state was in severe drought. Smoke from this fire reached parts of Ontario, Michigan, and Illinois.

Q & A

What new Minnesota weather record was set by the storm of October 26, 2010?

a. highest wind speed (150 mph)

b. lowest barometric pressure (28.21 inches)

c. most October snowfall (36 inches)

(answer on page 329)

. . . .

Haines Index

The Haines Index, or Lower Atmosphere Stability Index, indicates the potential for wildfire growth by measuring the air's stability and dryness. Used in fire management nationwide, the index is calculated by combining the lower atmosphere's stability and moisture content into a number that correlates with large fire growth. The stability term is determined by the temperature difference between two atmospheric layers; the moisture term by the temperature and dew point difference. The index has accurately described large fire growth when surface winds are not a dominant factor. The Haines Index is computed from morning radiosonde readings and can range between two and six: the drier and more unstable the lower atmosphere, the higher the index.

. . . .

Of the fires during the 1930s, probably the largest occurred at Red Lake (Beltrami County) in 1931, when 900,000 acres burned. In the years since those dramatic and tragic fires, technologies and methodologies have improved fire surveillance, fire prevention, fire mitigation, and fire fighting, all of which have helped prevent large-scale infernos like those of the past. In recent decades, too, foresters, climatologists, and meteorologists have worked together to develop better methods of forecasting wildfire risk. In the 1970s and early 1980s, Earl Kuehnast, Minnesota DNR state climatologist, and John Graff, meteorologist-in-charge at the National Weather Service Forecast Office in the Twin Cities, collaborated in the study of fire weather. After examining the state's fire history and analyzing the weather associated with major fires, they developed the concept of an "explosive fire day," defined by four conditions: drought or dry conditions across the area; maximum daytime temperatures from the low eighties to nineties; very low daytime relative humidity readings, typically less than 30 percent; and a relatively strong low-pressure system passing to the north and producing winds of 20 to 40 miles per hour. This work helped to advance the NWS's methods for routine fire weather forecasting. Today state and federal agencies with a mandate for fire prevention, management, and mitigation receive daily fire weather forecasts detailing expected sky conditions, temperatures, relative humidity, precipitation probabilities, and winds and atmospheric stability related to smoke dispersal. These valuable forecasts and the actions they inspire help reduce the risk of wildfires in Minnesota.

Q & A

What famous weather event happened on October 16, 1880?

a. flash flood in Rochester

b. earliest recorded blizzard in Minnesota

c. hottest October day in the Twin Cities

(answer on page 329)

Significant Days

Halloween

Following the end of summer, marked by Labor Day and the closure of the Minnesota State Fair, probably the most anticipated date on the fall calendar, at least for young people, is Halloween. Activities traditionally slated for October 31 include bonfires, hayrides, pumpkin-carving contests, costume balls, and of course trick-or-treating. With the loss of daylight saving time, the evening begins early as darkness sets in between 5:00 and 6:00 PM.

The weather is usually favorable, with temperatures in the forties and fifties and no rain or snow. In fact, the Twin Cities' climate record

Halloween traditions: costumes and a bonfire for warmth on a chilly fall night in Anoka, 1936

from 1872 to the present shows that measurable precipitation on Halloween has occurred in 37 years, or about 26 percent of the time. Snowfall has occurred in only 16 years, or about 11 percent of the time, and just six times was it a measurable amount. The most rain fell in 1979, when 0.78 inches was measured. Of course the most memorable Halloween snowfall was in 1991, when a three-day blizzard deposited 28.4 inches in the Twin Cities and 36.9 inches in Duluth. That weather system also produced an ice storm in southern counties, especially around Albert Lea and Austin, where some roads were closed for days. A storm unique in Minnesota history, this blizzard is the only one known to have taken place on Halloween for as far back as records are available.

The most interesting back-to-back Halloween weather conditions occurred in 1950 and 1951. Halloween 1950 was the warmest ever, with many communities reporting record highs in the eighties: 86°F at Worthington (Nobles County), 85°F at Farmington (Dakota County), Pipestone (Pipestone County), and Winona (Winona County), 83°F in the Twin Cities. In contrast, the following year's holiday was arguably the coldest ever, with snow on the ground from an October 30 storm and daytime temperatures ranging in the twenties. Brisk northerly winds pulled the wind chill into the teens, and some northern communities saw evening temperatures fall below zero: -2°F at Park Rapids in Hubbard County, for example. Those in costume can be grateful that these high and low temperatures are extremes, uncharacteristic of Halloween's routinely pleasant weather.

Election Day

The weather is often blamed for undesirable outcomes: "housing starts were down this quarter due to bad weather in the east"; "food prices will be higher because of poor growing season weather"; "the space shuttle landing was postponed due to foul weather"; "construction costs were higher because of weather delays." All of these assessments may well be accurate, but one that does not ring true, especially in Minnesota, is "voter turnout was low due to poor weather conditions." In *The Weather Factor,* historian David M. Ludlum suggests that weather may have played a significant role in past presidential elections, particularly in close contests where only a few thousand votes in key states would have reversed the outcome.

With the possible exception of parenting one's children, what is more important than exercising the right to vote? If weather conditions are dangerous—a blizzard, for example—voters may be justified in staying home, but "uncomfortable" or "inclement" weather is hardly an acceptable excuse. In truth, foul weather appears to have little effect on the turnout of steadfast Minnesotans, who traditionally cherish their voting privileges.

Since 1845 the Tuesday following the first Monday of November has served as Election Day, varying from November 2 to 8. In Minnesota, high temperatures are typically in the forties, with lows in the twenties and thirties. Frequently there is little or no precipitation. Exceptions exist, of course: the years 1901, 1910, 1911, 1933, 1936, 1951, 1959, and 1991 were all notable for being quite cold and/or snowy during election week. Of these, 1936 was the only year with a major national election. Heavy snows and glaze occurred in northern, western, and central counties during the first three days of November, accompanied by near-record cold temperatures. The lows on election morning—November 3—were in the single digits to the teens. The streets and sidewalks were icy from recent snows, and the wind-chill index was well below zero all day, registering in the dangerous category—colder than -25°F—in some places. Yet Minnesota's voter turnout was above 1.1 million, more than 70 percent of those eligible.

Election Day 1991, November 5, presented some weather challenges. Minnesota reported a relatively high voter turnout at the local and state election despite an all-day snow with blowing and drifting that made travel slow and cumbersome. As temperatures hovered in the teens to low twenties, many polling locations served weather-weary citizens hot

Q & A

On November 1, 2000, what unanticipated and rare weather event visited Minnesota?
a. freezing fog
b. a tornado
c. ball lightning
(answer on page 329)

tea and coffee as a reward for making the effort to vote. Conversely, Election Day 2008 was practically speaking a retro summer day, with daytime highs in the seventies across Minnesota (78°F at Redwood Falls in Redwood County and at Springfield in Brown County), and the state's voter turnout was once again the highest in the nation at 77.8 percent. The most recent national Election Day in Minnesota (November 6, 2012) brought unsettled weather with mixed rain and snow and generally seasonable temperatures. Minnesota led the nation in voter turnout again with 76.1 percent.

Thanksgiving

Early American communities set aside a weekday each autumn either to celebrate abundant and timely harvests—a Thanksgiving Day—or to pray for assistance during times of hardship and scarcity—a Fasting Day. By the middle of the seventeenth century, an annual Thanksgiving feast was pretty well established in many American colonies. The Continental Congress and early presidents declared periodic Thanksgivings, often in the month of December, but in 1863 President Abraham Lincoln selected the last Thursday of November as Thanksgiving Day. The timing of this national holiday stood until President Franklin Roosevelt signed a bill in 1941 to make Thanksgiving the fourth Thursday of November. This date, which varies from November 22 to November 28, adheres to the tradition of celebrating the agricultural harvest and hunting seasons' bounty; however, it also coincides with the highly volatile climate transition from fall to winter. For this reason, Thanksgiving is laden with more weather-related memories than any other holiday, especially in Minnesota.

By Thanksgiving week it is not uncommon to have some snow on the ground in the Twin Cities. In fact, the date of this holiday is often very close to the average date for the onset of snow cover across much of Minnesota. In the Twin Cities metropolitan area, snow cover of at least a trace occurs by Thanksgiving about 40 percent of all years. In 1921, 1983, and 1991, snow cover was 10 inches or more in many places. Statewide, 1880, 1896, and 1952 are the snowiest and coldest Thanksgivings on record. Thanksgiving 1880, on November 26, followed an early start

Q & A

What is a Turkey Tower?

a. a cagelike structure designed to house weather instruments

b. a shortlived tall, narrow cumulus cloud

c. a formation of perched wild turkeys

(answer on page 329)

to winter, which began with the worst October blizzard in Minnesota history, one recounted by Laura Ingalls Wilder in *The Long Winter*. All of November was cold and snowy, with temperatures as low as -18 to -22°F in the Twin Cities area. A severe blizzard on November 25–26, 1896, brought 10 to 15 inches of snow to western and northern communities and piled drifts to a height of 10 to 12 feet. The Ada weather observer, Fred Andrist, stationed in Norman County, noted that wagon and rail traffic was at a standstill for 60 hours. The storm brought such a strong cold front that temperatures plunged from the forties to below zero—with wind-chill values of -10 to -20°F—by Thanksgiving Day, November 27. Minneapolis observer J. H. Aschenbeck reported that the temperature never reached higher than 7°F. Similarly, in 1952 a strong winter storm preceded Thanksgiving Day, November 28, bringing several inches of snow with blizzard-like conditions. Wind-chill values again ranged from -10 to -20°F, and many people could not travel to see relatives for the holiday.

The weather was mild and the turkeys were wild at this Thanksgiving Day race, 1955.

The Wednesday and Sunday on either side of the holiday have evolved into two of the year's heaviest travel days. Climate records for the Twin Cities since 1891 show that 35 percent of the Wednesdays preceding Thanksgiving have seen some form of measurable precipitation, suggesting a better than 1 in 3 chance of snow or rain. Twenty-seven percent of these Wednesdays show measurable snowfall, the greatest being 11.4 inches in 1983. On the other hand, 26 percent of the Sundays following Thanksgiving have seen measurable precipitation. Twenty-one percent had measurable snowfall, the greatest being 8.4 inches in 1985. In 1939 not snow but dense fog caused numerous traffic accidents. Despite this example, statistics suggest a higher probability of having reasonable travel weather on the Sunday after Thanksgiving than on the Wednesday before.

Some rather extreme conditions have been recorded on Thanksgiving Day. Consider the Twin Cities climate record: in 1930, the afternoon high was only 7°F, with wind-chill factors between -30 and -33°F, while in 1914 and 1922 the temperature reached 62°F under sunny skies. The record snowfall in the Twin Cities area was 5.0 inches in 1970. Thanksgiving 1988 was remarkably pleasant: the mercury hit 52°F and families picnicked in city parks. A decade later (1998), golf courses were open on a sunny Thanksgiving Day and indeed for the entire four-day weekend, as daily afternoon highs ranged from 56 to 64°F in the Twin Cities area. Similarly, in 2012 temperatures reached the sixties (71°F at Winnebago, Faribault County), and many families celebrated the holiday with outdoor activities. The most recent cold Thanksgiving was in 2014, when most locations reported subzero morning lows and afternoon highs in the single digits, with wind chill values in the negative twenties to negative thirties.

The Friday and Saturday after the holiday have become two of the year's busiest shopping days. Mother Nature has generally been kind to shoppers: records show that there is a less than 1 in 4 chance of snowfall on Friday and a less than 1 in 5 chance on Saturday. Friday, November 29, 1991, was not a pleasant shopping day, however: 12 to 14 inches of snow blanketed the Twin Cities metropolitan area. Neither was Friday, November 29, 1929, which saw a high of only 4°F and daytime wind chills of -35 to -40°F. Blowing snow drastically reduced visibility, and cars had to negotiate icy streets. Then again, 1929 was long before shopping became such a popular post-Thanksgiving activity.

Taken as a whole, the holiday weekend in the Twin Cities area sees snowfall on at least one day 57 percent of the time and snow cover over 50 percent of the time. Interestingly, from 1944 to 1958 there was measurable snowfall every Thanksgiving weekend, while from 1963 to 1974 no significant snowfalls were recorded on the days surrounding the great turkey feast. Further, climate statistics show that when the Thanksgiving period comes white with snow, there is more than a 75 percent probability that Christmas will be white as well.

Telling Minnesota's Weather Story

Hundreds of people have contributed in countless ways to establishing and documenting Minnesota's climate history, improving weather forecasting, and disseminating essential information to the public. Some of these individuals dedicated much of their professional lives to expanding our understanding of weather and climate in the North Star State, for which they deserve special mention.

■ The journals of Alexander Henry the Younger of the North West Fur Company (d. 1814) provide the first documentation of climate in northwestern Minnesota's Red River Valley, along the North Dakota border near Pembina. Henry was one of the few literate men of his place and time, and his written records provide a rich source of daily measurements and observations of both nature and the weather in a landscape relatively untouched by humans. Though fragile documents after more than two centuries, his records and comments are available through the DNR–State Climatology Office.

■ Reverend A. B. Paterson, rector of St. Paul's Episcopal Church, was a Smithsonian observer at Ninth Street and Wacouta in St. Paul from January 1, 1859, until December 31, 1872. The Smithsonian Institution sent weather instruments to Paterson, and he provided monthly reports, recording temperatures—three times each day—sky conditions, precipitation, and snowfall. In addition, Paterson kept a journal of daily observations until March 17, 1876, two days before his death. His observations help establish continuity in the St. Paul climate record, for daily recordings at Fort Snelling ended when it was decommissioned in 1858 and did not resume until April 1, 1867, leaving a rather large gap. St. Paul's U.S. Army Signal Corps weather station, established in October 1870, made initial observations from a building at Third and Wabasha Streets and overlapped with Paterson's records until his death. Observations by the signal corps and, later, St. Paul's Weather Bureau Office continued until July 1933.

A committed and conscientious observer, Paterson kept meticulous records. When he traveled for business or pleasure, usually in late summer or early fall, family members filled in for him. For the Twin Cities area, Paterson's data provide the only reliable and consistent record of precipitation and snowfall from 1859 to 1872, making them quite important for hydrological application. He was the first to use a land-based Beaufort-type scale the Smithsonian had modified so that mean velocity of air could be estimated by the movement of trees or flags or other objects.

Paterson observed some unusual and extreme weather during his career. Among his records: St. Paul's coldest day: January 1, 1864, a high of -24°F and a low of -38°F, yielding a daily mean of -31°F; one of the coldest February minimum temperatures: -36°F on Valentine's Day 1862; coldest February: average daily temperature -2.6°F in 1875; coldest April reading: 3°F on April 8, 1865; wettest June: 11.67 inches in 1874, measurable rainfall on 17 days; coldest August temperature: a morning low of 33°F on August 29, 1863; and St. Paul's snowiest October: 14 inches in 1873, several inches more than the second-snowiest Twin Cities October—8.2 inches from the Halloween Blizzard of 1991. Additionally, from 9:00 PM on December 30, 1863, to 7:00 AM on January 9, 1864, the air temperature was continuously below zero degrees, the longest such period in St. Paul, 226 hours according to weather historian David M. Ludlum in *Early American Winters.*

■ William Cheney, Minneapolis city coroner and son of prominent nineteenth-century businessman William H. Cheney of Rochester, was a Smithsonian observer in St. Anthony Village beginning in December 1864. When the Smithsonian network ended in the early 1870s, Cheney continued as observer for the U.S. Army Signal Corps; then, in the 1890s, he signed on with the Weather Bureau, making observations until June 1901. Like Paterson's, Cheney's careful and consistent records filled gaps left by the absence of data from Fort Snelling and other locations. After 1867, his monthly climate summary—including personal interpretations of the weather's effects—was often published in the *Minneapolis Tribune* and had quite a following.

What is a pyranometer?

a. an instrument used to measure ozone

b. an instrument used to measure solar radiation

c. an instrument used to determine visibility

(answer on page 329)

Among the notable climate extremes Cheney recorded was a high temperature of 103°F on July 16, 1868, which stood as a Twin Cities record until July 1931. Also of note was the wet year of 1869, when 40.38 inches of precipitation fell, more than the modern-era record of 40.15 inches in 1911. Excessive rains in July (11.64 inches) and August (11.46 inches) were the biggest contributors, helping ensure the state's largest corn crop of the decade. Cheney remarked, "the weather seems to delight in extremes this year." In 1872 he recorded the coldest December in Twin Cities history, logging a monthly mean temperature of only 2.2°F, with 13 consecutive days of mean temperatures at or below zero. For comparison, the coldest December of the twentieth century occurred in 1983—its mean monthly value 3.7°F. Cheney's comment: "sleighing was excellent."

Cheney's record also quantified the mildest winter in Minnesota history: 1877–78, when the mean temperature for December through March was a whopping 31.1°F. St. Paul weather historian Charles Fisk called it "the year without a winter." During the warmest December in history, the terrain was especially muddy because the soil did not freeze. Eleven days reached the forties, and many nights never cooled off below freezing. The mild winter produced some unusual observations, including honeybee activity, small grain planting, and green grass in February, and the ice went out on Lake Minnetonka on March 11, the earliest date ever. Later climate researchers noted that the winter of 1877–78 coincided with a strong El Niño in the Pacific Ocean, a climate feature that brings mild weather to the western Great Lakes region.

■ Professor William Payne of Carleton College established Minnesota's first astronomical observatory after joining the faculty in Northfield (Dakota and Rice counties) in 1871. Revamped in 1890, the facility today is known as the Goodsell Observatory. With its highly accurate instruments for tracking the motions of stars and planets, Payne's observatory became the region's official time service, subscribed to by local railroads. The observatory took on the responsibility of monitoring and measuring weather for the U.S. Army Signal Corps in 1881, and Payne became the first director of the Minnesota Weather Service in 1883, assuming the duties of collecting, analyzing, and reporting on data from 35 climate stations across the state. He wrote monthly reports for the signal corps and provided "indications," or forecasts, and cold-wave warnings to the public.

One of Payne's subordinates, assistant state weather service director Private David R. McGinnis, in 1884 secretly negotiated with the University of Minnesota to take control of the Minnesota Weather Service and move it to the Twin Cities, apparently in exchange for a professorship at the university. A conference between the Regents of the University of Minnesota and the Trustees of Carleton College resulted in no change to the state weather service's organizational structure, which remained under Payne's control.

A strong-minded individual, Payne was accustomed to having things his own way. During the 1880s he bickered with the officer in charge of the U.S. Signal Corps in St. Paul, Thomas Woodruff, whom he considered his subordinate. In one dispute, Payne negotiated funding for the Minnesota Weather Service through the St. Paul Chamber of Commerce, an idea Woodruff deplored, considering the weather service to be a strictly government-supported entity. Each man wanted to operate the weather observation network in his own way, and behind the scenes they attempted to discredit each other with the signal corps administration and with the public. The lack of a timely cold-wave warning for the famous January 12, 1888, "Children's Blizzard" created quite a stir mostly because more than 200 lives were lost. Both Payne and Woodruff cast blame on each other. After a signal corps investigation, Woodruff was transferred from the St. Paul office, leaving the Minnesota Weather Service in Payne's control until 1889, when lack of funding forced its closure.

One of Payne's legacies is the monthly weather report, published in the government's *Monthly Weather Review.* Payne composed a report that not only considered the statistical description of the month's weather but also included interpretation of its impacts, with particular attention to agriculture and transportation, a format that prevails in a variety of government publications today.

Charting the weather the old-fashioned way at the Minneapolis Weather Bureau, 1940

▦ Edward A. Beals served as section director of the Minneapolis office of the U.S. Weather Bureau, later to be called the National Weather Service (NWS), from 1893 to 1896, extending a weather career begun when he enlisted in the signal corps in 1880. He was also director of the Minnesota Weather Service, which managed the state's volunteer observation network, and editor of the *Minnesota Weather and Crop Review,* later the *Northwest Weather and Crops Report,* a widely read monthly publication. Knowledgeable in both climate and agriculture, Beals wrote extensively on these topics and published articles by other experts—on soil tillage, soil fertility, irrigation, plant pests, and livestock management. He occasionally published poems in the *Report,* which was supported by advertisers that included many leading Twin Cities–area businesses. Well known and well regarded in the agricultural community, Beals promoted attendance at the Minnesota State Fair, where farmers could share new information with each other. He took a promotion within the NWS to be director at the Portland, Oregon, office and left the Minneapolis office in 1896. He served for 46 years, retiring in 1926.

▦ Thomas S. Outram began his weather career with the signal corps in 1879. He was section director of the Minneapolis NWS office from 1896 to 1906. Like Beals, he wrote a monthly report for the Climate and Crop Service in Minnesota, though he dropped the advertisers in lieu of government funding. Outram continued to publish educational materials on new climate research and farming practices, such as tillage strategies like subsoiling. He also provided a section called "progress of the crop season," describing the relative health of the crop and how well it was advancing toward maturation and harvest. He began writing annual climate summaries, which included narratives on major weather events such as the Red River Valley spring floods of 1897, the Morristown (Rice County) tornado of September 24, 1900, and the Twin Cities tornadoes of August 20, 1904. Mr. Outram died at his post of duty in Minneapolis on December 5, 1906, and many mourned his passing.

▦ Ulysses G. Pursell took over for Outram in late 1906, continuing a career started with the signal corps in 1885 and serving as Minneapolis NWS office section director until 1934. As local climate records continued to lengthen, Pursell published a station-by-station climatology for each community that had kept a daily record over several decades. These reports illustrated great variability in climate across the state, and under Pursell's administration the number of climate stations expanded. The NWS's monthly publication began to focus entirely on weather and climate and less on agriculture. Like Outram, Pursell reported on major weather events and episodes, describing the health implications of the July 1917 heat wave, the deadly results of the June 1919 Fergus Falls (Otter Tail County) tornado, and the damages inflicted by the long-lived F4 tornado that crossed the Twin Cities area on April 5, 1929. In contrast to earlier documentation, Pursell made a concerted effort to include damage estimates in his severe weather reports.

What is bloxam?

a. the residue left in a washing machine

b. a position on the rugby field

c. a smoothing technique used by climatologists to calculate average values

(answer on page 329)

▦ Martin Hovde served as NWS section chief of the Minneapolis office from 1934 to 1954—one of the longest tenures of the twentieth century— after previously holding a post at the bureau office in Huron, South Dakota. His monthly reports on Minnesota climate were published by the U.S. Department of Agriculture, and his observations of one of the most variable and traumatic decades of the century—the 1930s—have influenced historical perceptions of climate during this period. Hovde wrote about dust storms, noting that the dust-laden air resulted in zero visibility and penetrated windows and doorways to deposit dust inside homes and offices. About one event Hovde remarked, "spectacular becomes a weak adjective when applied to this storm. It was awful, terrifying—the howling winds and a darkness that turned the midday into night, will never be forgotten." He also reported on winter 1936, one of the coldest in the state's history, during which six communities reported lows of -50°F or colder. In extreme northwestern Minnesota, low temperatures below zero were reported for 50 consecutive days. Perhaps looking for a silver lining, Hovde remarked, "a good crop of ice was harvested" from area lakes and rivers, filling numerous iceboxes and meeting refrigeration needs for months. Hovde also reported on the famous July heat wave of 1936, when many communities set all-time high temperature records, including Moorhead in Clay County, 114°F on July 6. To this heat wave he attributed 759 deaths, as well as significant livestock losses.

Like his predecessors, Hovde wrote reports on severe weather, but he provided more details on the storms' meteorology and the damages found in aftermath surveys. He investigated the F4 tornado that tore through Anoka, Champlin, and Maple Grove in Anoka and Hennepin counties on June 18, 1939, killing nine and injuring 222 people. Hovde plotted the tornado's 25-mile path and collected eyewitness accounts to publish in his report. After striking Champlin, the tornado crossed the Mississippi River: "many witnesses saw the river dry up as the waters were lifted and wind velocities were so great as to prevent the flow and return of water until the funnel cloud had reached the opposite bank." Hovde also analyzed the meteorological aspects of the F4 tornadoes that passed over Blue Earth and Faribault counties on August 17, 1946, killing 11 and injuring 130 people. His detailed analysis of tornadoes and other storms, particularly the meteorological aspects that produced them, provided a model for future NWS employees. Hovde retired from the NWS in 1954 after a long and eventful career.

▨ During World War II many NWS employees were called to military duty, leaving forecast offices with vacancies to fill. The Minneapolis and Wold-Chamberlain Field offices hired a number of women as observers and to disseminate forecasts. Because weather reports were considered especially valuable information during the war, security was unusually tight: after maps, charts, and observations were used for briefings and public bulletins, they were kept under lock and key. Many women learned meteorology on the job; virtually none had taken courses in atmospheric science, particularly at the college level. At war's end many former NWS employees returned to their jobs, and most of the women left the weather offices. One exception was Lucille Sjostrom, who trained to become an observer-briefer meteorologist. She later married another meteorologist, Bob Nicholson, and retired in 1955 to raise a family. She was the first woman the Minnesota media recognized for meteorological expertise.

Radiosonde

Twice daily, at 12-hour intervals, weather services around the world probe the atmosphere with balloon-launched instruments called radiosondes. Their measurements of temperature, pressure, humidity, and wind throughout the atmosphere provide data to a number of forecast models, which require numerical input from certain constant pressure levels in the atmosphere, called mandatory levels (e.g., 1,000 mb, 850 mb, 700 mb, 500 mb, and others). Radiosondes are generally designed to sample the atmosphere up to an elevation of 19 miles—which includes nearly 99 percent of the Earth's atmosphere—at which point the balloon may burst and the instrument package will parachute back to Earth. The radiosonde balloons ascend at a nearly constant rate of 300 meters/minute, and numerous precautions are taken to ensure high-quality data. For example, if the balloon does not ascend to at least the 400 mb level, approximately 4.5 miles, then a second attempt is made. And even if the balloon reaches the 19-mile elevation, if for any reason it fails to transmit data for an interval of ten or more minutes, a second balloon is launched.

Weather balloons (radiosondes) help observers measure winds aloft. This launch took place at the University of Minnesota's Memorial Stadium in 1936.

■ Joseph H. Strub, Jr., began his career in the Air Force Weather Service during World War II and took meteorology training at the University of Chicago in the early 1940s. Strub served as section director of the Minneapolis NWS office from 1955 until its closure in 1965 and was meteorologist-in-charge of the forecast office at Minneapolis–St. Paul airport from 1968 to 1976. His illustrious career was cut short by cancer, but not before he had attained a reputation as one of the NWS's best flood forecasters, earning the prestigious

Department of Commerce Gold Medal in 1969. Strub accurately forecast the great Mississippi River flood in 1965; based on this experience and the study of other historical floods in Minnesota, he derived a method for estimating spring snowmelt flooding based on five key factors.

Strub continued to make storm surveys—of tornadoes, straight-line winds, blizzards, and hailstorms—and to write reports detailing their meteorological aspects and the resulting damage. He accurately forecast the so-called "Storm of the Century" on January 10–12, 1975, and later wrote about it. Much documentation of the famous Twin Cities tornado outbreak of May 6, 1965, rests on the survey reports of Strub and his staff.

During the 1950s and 1960s, while Strub headed the downtown Minneapolis office, Phil Kenworthy took charge of the NWS Forecast Office at Wold-Chamberlain Field (later Minneapolis–St. Paul International Airport). Both deserve credit for developing a strong working relationship with Twin Cities media and local civil defense personnel, which cooperatively provided weather warnings and storm reports to the public. As a working partner with the NWS, WCCO Radio became known as the "good neighbor of the north," the source for the most up-to-date information on threatening weather and school and business closings. Strub and Kenworthy nurtured this relationship, realizing that the public was better served through such cooperation.

In 1976, following the death of Joe Strub, John Graff took over as meteorologist-in-charge of the National Weather Service Forecast Office (WSFO) at Minneapolis–St. Paul International Airport, a post he held until 1986. Following closure of the downtown Minneapolis Weather Service Office in 1965, the airport WSFO became the primary forecast center for most of the state, coordinating with offices in Duluth (St. Louis County), International Falls (Koochiching County), Rochester (Olmsted County), and St. Cloud (Sherburne County). Graff was instrumental in encouraging development of a fire weather program to help state and federal foresters cope with wildfire threats. He also coordinated severe storm surveys with the Minnesota State Climatology Office and wrote detailed reports on all major storms in the state. Under Graff the system of weather warning alerts communicated through civil defense sirens became decentralized, as county-level emergency managers gained responsibility for tripping warning sirens. NOAA weather radio broadcasts of current weather conditions and forecasts also expanded, bringing coverage to more areas of the state. Graff began to establish working relationships with local television meteorologists, more of whom were becoming certified broadcast meteorologists rather than weather presenters. Graff also developed good relationships with the state emergency managers and nurtured development of several storm-spotter networks.

Hovde, Kenworthy, Strub, and Graff supported a novel idea to disseminate the daily weather forecast in the Twin Cities. In October 1949 a tower erected by Northwestern Bank of Minneapolis at 600 Marquette Avenue was fitted with a large ball on its roof—the "Weatherball." The ball was illuminated by neon tubes, and its color and whether or not it was blinking conveyed the weather forecast to the public. Forecast information was initially provided to bank employees who controlled the Weatherball, but later the airport WSFO forecasters themselves manipulated it. The ball's code was summarized in a popular jingle:

> *When the Weatherball is red, warmer weather is ahead,*
> *When the Weatherball is green, no change in weather is foreseen,*
> *When the Weatherball is white, colder weather is in sight,*
> *If colors blink by night or day, precipitation's on the way.*

The Weatherball system remained in place until November 1982, when it was disassembled and stored in buildings on the Minnesota State Fairgrounds.

After receiving training as an air force meteorologist, Earl L. Kuehnast served as Minnesota state climatologist from 1968 to 1986, his expertise in meteorology and climatology complemented by his vast knowledge of agriculture from years spent on a farm in northern Iowa. He developed a number of educational programs and coordinated a variety of forecast and outlook products with the NWS, including energy outlooks for the winter heating season (with John Graff of the WSFO), fire weather forecasts, drought assessments and contingency plans, and even mosquito outlooks. Kuehnast was an employee of the

Minnesota Department of Natural Resources, but he was located with the climate group at the University of Minnesota–St. Paul. Within the Department of Soil, Water, and Climate he used his skills and expertise to better serve agriculture, particularly through studies of soil moisture cycles, prediction of crop pest migrations, evaluation of weather for plant diseases, and analysis of energy requirements for crop drying and storage. Many of his published studies on snow, precipitation, and flash floods are still utilized today. He was so highly respected that, following his death in 1990, the university set up the Kuehnast Endowment Fund to support public engagement in the atmospheric and climate sciences, including an annual lectureship.

The beloved Weatherball topped the Northwestern Bank Building in Minneapolis for decades.

◼ When Earl Kuehnast retired as state climatologist in 1986, his assistant Jim Zandlo assumed the role. Zandlo was adept in the application of technology and modern database management. He emphasized the need to digitize and store, in a geo-referenced retrieval format, all of the state's printed climatic records dating back to the nineteenth century. Under his leadership, this task was accomplished throughout the 1980s and 1990s. He also focused on developing tools for analyzing and displaying the data in usable formats through the Internet and web-based tools; and he developed methods to eliminate some "observation bias" in the records due to landscape change. Zandlo expanded the networks of volunteer weather observers in the state, increasing the number to more than 1,500. He especially enhanced the snow observational network along the north shore of Lake Superior. Under his leadership, the Minnesota State Climatology Office expanded, hiring two more climatologists to assist in data management, web-based tools, public inquiries, and interactions with other state agencies. Finally, the Minnesota Climatology Working Group took form under Zandlo's leadership. This group, which functioned in the roles of research, teaching, and public outreach, included Zandlo's assistants Greg Spoden and Pete Boulay as well as Dave Ruschy, Don Baker, Dick Skaggs, and Mark Seeley. Jim Zandlo retired as state climatologist in 2011, and since that time Greg Spoden has served in the role, with assistant Pete Boulay handling much of the observer network management and a high volume of public inquiries. They have built a national reputation as one of the best-run state climate offices in the country and provide the public with nearly complete access to the state database through the DNR–State Climatology website.

▨ In the late 1970s, Kathy Ericksen, a secretary for John Graff, trained to become a meteorological technician and observer. During the 1980s and 1990s, she recorded most of NOAA weather radio's broadcasts, her voice becoming widely known and trusted throughout the state, especially as the use of portable NOAA weather radios became more popular with campers, fishermen, and hunters. Many Minnesotans were disappointed when technology took over the weather radio airwaves in the new millennium, broadcasting announcements, warnings, and forecasts using synthesized computer voices from an automated system. Kathy Ericksen retired in 2011, having recorded more NOAA weather radio messages than any other person in the region.

What is *the going forecast?*

a. the most recent forecast

b. a travel forecast

c. a forecast for the horse racing track

(answer on page 329)

▨ James Campbell became meteorologist-in-charge of the Twin Cities WSFO in 1986 and served in this role until October 1991. During his tenure many changes in the NWS were initiated, including a $4 billion system-wide modernization. This modernization included deployment of new Automated Surface Observation Systems (ASOS), Wind Profilers (vertically pointed Doppler systems to detect winds aloft), Doppler radar systems (WSR-88D), and weather satellites. In order to accommodate the new radar technology, the WSFO would need to be relocated, and Campbell negotiated a move to the Twin Cities suburb of Chanhassen. The year 1990 marked the NWS Cooperative Weather Observer Program's centennial, and Campbell helped organize a celebration at Fort Snelling. At the event NWS director Dr. Joe Friday recognized Minnesota communities that had kept more than a hundred years of continuous daily weather observations. Citizens from 15 Minnesota communities proudly accepted the awards and were interviewed by many local media.

▪ Dr. Donald Baker, professor of climatology at the University of Minnesota from 1958 to 1994, founded the climatology academic program and was instrumental in bringing the Minnesota State Climatology Office to the St. Paul campus. In October 1960 Baker established the St. Paul Climate Observatory. This facility maintains the most complete set of climate observation of any location in the state, including solar radiation, wind speed and direction, soil temperature, air temperature and humidity, pan evaporation, frost depth, snow, and precipitation. These data have appeared in hundreds of publications, and they have been analyzed by many researchers. Dave Ruschy of the Department of Soil, Water, and Climate was the chief architect for the development of the observatory database and the interface that allows display of near-real-time weather data at the DNR–State Climatology website. Ruschy and Baker collaborated to write many climate publications based on the

observatory data. For many years Don Baker also teamed with Dr. Richard Skaggs of the Geography Department to study some of the earliest climate records in the state. They published work on a variety of climatic trends. Baker and Skaggs also conducted a very popular introductory meteorology class at the university; they taught weather to literally thousands of students over parts of three decades. Many of their students went on to productive science and academic careers of their own. Baker and state climatologist Earl Kuehnast collaborated on many studies of Minnesota's climate, some of which are still used today. Baker's early research on Minnesota's wind climatology supported the Wind Resource Assessment Pro-

Don Baker giving Governor Arne Carlson a lesson on making weather measurements on Earth Day, 1995.

gram, which led to deployment of wind turbine generators around the state. Don Baker passed away in 2014, but much of his work lives on through the graduate program in Land and Atmospheric Science at the University of Minnesota.

▪ Craig Edwards became meteorologist-in-charge at the Twin Cities WSFO in 1991, arriving on the scene just in time for the famous Halloween

Blizzard. He oversaw the station's move from the Minneapolis–St. Paul International Airport to new offices in Chanhassen during 1995. Many changes occurred during his tenure: NWS offices were restructured and relocated and new procedures for coordinating forecasts instituted. In addition, NOAA weather radio broadcasts were automated and expanded: computerized and synthesized voices broadcast messages as they are typed on the console by meteorologists, saving time in issuing weather warnings; nearly all parts of the state now have access to continuous weather broadcasts. Like his predecessors, Edwards coordinated with broadcast meteorologists in the state and was very involved in storm survey assessment and reporting. He also emphasized weather education by being regularly engaged in public radio and public television events and in community meetings. Edwards retired from the NWS in 2006 but has remained quite active.

He is the Minnesota Twins baseball game weather forecaster for all home games at Target Field, provides occasional on-air forecasts and commentary for Minnesota Public Radio, and does a number of public lectures each year. Following Edwards's retirement, Dan Luna took over as meteorologist-in-charge at the National Weather Service Forecast Office in Chanhassen. He is widely known throughout the Twin Cities area.

In 1983 Gary McDevitt joined the Twin Cities NWS-WSFO as a service hydrologist, responsible for coordinating all hydrological—that is, water—products released by the office. He worked closely with Dean Braatz, director of the North Central River Forecast Center, which is co-located with the WSFO. As all flood outlook and forecast products underwent transition, a new public education program was needed to ensure that city managers and other government officials who respond to flood threats would use the information properly. During the 1980s and 1990s McDevitt and Braatz contributed significantly to efforts that resulted in a number of flood contingency plans and educational programs, working with partners in the U.S. Geological Survey, U.S. Army Corps of Engineers, Department of Emergency Management, and other agencies. Following the floods of 1989, 1993, and 1997, public meetings

Q & A

When the Scottish weather forecaster predicts "shoosh conditions," what should you expect?

a. snow squalls

b. absence of wind/calm conditions

c. pleasant, sunny weather

(answer on page 329)

were held to evaluate the flood forecasts and determine how they might be improved. As a result, a new, more numerically intensive forecasting system, the Advanced Hydrologic Prediction System (AHPS), was deployed. Following McDevitt's retirement in 2003, Steve Buan served as service hydrologist and continued many of the public engagement programs on flood awareness and education.

. . . .

Long-range outlooks

Weather statements, maps, and graphics that look beyond seven days are described as long-range outlooks. They provide information about expected departures from normal temperatures and precipitation. Outlooks to 14 days are released once each day; those that look ahead one to 12 months are released once per month.

Long-range forecasting

Weather statements, maps, and graphics that look beyond 48 hours are called long-range forecasts. Predictive models assimilate radiosonde (instrumented balloon) data every 12 hours and depict conditions for hourly, 3-, 6-, and 12-hourly time steps for from three to seven days ahead. The National Weather Service uses the Global Forecast System (GFS), the European Centre for Medium-Range Weather Forecasts (ECMWF), and Canada's Global Environmental Multi-scale Model (GEM) to create its models.

Short-range forecasting

Weather statements, maps, and graphics that look ahead 12 to 48 hours are called short-range forecasts. All of the models, including GFS, ECMWF, and GEM, depict detailed hour-by-hour forecasts for grid points on a map.

Nowcasting

Weather statements, maps, and graphics produced throughout the day can be described as nowcasts. They often refer to the next one to 12 hours and are highly dependent on the changing nature of weather conditions. When conditions vary outside the range of what was forecast, nowcasting uses information provided by radar, satellite, automated surface observing systems (ASOS), or even individual observers.

. . . .

▨ Bruce Watson was one of the state's leading private meteorologists from the 1960s until his untimely death in 2004. He published the *WCCO*

Radio Weather Almanac in 1975 and created the *Minnesota Weather-Guide Calendar* in 1977. He collaborated with Donald Baker, Richard Skaggs, and Earl Kuehnast in publishing the eastern Minnesota climate record back to 1820, a compilation of Fort Snelling and other cooperative observer data and NWS records; the result is one of the longest published climate records in America. Well known for his public speaking, published articles, and radio and television appearances, Watson wrote extensively about Minnesota's weather history and served as a consultant to many organizations, including the Science Museum of Minnesota. He and his son, Frank Watson, made daily climate observations in Roseville in Ramsey County; the *Watson Weather Watch Calendar* is still published annually by his son, Frank, who serves as a consulting meteorologist in the region.

As a graduate student in the 1980s, Charles Fisk began to reconstruct Fort Snelling's entire climatic record, which spans the years from 1820 to 1872. Because the old temperature measurements, taken at fixed times during the day, did not conform to modern-day observations, Fisk used statistical methods to transform the older readings into equivalent daily maximum and minimum temperature values. He subsequently developed a time series of precipitation measurements as well. He has published the complete set of climate measurements, along with narratives about significant weather events in the Twin Cities area from 1820 to the present. One of the longest climate records in the Midwest, Fisk's work has been used extensively to study variations and extremes.

Twin Cities weather historian Tom St. Martin has reconstructed climate data and observations from Fort Snelling records, from pioneer-era Smithsonian observations taken in St. Paul and Minneapolis, and from early signal corps records that predate the NWS office, established in 1891. He has also studied weather descriptions in pioneer newspaper accounts (most notably the *Pioneer Press*) and in other Minnesota Historical Society documents. As a result of this work, St. Martin has published more than 20 volumes of data and narratives, all of which are cataloged in the Kuehnast Library at the University of Minnesota. He has proven particularly adept at identifying faulty records and errors in historical climate data and at unearthing interesting stories about extreme weather in the state. In October 2014 St. Martin was given a State Appreciation Award by NOAA-NWS, the MN-DNR State Climatology Office, and the University of Minnesota for decades of service in preserving state climate records.

■ Much can be said about the work of Minnesota's broadcast meteorologists. Beginning in the 1970s, television evening news programs began to shift from employing weather presenters to hiring professional meteorologists. Both television and radio stations developed their own weather departments with qualified and experienced staff. These meteorologists have served the public well, presenting new technology and information about the weather and collaborating with the NWS during weather threats. In many respects they have nurtured a renewed interest in the weather among the state's citizens, many of whom volunteer as weather observers or severe-weather spotters. The broadcasters of the past four decades—too numerous to mention here—are most deserving of gratitude and respect for developing weather literacy among Minnesotans as well as for care and respect for the state's climatic records.

■ Lastly, it is worth mentioning that community-wide interest in weather and climate is a cultural trait conducive to the use of social media. In this regard, our state's cultural landscape now includes many online blogs dedicated to weather and climate matters. Some present real-time data, forecasts, and analysis, while others include research findings, public policy, educational issues, or interviews with prominent scientists or government officials. Some personalities with an avid following include Paul Douglas of the *Star Tribune,* Paul Huttner of Minnesota Public Radio (*Updraft* blog), Dr. Kenny Blumenfeld (*Weather and BS*), and severe weather expert and former student of Richard Skaggs—and this book's author—Mark Seeley (*Minnesota WeatherTalk,* available through the University of Minnesota Extension website).

Enthusiasm for telling Minnesota's weather story is evident not only in such forums but as a common theme in thousands of Facebook entries and tweets over the Internet. There is no question that weather observations are one of the most commonly shared topics in the social media world. Regardless of platform, that will likely continue to be the case.

Climate Variability, Climate Change, and Minnesota's Future

Living in the middle of a continent and at a middle-latitude position within the northern hemisphere affords many opportunities to observe erratic climate behaviors and to gain an appreciation for the sometimes beautiful, frightful, dramatic, and traumatic weather that Mother Nature can deal from her deck of cards. Thousands of different weather scenarios can occur at each point on the Minnesota landscape, on any given day of the year.

Unlike those who live in equatorial or coastal environments, Minnesotans recognize that climate behavior is not stable or reliable but highly dynamic and ever changing. There have been periods of great climatic variability, when each year was distinctly different from those that immediately preceded or followed. There have also been periods when each year brought somewhat similar patterns of temperature, moisture, or storminess. Today's Minnesotans do not live in precisely the same climate that their parents or grandparents did. Rather, each generation seems to experience different patterns and extremes of the various climatic elements. Each generation has memorable weather-related stories to tell.

In the broadest context, we should remember that our Minnesota landscape lies nearly in the middle of the North American continent, about halfway between the equator and the North Pole and about halfway between the Pacific Ocean and the Atlantic Ocean. From this position, we have always recorded a high degree of variability in our climate, and in recent decades it has become clear that our climate is changing as much or more than all other areas of the United States. During the three most recent decades, the Minnesota climate has shown some very significant trends, all of which have had many observable impacts. These trends are statistically detectable in measured climate attributes kept by our state observer networks, but they are also evident in character changes, such as seasonality and even types of precipitation. Among the detectable measured quantity changes are: (1) warmer temperatures, especially daily minimum temperatures, more weighted to winter than any other season; (2) increased frequency of high dew points, especially notable in mid- to late summer as they push the Heat Index values beyond 100°F; and (3) greater annual precipitation, with a profound increase in the contribution from intense thunderstorms. These trends and their implications deserve a closer look.

DID YOU KNOW?

A kytoon is a kite-shaped tethered balloon for making meteorological measurements. In the nineteenth century, kytoons measured temperature and humidity aloft; today they measure carbon dioxide and ozone profiles.

The statistical signal for warmer temperatures appears in a number of ways when one examines the state's mean monthly temperature values since 1895. Statewide ranking of the warmest November-through-March periods—a calendar window encompassing nearly all of the snow season for most communities as well as the core of the heating season—reveals that 14 of the warmest 16 such periods have occurred since 1980 and the warmest 5 have occurred since 1997. In the Twin Cities climate record, the fewest Heating Degree Days (HDD) occurred in the heating season of 2011–12, with just 4,792 HDD, a total that is only 78 percent of normal. In fact, the upward trend in mean annual temperature for all Minnesota communities is in large part due to these contributions from the heating season months (November-March).

Further, ranking meteorological winter (December-February), a narrower time frame, shows that 9 of the warmest 11 winters in state history have occurred since 1982, with the warmest three coming in 1997–98, 2001–02, and 2011–12. Figure 1 illustrates the long-term statewide average winter (December-February) minimum temperature rising by 4.5 degrees per century despite the most recent harsh winter of 2013–14 (ranked among the 5 coldest since 1900). At individual climate stations within the state, some of the changes in average minimum temperature have been even more amplified than shown in Figure 1. For example, at Milan (Chippewa County), regarded as one of the best-quality long-term records in the state, the average January minimum temperature has changed from -4.3°F (1951–80) to 3.7°F (1981–2010), a remarkable 8.0 degrees warmer. Similarly, the average February minimum temperature at Milan has changed from 2.3°F (1951–80) to 9.3°F, a rise of 7.0 degrees.

Since the back-to-back cold winters of 1995–96 and 1996–97, 68 percent of all winter months have recorded above-normal temperatures on a statewide basis. The distribution of daily temperature measurements across the state has been even more heavily skewed toward a higher fraction of above-normal readings, especially in the past two decades. In some meteorological winters (as in 2011–12), over 70 percent of all daily temperatures were warmer than normal. Of further significance is that some of the individual daily temperature departures have been remarkable, as in January 2003 and December 2014, when daytime temperatures were in the fifties, 25 to 30 degrees above the historical average.

Another important detail about the warming trend in the state of Minnesota is that the minimum daily temperature values are warming at roughly twice the rate of the maximum daily temperature values. This disparity in the rate of warming between the maximum temperature (usually daytime values) and the minimum temperature (usually nighttime values) is not unique to Minnesota. The data show similar trends at a number of other geographic locations, as reported by the United Nations Intergovernmental Program on Climate Change. In Minnesota, a greater number of climate station records for warmest daily minimum temperature values have been established over the past 30 years than new records for warmest daily maximum temperature value. Figure 2 illustrates the average rate of change in state-averaged daily maximum temperature versus state-averaged daily minimum temperature across each month in the period from 1895 to 2013. The average rate of change in minimum temperatures is about twice that of maximum temperatures.

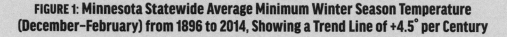

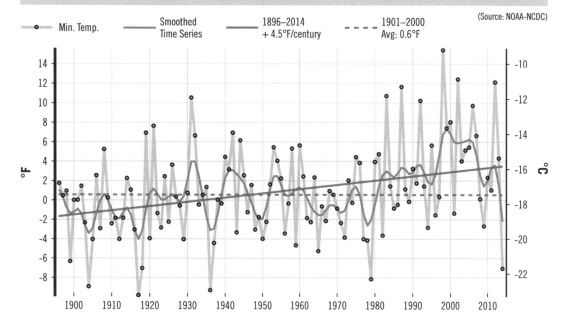

FIGURE 1: Minnesota Statewide Average Minimum Winter Season Temperature (December–February) from 1896 to 2014, Showing a Trend Line of +4.5° per Century

In addition, it is clear that the greatest magnitude of change has occurred in the months of January, February, and March. The rate of temperature change clearly varies by month. No change has occurred in July maximum temperature.

Impacts of this warm winter trend are evident even to the casual observer. Golf courses have been open for business in every winter month during some years, though this example depends on the absence of snow cover as well as warm temperatures. Outdoor construction season has lingered deep into the winter months, with the weather offering relatively minor obstructions to tasks like pouring concrete, welding, excavating, or re-roofing. There has been a detectable change in the depth and duration of soil freezing as well as in the onset of ice cover on area lakes, the thickness of the ice, and the loss of ice cover in the spring. In other words, Minnesota soils are frozen for fewer days, and short ice-fishing seasons have become more common. From a public health perspective, the milder temperature trends in spring and autumn have extended the mold and allergy seasons for people who suffer maladies

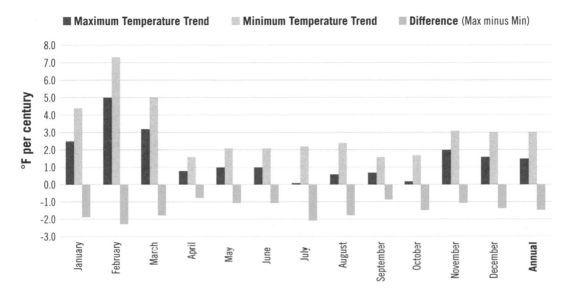

FIGURE 2: Minnesota State-Averaged Trends in Maximum and Minimum Temperature over the Period from 1895 to 2013 in °F

(Source: NOAA-NCDC and MN-DNR State Climatology Office)

from those elements. One positive economic impact has been reduced consumption of energy for residential and commercial heating, as Heating Degree Days (HDD) have more frequently averaged fewer than normal during the winter months.

The effects of high minimum temperatures are perhaps more subtle than those of higher winter temperatures. Many organisms exhibit population dynamics and behaviors associated with minimum temperatures. Additionally, important plant processes such as growth and respiration are affected—that is, increased—by higher minimum temperatures. Many soil microbes, plant pathogens, and insects survive better and even thrive when minimum temperatures are higher. Wildlife biologists find that animal migration, hibernation, and winter foraging behaviors can be altered by warmer minimum temperatures as well. One result of warmer winters and higher minimum temperatures has been a change in the freeze-free growing season. Recent studies have shown that the average last spring freeze date (32°F or colder) for most Minnesota climate stations has arrived earlier, in some cases by as many as nine to 17 days. The average first fall freeze date has also changed, but to a lesser degree, becoming both earlier and later by a few days at various stations. The overall result is a lengthening of the freeze-free growing season by five to 15 days for a number of climate stations. This extension, particularly in the first fall freeze date, has benefited farmers, who sometimes encounter wet spring soil conditions that delay planting or endure persistent bad weather that requires them to replant fields following floods or hailstorms. Yet another benefit from the change in minimum temperatures has been an expansion in Plant Hardiness Zones, especially Zone 4 and Zone 5a, in Minnesota's landscape. Today landscape nursery association businesses can market a much broader range of plant species for landscaping thanks to these changes in plant hardiness zones. Plants formerly not winter hardy enough to survive in Minnesota can now thrive.

· ·

Psychrometric tables

Psychrometric tables are derived from empirical calculations related to atmospheric water vapor content; they provide values for dew point, vapor pressure, and relative humidity based on observed dry-bulb and wet-bulb temperatures taken with a psychrometer. Today's electronic instrumentation is often programmed to give these values. The term is composed of the Greek psychro, "cooling," and meter, "to measure." Air's cooling power is related to water vapor content: the drier the air, the more rapidly evaporation, or cooling, will occur.

· ·

Another climatic trend of recent decades involves dew point, a measure of atmospheric water vapor that corresponds to the temperature of the air (at constant pressure) when it reaches saturation and the water vapor condenses into droplets. Because water vapor is the primary greenhouse gas in the Earth's atmosphere, it has a significant effect on air temperature, particularly at night, when long-wave or infrared radiation is emitted from the ground and absorbed by water vapor molecules in the air, then released in all directions, the heat in part directed back toward the surface. The summer months' highly erratic pattern of dew points trends slightly upward in the Twin Cities record since 1900 for June, July, and August. There are very few other long-term dew point records to draw on for comparison, since the modernization of the National Weather Service and the deployment of electronic hygrometers did not occur until the mid-1990s. Most disturbing about the Twin Cities record is the increased frequency of very high, tropical-like dew points during the summer months.

Dew points of 70°F are common in the Persian Gulf and in low-latitude coastal and tropical rainforest climates, but they are rare in Minnesota's historical records. Dew points of 70°F serve as one criterion used in meteorology to identify an air mass of tropical origin. The northward migration of such air masses infrequently reaches the latitude of Minnesota. From 1945 to 2013 the Twin Cities record shows dew points of 70°F or higher for approximately 170 hours per year. Figure 3 illustrates the year-by-year variation in the hours with a dew point of 70°F or higher in the Twin Cities. Since 1995 the record shows a frequency of 209 hours per year,

Statewide, what was the wettest decade of the twentieth century?

a. the forties

b. the sixties

c. the nineties

(answer on page 329)

an increase of 23 percent. In 2002 a record 512 hours with dew points of 70°F occurred in the Twin Cities area. Of course, the vast majority of these hours come in the summer months of June, July, and August, sometimes lasting over a period of several days before a new weather frontal system brings a change in the air mass. Dew points of 70°F or higher have been measured as early as May 4 and as late as October 14. These high dew points also set the floor for the daily temperature range because as a dropping overnight temperature reaches the dew point, condensation occurs and latent heat is released, often stabilizing the air temperature at the dew point. As a result, it is not surprising that there has been an increase in the frequency of overnight low temperatures remaining at 70°F or higher. For the Twin Cities, the long-term average (1872–2014) number of days when the minimum temperature does not drop below 70°F is 9, but over the last two decades it has been 15. As recently as 2013, there were 27 nights when the low temperature in the Twin Cities never dropped below 70°F, and three of those nights the temperature did not drop below 80°F. Such conditions are particularly stressful for people living without the convenience of air conditioning, and they amplify residential and commercial use of water for drinking as well as for cooling.

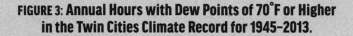

FIGURE 3: Annual Hours with Dew Points of 70°F or Higher in the Twin Cities Climate Record for 1945–2013.

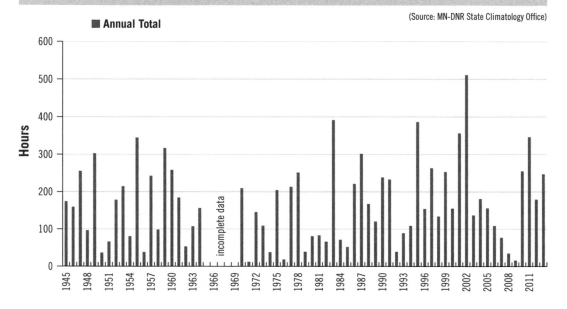

(Source: MN-DNR State Climatology Office)

Using the statewide automated weather station network operated by NOAA, some years have shown spells of extremely humid weather, with dew points rising to 80°F or higher for ten hours or more. These years include 1997, 1998, 1999, 2001, 2002, 2005, 2006, 2007, 2010, 2011, 2012, and 2014. These extreme summer dew point values invariably lead to the National Weather Service issuing a Heat Advisory or an Excessive Heat Warning, as the Heat Index Value rises to 100°F or higher, levels that cause stress and even health risk to large segments of the Minnesota population. For example, an air temperature of 97°F is uncomfortable but manageable for most healthy people who have enough water and shade. But when that temperature is combined with an 84°F dew point— as it was at Red Wing on July 30, 1999—the Heat Index becomes 127°F, a value that makes heat exhaustion likely and heatstroke a possibility for those exposed to such an environment for even short periods of time. Few Minnesota citizens realize that the all-time highest Heat Index measured in the United States comes from July 19, 2011, at Moorhead. At 6:00 that evening the air temperature was 95°F and the dew point was 88°F (an all-time high for the state), giving a Heat Index Value of 134°F, just like being in a sauna. That value was thought to be the record value for the nation. Who would have thought that such a record would come from Minnesota?

Since the heat wave and drought of 1988, every heat advisory and excessive heat warning issued by the NWS in Minnesota has been associated with high dew points combined with temperatures in the nineties. This record differs distinctly from many other historical heat waves, especially those of the 1930s, when detrimental and even dangerous conditions were prompted only by extremely high daytime maximum temperatures of 100°F or greater. This trend toward high dew point spells during the summer months yields wide-ranging, observable impacts. Air conditioning installations, both in new construction and in retrofits to older buildings, have increased. The health care industry has seen more frequent spikes in daily workloads at walk-in and emergency clinics on days when the Heat Index has topped 100°F. Almost invariably the elderly who are without air conditioning or people with respiratory problems or otherwise frail health need to seek relief and care. In fact, recent statistics from the National Oceanic and Atmospheric Administration show more annual human fatalities associated with heat waves than with floods, lightning, tornadoes, and winter storms. Many cities have responded to this worrisome situation by investing in heat-wave response plans to ensure that those who may suffer most (the homeless or those in marginal health without air conditioning) can obtain the relief and care they need. The Twin Cities introduced such a plan through the Heat Emergency Response Task Force (2005). The Minnesota Department of Health developed the Extreme Heat Toolkit and associated publications to more broadly educate at-risk populations about how to take care of themselves and reduce risks associated with these heat waves. Finally, in recent years the Minnesota farmer has responded to higher dew points by redesigning and retrofitting dairy, hog, and poultry barns with better watering, feeding, and ventilation systems to minimize the detrimental effects of heat stress on animal metabolism and weight gain.

Perhaps the most important significant Minnesota climate trend is greater precipitation. Map 9 shows how Minnesota's average annual precipitation has varied geographically over time. The area of the state that records an average annual precipitation of 29 inches or greater has waxed and waned over the years, but for the current averaging period of 1981–2010, over half of the state landscape hit this mark, for the first time in the state's documented climate history.

The statistical details are revealing both about the quantity changes in precipitation and the character changes in precipitation. Most of the state's climate stations show an increase in average annual precipitation that ranges from 1 to 4 inches over the past 60 years. Some cases are more extreme than others. For example, at Waseca (Waseca County), the mean annual precipitation for 1921–50 is 27.55 inches, while for the most recent averaging period (or current normal) of 1981–2010 the mean is 35.72 inches, an increase of more than 7 inches. The same comparison of mean annual precipitation for the Twin Cities shows an increase of 5.88 inches; and for Grand Rapids (Itasca County), the same comparison reveals an increase of 4.58 inches. Thus, nearly all of the annual precipitation "normals" expressed in Table 4 (page 26) are essentially greater than what was depicted when the 1971–2000 values were used.

Embedded in this variable rate of increase in total annual precipitation is a trend for greater seasonal snowfall, at least at a majority of Minnesota climate stations, and higher water content of the snow. A study of the seasonal snowfall records from 46 locations across Minnesota covering the period 1890–2000 revealed that 41 of these climate stations show an increase in average annual snowfall, as much as 10 inches in some cases. Since snow density, or water content, varies so greatly, one might wonder whether additional snowfall translates into more water on the landscape. Indeed, for the most part it does, but even where there is little or no change in snowfall, there is an increase in precipitation quantity. For example, examining month-by-month precipitation trends for liquid equivalent values at Willmar (Kandiyohi County), where mean seasonal snowfall is about 46 inches, shows that the average precipitation for November through March has increased by roughly 25 percent since 1921: from 4.08 inches to 5.09 inches. A similar comparison at Two Harbors in Lake County, where mean seasonal snowfall is about 56 inches, shows a gain of more than 29 percent in November-to-March precipitation since 1921: from 5.31 inches to 6.85 inches. The added snowfall and its liquid equivalent mean greater runoff potential during spring snow melt; indeed, many Minnesota watersheds have shown more consistent measures of high-volume flows on stream gauges in the spring, often at or above flood stage.

MAP 9: Depiction of Average Annual Precipitation for Minnesota over Four Distinct Periods

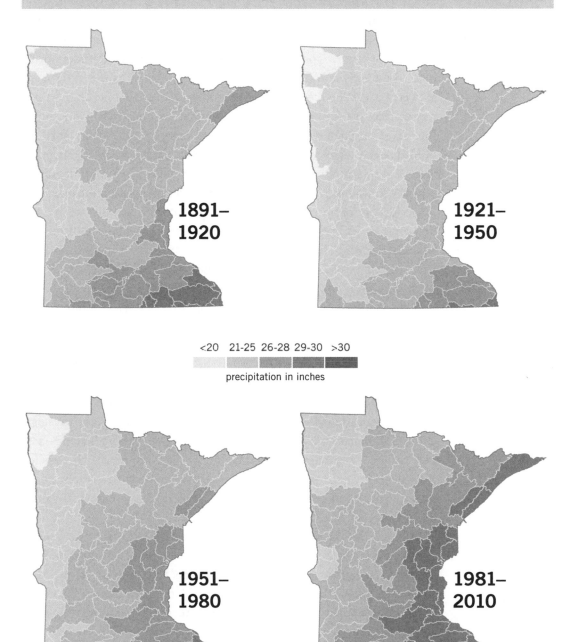

1891–
1920

1921–
1950

<20 21-25 26-28 29-30 >30
precipitation in inches

1951–
1980

1981–
2010

For most climate stations in the state, another character change in precipitation is very significant: a larger fraction of annual precipitation is coming from intense thunderstorms. Associated with this character change is higher spatial variation of precipitation across the state landscape, which proves problematic. Areas that experience thunderstorms tend to get a great deal of rainfall, while areas missed by thunderstorms get very little. When repeated thunderstorm tracks occur in summertime, some places get inundated while others that are consistently missed go into drought. Such was the case in the summer of 2012, when portions of Minnesota were declared federal drought disaster areas, while other areas even within the same counties were declared federal flood disasters, the first time in state history that such a climatic situation emerged.

Evidence for an increased frequency of intense thunderstorms can be found at many climate stations. According to an older hydrological frame of reference depicted in U.S. Weather Bureau TP 40, the hundred-year rainfall for places in southern Minnesota like Austin and Albert Lea during a single day was 6.3 inches. Stated differently, there was a one percent chance for a rainfall of that amount to occur in a 24-hour period. Using compiled climate data from more recent decades, the NOAA–NWS Hydrometeorological Design Center reworked the data series and published the new NOAA Atlas 14 in 2013. The new value for a hundred-year 24-hour rainfall event at Austin or Albert Lea is 7.8 inches, representing an increase of 24 percent. Further north in Moorhead, a similar comparison of calculations for hundred-year 24-hour rainfall events reveals a TP-40 value of 5.3 inches versus a NOAA Atlas 14 value of 6.5 inches, a 23 percent increase. These figures show how the computed estimates for historical extreme rainfalls have changed over time with so many heavy thunderstorms in recent decades that exceeded any historical values previously measured.

Thunderstorm rainfalls come with a variety of intensities, many times exceeding the capacities of storm sewer runoff systems or agricultural drainage systems. As a consequence, flash floods occur. Oftentimes this happens with a daily rainfall of 2 or more inches. Earlier in our state climate history, a 2-inch thunderstorm rainfall occurred across southern Minnesota counties about once per year. However, since 1991 this frequency has doubled to twice per year, as depicted in the climate data of places like Waseca, Winnebago (Faribault County), and Lake City

(Goodhue County). Similarly, in the north at locations such as Baudette (Lake of the Woods County), historical frequency of thunderstorm rainfalls of 2 inches or greater was about once every two years. Again, since 1991 this frequency has doubled to once per year. In fact, Baudette, with a climate history back to 1908, has reported daily thunderstorm rainfalls greater than 5 inches only twice in history, both times since 2001.

The Minnesota DNR State Climatology Office has documented the history of mega-rainfall events back to 1866. The defining characteristics of these events include a 6-inch one-day rainfall that encompasses more than a thousand square miles and a peak rainfall measurement of 8 inches somewhere within that area. Only 12 such events have been documented in Minnesota's post-statehood history. These storms are described elsewhere in the book but are chronologically listed here:

▦ August 6, 1866, Southern Minnesota
Known as the Wisel Flood, this event killed 16 people, including three members of the Wisel family in Fillmore County. A total of 10.30 inches of rain fell at the Sibley Indian Agency, located in Sibley County. The story of the imperiled Wisel family appeared in the Harmony/Mabel/Canton *News Record* from December 2011 to January 2012.

▦ July 17–19, 1867, Central Minnesota
Rainfall was estimated at 30–36 inches in 36 hours, with at least 8 inches over a multi-county area in west-central Minnesota. This storm is known as Minnesota's greatest flash flood. Most of what is known about this event is from an account written by George Wright and read before the Minnesota Academy of Sciences on March 7, 1876. Climate historian Tom St. Martin also summarized this event from old newspapers articles.

▦ July 20–22, 1909, Northern Minnesota
This very extensive flash flooding event extended from northwestern Minnesota across to upper Michigan. The highest one-day rainfall total was 10.75 inches at Beaulieu in Mahnomen County (11.10 inches for the three-day total). This storm also did extensive damage in Duluth and killed two children in the city when they were swept out of their mother's arms.

▨ September 9–10, 1947, Iron Range

For this storm, 24-hour totals of 6 inches or more were reported at Hibbing, Ely, and Winton. An unofficial report of 8.60 inches in five hours came from Hibbing. There was extensive flood damage over the Iron Range district.

▨ July 21–22, 1972, Grand Daddy Flash Flood

This storm delivered 10.84 inches in 24 hours to Fort Ripley, setting a new statewide rainfall record for a single date. Most roads and highways across central Minnesota counties were flooded. This record rainfall amount was finally broken in Hokah during the 2007 flash flood (see page 311).

▨ June 28–29 and July 1–2, 1975, Northwestern Minnesota

This geographically extensive storm event encompassed parts of North Dakota and northwestern Minnesota. Many observers reported rainfall totals of 6 to 8 inches, and a number of rural roads were washed out. Two separate thunderstorm complexes pounded the region with rain and hail.

▨ July 23–24, 1987, Twin Cities Superstorm

This storm brought the biggest calendar-day rainfall ever to the Twin Cities, with 9.15 inches before midnight on July 23 and then another 0.85 inches after midnight, for a storm total of 10 inches in about 6.5 hours. Most metro area counties were affected with widespread flooding. Minneapolis–St. Paul airport reported 17.90 inches of rain that month.

▨ June 9–10, 2002, Northern Minnesota

This storm delivered 12 or more inches to portions of Roseau, Lake of the Woods, and Koochiching counties, with an observer near Lake of the Woods reporting 14.55 inches. It produced an all-time record flood crest on the Roseau River.

▨ September 14–15, 2004, Southern Minnesota

This storm had a total duration of 36 hours and was centered over Faribault, Freeborn, and Dodge counties, where it delivered a foot or more of rain. The storm sent an all-time flood crest down the Cedar

River at Austin. Widespread flooding was reported in several southern communities.

▧ August 18–20, 2007, Southern Minnesota
The 15.10 inches of rain measured one mile south of Hokah stands as the all-time state 24-hour record rainfall. The three-day total for this station was 16.27 inches. The town of Rushford (Fillmore County) was almost entirely under water.

▧ September 22–23, 2010, Southern Minnesota
The National Weather Service observer in Amboy (Blue Earth County) reported 9.48 inches from this storm on September 23, with 10.68 inches in total over the two days. The area receiving 6 or more inches of rain encompassed more than 5,000 square miles, and there was a great deal of street and basement flooding in southern Minnesota cities.

▧ June 19–20, 2012, Northeastern Minnesota
This storm encompassed portions of Carlton, St. Louis, and Lake counties. The two-day total rainfall was more than 10 inches in Cloquet, more than 11 inches in Duluth, and more than 12 inches near Floodwood. Jay Cooke State Park was flooded out, as were many roads and highways. The St. Louis River reached an all-time flood crest.

Of particular note is that five of these twelve mega-rainfall events happened since the year 2001. Further, the NOAA National Climatic Data Center reports that these types of storms are increasing in frequency across midwestern states.

These precipitation trends yield widespread impacts. Many communities have redesigned storm sewer runoff systems to handle larger rainfall volumes. Farmers have decreased the spacing in their tile lines, called pattern tile drainage, to more quickly shed surplus water from their croplands. Many farmers are also using conservation tillage practices or planting field border areas with perennial grasses to reduce erosion and to slow the rapid runoff intense thunderstorms can generate. Several communities and local watershed districts have conducted studies to find ways of mitigating the effects of flash or spring snowmelt floods. In the case of winter precipitation, the Minnesota Department of Transportation has embarked on a statewide program to make better use of snow fences, both constructed and vegetative, to reduce blowing and drifting snow that forces repeated road closures.

The governing leadership in our state certainly has plenty to think about when it comes to climate. Our state historical climate database is of significant value for this purpose. The evidence for climate change in Minnesota is pretty overwhelming, and the evidence of impacts on both our natural resources and our societal infrastructure abounds. Further, it appears these climate trends are not about to reverse themselves. Their statistical signals and impacts, though not uniform across the state, have been fairly consistent, and residents have reacted in predictable ways. These trends have increased pressure to better manage natural resources, particularly soil and water, and to diminish risks posed by various climate elements, including heat waves, floods, and blowing snow. These trends also pose a challenge to our societal infrastructure, which supports our quality of life and encompasses such features as fresh water supply, power usage, transportation efficiency, insurance options, and public health. But the old adage "what goes around comes around" holds true.

A look through Minnesota's climate history shows that drought repeats itself (1894, 1930s, 1976, 1988, 2012), as do extreme winters (1872–73, 1874–75, 1882–83, 1884–85, 1886–87, 1903–04, 1916–17, 1935–36, 1978–79, 2013–14), severe spring floods (1826, 1851, 1881, 1897, 1952, 1965, 1997, 2001, 2009), and torrential flash floods (1866, 1867, 1909, 1947, 1975, 1987, 2002, 2005, 2007, 2010, 2012). Minnesotans' vulnerability to such events and episodes is no less than it once was; in fact, some argue that it has increased, if not in threats to health, then surely in heightened economic consequences, which more frequently run in the millions or even billions of dollars. Attending to this vulnerability—reducing or mitigating it—seems a worthwhile endeavor.

Another feature of our society, particularly evident in young people, is the adoption of an environmental ethic, a perspective that acknowledges how greatly our choices affect the environment, particularly in relation to food production, waste disposal, water use, and energy consumption. This shared value system, which prioritizes minimal impacts on the environment, has gained political strength and is likely to set yet a new framework on how we view climate behavior and trends. We have acknowledged for generations that Minnesota's climate has helped shape who we are. This understanding combined with an evolving environmental ethic and knowledge of climate behaviors may enlighten us to forge a direction that sustains Minnesota's land, air, and waters for future generations. This perspective has inspired a number of nongovernmental organizations, church denominations, and even private companies to adopt a mission statement to promote products, procedures, and protocols that reduce the vulnerability of individuals and communities to natural hazards, improve the efficient use of limited resources, encourage the development and utilization of sustainable resources, and protect the Minnesota environment. Just as the weather has shaped us, this admirable mission statement says a good deal about who we are. And what we can become.

TABLE 25: Statewide Daily Climate Records (1850–2014)

Month Day	MAXIMUM TEMPERATURE (°F)			MINIMUM TEMPERATURE (°F)			PRECIPITATION (INCHES)			SNOWFALL (INCHES)		
	Value	Year	Location	Value	Year	Location	Value	Year	Location	Value	Year	Location
Jan 1	56	1998	Luverne	-46	1885	St. Vincent	1.75	2007	Elk River	18.0	1921	Ada
Jan 2	53	1944	Austin 3S	-50	1904	Pokegama Dam	1.25	1999	Luverne	17.0	1941	Pigeon River
Jan 3	53	1998	Canby	-54	1904	Pokegama Dam	1.90	1897	St. Cloud	15.5	1943	Willmar
Jan 4	52	2012	Milan 1NW	-48	1896	Leech Lake	2.13	1886	Red Wing	24.0	1997	Wheaton
Jan 5	62	2012	Granite Falls	-49	1896	International Falls	1.92	1997	Byron	23.0	1997	Remer
Jan 6	62	2012	Marshall	-55	1909	International Falls	1.85	1997	Orwell Dam	19.0	1997	Hinckley
Jan 7	59	2003	Amboy	-54	1909	International Falls	1.60	1861	Beaver Bay	36.0	1994	Wolf Ridge (Finland)
Jan 8	60	2003	Fairmont	-48	1887	Moorhead	1.14	1989	Tamarac Refuge	17.0	1969	Isabella
Jan 9	60	2002	Amboy	-49	1930	Warroad	1.70	1873	Beaver Bay	17.0	1873	Beaver Bay
Jan 10	58	1990, 2012	Madison SE, Granite Falls	-52	1905	Pokegama Dam	2.12	1975	Grand Portage	15.0	1983	Brainerd
Jan 11	59	2012	Canby	-53	1888	St. Vincent	2.70	1866	Beaver Bay	24.0	1975	Riverton
Jan 12	58	1987	Browns Valley	-53	1912	Pine River	1.80	1975	Remer	18.0	1975	Remer
Jan 13	60	1987	Lamberton	-50	1916	Bagley	1.41	2008	Grand Marais	15.0	2008	Grand Marais
Jan 14	57	1987	Browns Valley	-50	1965	Cook 18W	1.07	1950	Grand Meadow	12.0	1923	Campbell
Jan 15	57	1914	Winnebago	-53	1972	Moose Lake	1.28	1952	Campbell	20.0	1982	Winsted
Jan 16	54	1974	New Ulm 2S	-48	2005	Embarrass	1.65	1870	Beaver Bay	16.0	1870	Beaver Bay
Jan 17	58	1889	Winona	-54	2005	Embarrass	2.20	1996	Byron 4N	15.0	1870	Fort Ripley
Jan 18	57	1996	Caledonia	-51	2005	Babbitt	2.28	1996	Jordan	18.0	1866	Sibley
Jan 19	61	1900	Milan	-47	1963	Bigfork 5E	2.03	1982	Lake City	18.0	1988	St. James
Jan 20	61	1944	Madison SE	-57	1996	Embarrass	1.76	1988	Preston	17.1	1982	MSP Airport
Jan 21	62	1942	Canby	-57	1996	Embarrass	2.20	1917	Redwood Falls	24.0	1917	Tracy
Jan 22	59	1900	Lynd	-51	1922	Itasca State Park	2.53	1973	Austin 3S	22.0	1917	Lynd
Jan 23	62	1942	Itasca State Park	-55	1936	Warroad	1.43	1982	Stillwater	17.0	1982	Stillwater
Jan 24	69	1981	Montevideo	-57	1904	Pokegama Dam	2.05	1858	Princeton	14.0	1982	Blue Earth
Jan 25	67	1981	Springfield	-55	1904	Pokegama Dam	1.78	1967	Grand Meadow	16.5	1982	Isabella
Jan 26	63	1944	Winnebago	-55	1904	Pokegama Dam	1.00	2004	Grand Marais	24.0	2004	Tamarac Refuge
Jan 27	61	2002	Lakefield	-54	1904	Pokegama Dam	1.55	1944	Canby	18.0	1996	Hokah 1SW
Jan 28	63	1946	Chaska	-50	1966	Baudette	1.55	1909	Bird Island	12.0	1949	Caledonia
Jan 29	60	1931	Canby	-54	1899	Pokegama Dam	1.84	1909	Windom	19.0	1996	Lutsen
Jan 30	56	1989, 2012	St. Peter, Granite Falls	-52	1899	Duluth Harbor	2.00	1927	Crane Lake	16.0	1947	Worthington
Jan 31	57	1989	Springfield	-55	1996	Embarrass	1.70	1986	Glenwood 2	14.8	1858	Burlington
Feb 1	60	1931	St. Peter	-58	1996	Tower 3S	1.85	1915	Fairmont	18.5	1915	Fairmont
Feb 2	66	1991	Wheaton	-60	1996	Tower 3S	2.00	1915	Caledonia	20.0	1915	Caledonia
Feb 3	65	1991	Browns Valley	-52	1996	Itasca State Park	1.50	2000	Red Lake Indian Agency	12.0	1983	Caledonia
Feb 4	65	2005	Tracy	-52	1907	Detroit Lakes	2.15	1865	Fort Ridgley	14.0	2008	Halstad

Month Day	MAXIMUM TEMPERATURE (°F)			MINIMUM TEMPERATURE (°F)			PRECIPITATION (INCHES)			SNOWFALL (INCHES)		
	Value	Year	Location	Value	Year	Location	Value	Year	Location	Value	Year	Location
Feb 5	68	2005	Lamberton SW exp S	-54	1895	Leech Lake	1.70	1975	Bird Island	24.0	1915	Lynd
Feb 6	59	1963	Madison SE	-50	1907	Detroit Lakes	1.78	1941	High Landing	16.0	1946	Beardsley
Feb 7	62	1987	Browns Valley	-53	1899	Leech Lake	1.75	1928	Lynd	14.0	1946	Campbell
Feb 8	66	1991	Madison SE	-55	1933	Warroad	1.35	1947	St. James	15.0	1937	Grand Marais
Feb 9	63	1991	Canby	-59	1899	Leech Lake	1.75	1909	Collegeville	15.0	1909	Albert Lea
Feb 10	57	1977	Luverne	-49	1899	Tower 3S	1.69	1965	Montevideo	20.0	1939	Pigeon River
Feb 11	61	1977	Luverne	-55	1899	Leech Lake	2.05	2013	Detroit Lakes 1NN	19.0	2013	Rothsay
Feb 12	63	2005	Windom	-50	1914	Roseau 1E	1.86	1922	Brainerd	17.0	1965, 2013	Fairmont, Orwell Dam
Feb 13	63	1990	Windom	-46	1916	Detroit Lakes	1.87	1915	Pipestone	20.0	1936	Pigeon River
Feb 14	66	1954	Windom	-47	1906	Bagley	2.15	1919	Lynd	22.0	1936	Grand Marais
Feb 15	67	1921	Winona	-53	1936	Ada	1.84	1967	Winsted	12.5	1945	Fergus Falls
Feb 16	67	1981	Canby	-59	1903	Pokegama Dam	2.20	1921	Brainerd	12.0	1990	Aitkin
Feb 17	66	1981	Luverne	-52	1903	Pokegama Dam	1.85	1984	Hokah 1SW	13.0	1870	Beaver Bay
Feb 18	66	1981	Pipestone	-48	1966	Roseau 1E	1.80	2004	Blackduck	16.0	1962	Worthington
Feb 19	68	1981	Winona	-52	1966	Baudette	2.50	1984	Montevideo 1 SW	23.0	1962	Luverne
Feb 20	65	1981	Canby	-50	1966	Baudette	1.60	1898	Caledonia	16.0	1952	Marshall
Feb 21	64	1943	Whitewater	-51	1966	Baudette	2.02	2014	Two Harbors	19.2	2011	Madison Sewage Pit
Feb 22	66	1943	Whitewater	-46	1939	Berridji	3.00	1922	Willmar ST	15.0	2011	Gaylord
Feb 23	70	2000	Lake Wilson	-43	1889	St. Vincent	2.35	1922	Detroit Lakes	25.0	1922	Detroit Lakes
Feb 24	67	1958	Pipestone	-46	1955	Red Lake Falls	2.10	1964	Tower 3S	19.0	1868	Beaver Bay
Feb 25	66	1958	Beardsley	-50	1897	Leech Lake	2.28	1930	Cass Lake	19.0	2001	Wolf Ridge (Finland)
Feb 26	73	1896	Pleasant Mound	-49	1897	Pokegama Dam	2.50	1964	Wannaska	15.0	2001	Sandy Lake
Feb 27	66	1896	Pleasant Mound	-40	1913	Warroad	2.35	1971	Comfrey 4E	18.0	1893	Blooming Prairie
Feb 28	66	1924	Pipestone	-50	1897	Pokegama Dam	2.21	1998	Isabella 1	16.5	1948	Gull Lake
Feb 29	65	2000	Forest Lake	-38	1916	Roseau 1E	2.23	2012	Faribault	12.0	2012	Ortonville 1N
Mar 1	76	1907	Ashby	-47	1962	Bigfork 5E	2.50	2007	Jordan 2E	18.8	1965	Collegeville
Mar 2	71	1992	Browns Valley	-50	1897	Pokegama Dam	2.45	1965	Young America	25.0	2007	Wolf Ridge (Finland)
Mar 3	71	1905	Milan	-44	2014	Embarrass	3.06	1985	Benson	18.7	1985	Benson
Mar 4	70	1905	Currie	-43	1917	Bagley	3.54	1966	Isabella	26.0	1966	Isabella
Mar 5	79	2000	Milan	-40	1890	Pokegama Dam	3.00	1896	Red Lake Indian Agency	28.5	1966	Karlstad
Mar 6	76	2000	Canby	-41	1996	Embarrass	2.50	2006	Austin Waste Water	16.0	1959	La Crescent
Mar 7	80	2000	Winona	-38	1913	Littlefork	1.56	1915	Lynd #2	22.7	1959	Caledonia
Mar 8	84	2000	Amboy	-38	1908	McIntosh	2.22	1878	Breckenridge	14.5	1961	Spring Grove
Mar 9	77	2000	Caledonia	-33	1984	Tower 3S	1.96	1992	St. James	16.0	1999	Red Wing
Mar 10	68	2012	Granite Falls	-44	1948	Itasca State Park	2.10	1892	Winnibigoshish	16.8	1892	Winnibigoshish
Mar 11	72	2012	St. James Filt Pit	-41	1948	Moose Lake	3.00	1918	Waseca EXP	16.0	1897	New London
Mar 12	70	1934	Beardsley	-35	2009	Babbitt	3.30	1977	Madison SE	18.0	1997	Elgin
Mar 13	80	1927	Waseca	-36	1896, 2009	Campbell, Embarrass	1.85	1997	Winona	18.0	1940	Cloquet

Month Day	MAXIMUM TEMPERATURE (°F)			MINIMUM TEMPERATURE (°F)			PRECIPITATION (INCHES)			SNOWFALL (INCHES)		
	Value	Year	Location	Value	Year	Location	Value	Year	Location	Value	Year	Location
Mar 14	74	2012	Rochester Intl AP	-40	1897	Detroit Lakes	2.50	2013	New York Mills	18.0	1917	Grand Marais
Mar 15	80	1927	Waseca	-49	1897	Pokegama Dam	3.22	1971	Brimson 1E	21.0	1971	Isabella
Mar 16	83	2012	Granite Falls	-34	1899	Pokegama Dam	2.44	1917	Lynd	24.0	1917	Lynd
Mar 17	83	2012	St. James Flt Plt	-33	1897	International Falls	2.00	1942	Whitewater	23.6	1965	Collegeville
Mar 18	84	1921	Canby	-48	1939	Sawbill Camp	3.12	2005	Hokah	20.0	1933	Albert Lea
Mar 19	84	2012	Madison Sewage Plt	-40	1923	Meadowland	2.90	2007	Littlefork 10SW	18.0	1933	Grand Meadow
Mar 20	80	2012	Cannon Falls 2SW	-37	1965	Bigfork 5E	2.12	1982	Browns Valley	15.0	1982	Browns Valley
Mar 21	81	1910	Montevideo	-33	1965	Cotton 3E	2.00	1893	Ortonville	15.0	2008	Milan 1NW
Mar 22	81	1939	Pipestone	-30	1888	Pokegama Dam	3.00	1865	Elk River	14.6	1952	Fairmont
Mar 23	88	1910	Montevideo	-37	1965	Baudette 2	3.87	1979	Isabella	18.0	1966	Montgomery
Mar 24	86	1910	New Ulm 2S	-41	1974	Thorhult	2.50	1966	Waseca	15.0	1937	Bird Island
Mar 25	83	1939	Canby	-31	1965	Bigfork 5E	3.31	1996	Halstad	14.0	1914	Bemidji
Mar 26	84	1907	Winnebago	-31	1996	Fosston	2.50	1902	Beardsley	13.0	1936	Redwood Falls
Mar 27	88	2007	Winona Dam 5A	-29	1955	Red Lake Falls	2.70	1975	Two Harbors	15.0	1975	Virginia
Mar 28	84	1946	Bemidji	-30	1923	Roseau 1E	2.60	1924	Canby	13.0	1924	Canby
Mar 29	83	1986	Gaylord	-23	1921	Itasca State Park	2.88	1998	Lake City	17.5	1924	Maple Plain
Mar 30	87	1968	New Ulm 2S	-28	1975	Thorhult	3.39	1933	Park Rapids	16.0	1934	Winona
Mar 31	84	1968	New Ulm 2S	-32	1975	Tower 3S	3.00	1896	New London	20.0	1896	St. Cloud
Apr 1	85	1986	Winona	-21	1975	Thorhult	2.52	2009	Beaver Bay 55W	18.0	2009	Frazee
Apr 2	88	2012	Pipestone	-21	1975	Tower 3S	3.17	1967	Luverne	12.0	1960	Canby
Apr 3	86	1929	Beardsley	-19	1954	Big Falls	3.05	1934	Farmington	14.0	1982	Kettle Falls
Apr 4	89	1929	Tracy	-17	1975	Tower 3S	2.57	1981	Hokah 1SW	18.0	1968	Meadowland
Apr 5	88	1991	Madison SE	-18	1936	Warroad	2.95	1933	Pigeon River	28.0	1933	Pigeon River
Apr 6	90	1991	Madison SE	-22	1979	Karlstad	2.67	1997	Dawson	18.0	1947	Fosston
Apr 7	91	1991	Canby	-16	1982	Tower 3S	3.52	2006	Blue Earth	24.0	2008	Tower DNR
Apr 8	88	1931	Canby	-12	1939	Sawbill Camp	2.45	1894	Dawson	13.2	1904	Moorhead S
Apr 9	91	1977	Redwood Falls	-19	1939	Sawbill Camp	3.19	1878	Breckenridge	27.5	2008	Tower 2S
Apr 10	92	1977	Browns Valley	-8	1989	Brimson 1E	3.14	2013	Harmony	14.0	1913	Lynd
Apr 11	92	1977	Madison SE	-4	1940	Baudette	3.75	2001	Rochester	12.0	1959	Minneota
Apr 12	90	1931	Beardsley	-7	1924	Warroad	3.74	2001	Grand Meadow	11.8	1995	Madison SE
Apr 13	90	2003	Wheaton	-11	1950	Roseau 1E	3.57	2010	Hutchinson 1W	13.0	1892	Kinbrae
Apr 14	94	2003	Benson	-5	1950	Roseau 1E	2.95	1886	Northfield	16.0	1983	Farmington
Apr 15	93	2003	Canby	-12	2014	Warroad	3.00	2013	Pipestone	13.0	1961	Mahoning Mine
Apr 16	93	2002	Fairmont	-18	2014	Warroad	3.40	2003	Belle Plain	13.0	1945	Itasca State Park
Apr 17	91	1914	Tracy	-5	1983	Gunflint Lake	4.00	1894	Belle Plain	13.0	1945	Detroit Lakes
Apr 18	94	1985	Marshall	2	1983	Gunflint Lake	4.80	1898	Bingham Lake	13.0	1869	Beaver Bay
Apr 19	95	1985	Canby	-3	1897	Tower 3S	2.87	1916	New Richland	25.0	2013	Isabella 14W
Apr 20	96	1980	Georgetown	-14	2013	Embarrass	3.08	1893	Collegeville	18.0	1893	Fort Ripley

Month Day	MAXIMUM TEMPERATURE (°F)			MINIMUM TEMPERATURE (°F)			PRECIPITATION (INCHES)			SNOWFALL (INCHES)		
	Value	Year	Location	Value	Year	Location	Value	Year	Location	Value	Year	Location
Apr 21	100	1980	Ada	-14	2013	Embarrass	2.49	1974	Warroad	15.0	1893	Lynd
Apr 22	101	1980	Hawley	0	1952	Canby	3.52	2001	St. Cloud	10.0	1902	Moorhead
Apr 23	94	2009	Granite Falls	-1	1918	Grand Rapids	4.22	2001	Marshall	12.0	1968	Meadowland
Apr 24	92	1962, 2009	Madison SE, Montevideo 1SW	3	1936	Sawtill Camp	5.44	1990	Winona Dam	15.0	1937	Fosston
Apr 25	96	1962	Madison SE	5	1909	Leech Lake	3.55	1994	Hokah 1SW	16.0	1950	Two Harbors
Apr 26	94	1962	Marshall	8	2002	Embarrass	6.90	1954	Morris	16.0	2008	Pelican Rapids
Apr 27	96	1952	Hallock	7	1996	Brimson 1E	3.76	1975	Cambridge	14.0	2008	Ottertail
Apr 28	95	1910	Lynd	-2	1892	St. Vincent	3.58	1975	Red Wing	11.0	1907	Stillwater
Apr 29	93	1910	New Ulm 2S	3	1958	Babbitt 2S	3.25	1940	Orr	14.0	1956	Windom
Apr 30	96	1934	Maple Plain	6	1877	Fort Ripley	3.33	1936	New London	13.7	1945	Fosston
May 1	100	1959	Beardsley	4	1909	Pine River	3.83	1936	Winona	8.0	1909	Hinckley
May 2	99	1959	Wheaton	4	1909	Pine River	3.05	1950	Trail 2N	15.4	2013	Dodge Ctr
May 3	97	1949	Bird Island	6	1967	Crookston	4.00	1902	Albert Lea	12.0	1954	Leonard 8N
May 4	96	1949	Montevideo 1SW, Wheaton	8	1911	Cloquet	4.00	1949	Blanchard	5.0	1890	Mankato
May 5	97	1926	Angus 1N	10	1951	Canby	4.38	1950	Two Harbors	4.6	1931	Virginia
May 6	98	1934	Grand Meadow	12	1944	Alborn	3.78	2012	Lake Wilson	10.0	1938	Roseau 1E
May 7	95	1934	Maple Plain	13	1907	Hallock	3.74	2012	Rosemount Agr Exp	5.0	1924	Gonvick 2W
May 8	102	1934	Beardsley	10	1907	Pine River	4.85	2014	Santiago 3 E	12.0	1938	Windom
May 9	99	1928	Milan	9	1966	Isabe la	3.22	1979	St. Cloud	4.0	1924	Farmington
May 10	97	1928	Beardsley	11	1905	Pine River	4.27	1920	Wheaton	6.0	1927	Pigeon River
May 11	104	2011	Blue Earth 1S	11	1946	Fosston	4.60	1922	Crookston	3.0	1966, 2008	Isle 8N, Tower 2S
May 12	98	1900	Hallock	11	1946	Hallock	4.48	2004	Warroad	2.5	1953	Babbitt 2S
May 13	95	1894, 2007	Beardsley, Granite Falls	10	1997	Tower 3S	4.63	1999	St. Francis	3.0	1924	Argyle
May 14	102	2013	Amboy	11	1945	Alborn	3.82	1941	Crookston	6.0	1974	Cook 18W
May 15	103	2013	Winnebago	16	1888	St. Vincent	4.20	1911	New London	8.0	1907	Mount Iron
May 16	100	1934	Artichoke Lake	12	1929	Meadowland	5.00	1894	St. Cloud	2.5	1932	Mahoning Mine
May 17	100	1934	New Ulm 2S	14	1888	St. Vincent	4.43	2000	Blue Earth	12.0	1890	St. Cloud
May 18	101	1934	Fairmont	16	1924	Duluth	5.01	2000	Lanesboro	3.0	1915	Minneapolis
May 19	104	1934	Redwood Falls	16	2002	Embarrass	4.17	1996	St. Francis	8.6	1971	Grand Rapids
May 20	100	1934	Fairmont	16	2002	Embarrass	4.83	1970	Winton	4.8	1931	Virginia
May 21	100	1964	Thief River Falls	17	2006	Embarrass	5.75	1960	Montgomery	2.5	1926	Pigeon River
May 22	100	1925	Fairmont	19	2006	Embarrass	5.84	1962	Collegeville	2.5	2001	Big Falls
May 23	97	1928	Fergus Falls	17	2008	Kelliher	4.54	1933	Park Rapids	0.7	1924	Fergus Falls
May 24	98	1928	Beardsley	18	1988	Mora	4.91	2012	St. Francis	1.0	1930	Pigeon River
May 25	100	1967	Luverne	19	1983	Tower 3S	4.32	1953	St. James	4.0	1970	Baudette
May 26	103	1914	Tracy	20	1961	Cook 18W	3.48	1978	Grand Meadow	2.0	1970	Kelliher
May 27	104	1934	Hallock	13	1895	Sandy Lake	4.22	2012	New York Mills	5.0	1932	Virginia
May 28	106	1934	Beardsley	17	1889	Pokegama Dam	6.15	1970	Wabasha	4.0	1947	Orr

Month Day	MAXIMUM TEMPERATURE (°F)			MINIMUM TEMPERATURE (°F)			PRECIPITATION (INCHES)			SNOWFALL (INCHES)		
	Value	Year	Location	Value	Year	Location	Value	Year	Location	Value	Year	Location
May 29	101	1934	Chaska	20	1965	Bigfork 5E	7.50	1949	Thief River Falls	2.6	1947	Spring Grove
May 30	108	1934	Pipestone	20	1889	Pokegama Dam	5.63	1980	Preston	0.1	1897	Bemidji
May 31	112	1934	Maple Plain	19	1889	Pokegama Dam	4.92	1888	Maple Plain	4.6	1946	Virginia
Jun 1	105	1934	Chaska	15	1964	Bigfork 5E	7.98	1967	Stillwater	0.0	1967	
Jun 2	105	1934	Fairmont	20	1947	Ely	7.02	2007	Wheaton	5.0	1945	Virginia
Jun 3	98	1940	Canby	19	1947	Ely	7.10	1898	Pine River			
Jun 4	100	1968	Lamberton	21	1964	Bigfork 5E	5.30	1958	Zumbrota	1.5	1935	Mizpah
Jun 5	99	1934	Fairmont	18	1985	Remer	5.47	1994	Rosemount			
Jun 6	106	1933	Pipestone	20	1985	Remer	6.51	1896	Luverne	0		
Jun 7	103	2011	Little Falls 1N	22	1897	Tower 3S	4.33	1962	Springfield	0		
Jun 8	102	1985, 2011	Chaska, Wild River SP	20	1935	Sawbill Camp	8.07	2001	Thief River Falls	0		
Jun 9	102	1985	Albert Lea	23	1877	Fort Ripley	7.20	1895	Willmar	0		
Jun 10	106	1933	Fairmont	22	1985	Remer 2	6.05	2002	Agassiz Refuge	0		
Jun 11	102	1933	Fairmont	24	1877	Fort Ripley	5.31	2002	Camp Norris	0		
Jun 12	102	1893	Crookston	23	1985	Remer	8.00	1899	Minnesota City	0		
Jun 13	104	1956	Redwood Falls	25	1969	Cotton 3E	6.08	1950	Red Wing	0		
Jun 14	105	1979	Montevideo	27	1927	Alborn	5.70	1921	Fosston	0		
Jun 15	106	1979	Lamberton	24	1917	Meadowland	8.83	2012	Cannon Falls 25W	0		
Jun 16	106	1933	Beardsley	23	1999	Embarrass	4.98	1967	Willmar	0		
Jun 17	102	1933	Campbell	22	2000	Tower 3S	8.67	1957	Minneota	0		
Jun 18	106	1933	Beardsley	26	2001	Kelliher	5.15	1956	Faribault	0		
Jun 19	108	1933	Beardsley	26	2001	Kelliher	5.13	2000	Moorhead	0		
Jun 20	104	1988	Olivia	23	1985	Remer	10.45	2012	Two Harbors	0		
Jun 21	107	1988	Browns Valley	25	1992	Brimson 1E	6.25	1941	West Union	0		
Jun 22	107	1988	Canby	25	2001	Kelliher	5.42	1957	Itasca State Park	0		
Jun 23	104	1937	Springfield	27	1917	Sandy Lake	5.47	1940	Reads Landing	0		
Jun 24	106	1988	Artichoke Lake	20	1985	Remer	7.60	2003	Browns Valley	0		
Jun 25	109	1933	Beardsley	25	2001	Kelliher	6.60	2003	Elk River	0		
Jun 26	106	1934	Faribault	30	1929	Fosston	5.20	1914	Morris	0		
Jun 27	108	1934	New London	20	1970	Baudette	6.46	1998	Zumbrota	0		
Jun 28	108	1931	Canby	24	1936	Sawbill Camp	3.60	1897	Moorhead	0		
Jun 29	110	1931	Canby	27	1925	Pine River	6.37	1969	Worthington	0		
Jun 30	109	1931	Canby	24	1925	Pine River	5.34	1901	Faribault	0		
Jul 1	103	1921	Fergus Falls	30	1988	Brimson 1E	8.00	1978	Theilman	0		
Jul 2	105	1925	Canby	27	2009	Kelliher	6.44	1901	Newfolden	0		
Jul 3	107	1949	Beardsley	29	1927	Meadowland	4.61	1983	Itasca State Park	0		
Jul 4	107	1936	Pipestone	27	1972	Tower 3S	9.78	1995	Milan	0		
Jul 5	108	1936	Pipestone	27	2001	Embarrass	6.25	1943	Albert Lea	0		

Month Day	MAXIMUM TEMPERATURE (°F)			MINIMUM TEMPERATURE (°F)			PRECIPITATION (INCHES)			SNOWFALL (INCHES)		
	Value	Year	Location	Value	Year	Location	Value	Year	Location	Value	Year	Location
Jul 6	114	1936	Moorhead	30	1969	Cotton 3E	5.30	1978	Minnesota City	0		
Jul 7	108	1988	Browns Valley	24	1997	Tower 3S	5.00	1990	Elgin	0		
Jul 8	110	1936	Fosston	25	2003	Kelliher	6.03	1950	White Rock	0		
Jul 9	110	1936	Beardsley	30	1900	Tower 3S	6.00	1858	Forest City	0		
Jul 10	112	1936	Wadena 3S	32	1978	Tower 3S	7.02	1954	Leech Lake	0		
Jul 11	111	1936	Ada	30	1985	Meadowland	7.47	1981	Rochester	0		
Jul 12	111	1936	Canby	27	1975	Tower 3S	5.45	1961	Buffalo	0		
Jul 13	111	1995	Minnesota City	24	2009	Kelliher	5.02	1999	Indus 3W	0		
Jul 14	111	1936	New Ulm 2S	30	1930	Albon	5.49	2011	Carlos 3 SW	0		
Jul 15	112	1931	Beardsley	30	1930	Albon	7.37	1916	New Ulm 2S	0		
Jul 16	113	1936	Wheaton	32	1940	Sawbill Camp	6.38	1993	Moorhead	0		
Jul 17	110	1936	Worthington	33	1971	Bigfork 5E	5.90	1952	Gull Lake	0		
Jul 18	109	1940	Beardsley	30	2003	Kelliher	7.50	1867	Fort Ripley	0		
Jul 19	108	1932	Canby	29	2000	Tower 3S	8.97	1909	Fosston	0		
Jul 20	110	1901	New London	27	2009	Kelliher	10.75	1909	Beaulieu	0		
Jul 21	113	1934	Milan	34	1947	Angus 1N	7.83	1987	Chaska	0		
Jul 22	111	1934	Beardsley	31	1985	Meadowland	10.84	1972	Fort Ripley	0		
Jul 23	108	1934	Milan	28	2002	Kelliher	9.15	1987	MSP Airport	0		
Jul 24	110	1940	Canby	29	2003	Kelliher	5.80	1987	Rosemount	0		
Jul 25	109	1931	Wheaton	30	1915	Cloquet	5.20	1993	Tamarac Refuge	0		
Jul 26	107	1931	Beardsley	27	2001	Kelliher	5.24	1949	Rochester	0		
Jul 27	111	1931	Beardsley	31	2009	Kelliher	6.67	1892	Minneapolis	0		
Jul 28	113	1917	Beardsley	31	1936	Hill City	5.01	1972	Pine River	0		
Jul 29	115	1917	Beardsley	17	1929	Littlefork	5.75	2011	Winona Dam 5A	0		
Jul 30	107	1933	Milan 1 NW	31	1984	Virginia	4.84	1950	Dodge Center	0		
Jul 31	110	1988	Madison SE	31	1937	Sawbill	6.70	1961	Albert Lea	0		
Aug 1	110	1988	Montevideo	31	1920, 2008	Cloquet, Kelliher	6.75	1906	Park Rapids	0		
Aug 2	106	1958	Beardsley	21	2002	Kelliher	7.03	1964	Agassiz Refuge	0		
Aug 3	109	1930	Fairmont	28	1972	Wannaska	5.03	1983	Pokegama Dam	0		
Aug 4	106	1947	Beardsley	29	1972	Tower 3S	6.00	2002	St. Francis	0		
Aug 5	105	1947	New Ulm 2 SE	31	1994	Brimson 1E	4.75	1945	Albert Lea	0		
Aug 6	107	2001	Vadnais Lake	28	1976	Tower 3S	10.30	1866	Sibley	0		
Aug 7	104	1983	Alexandria	29	1989	Brimson 1E	8.62	1968	St. Peter	0		
Aug 8	105	1936	Beardsley	33	1964	Thorhult	6.20	2009	Chaska 2NW	0		
Aug 9	104	1947	Canby	28	1994	Tower 3S	5.25	1939	Two Harbors	0		
Aug 10	110	1947	Beardsley	27	1923	Duluth	7.72	1948	Mankato	0		
Aug 11	105	1947	Wheaton	28	1997	Embarrass	8.06	1945	Hastings	0		
Aug 12	101	1988	Canby	29	2003	Kelliher	5.22	2008	Breckenridge	0		

Month Day	MAXIMUM TEMPERATURE (°F)			MINIMUM TEMPERATURE (°F)			PRECIPITATION (INCHES)			SNOWFALL (INCHES)		
	Value	Year	Location	Value	Year	Location	Value	Year	Location	Value	Year	Location
Aug 13	108	1965	Beardsley	25	2002	Kelliher	6.41	1911	Grand Meadow	0		
Aug 14	105	1984	Hawley	25	1977	Tower 3S	5.29	1981	Gaylord	0		
Aug 15	108	1937	Beardsley	26	1976	Tower 3S	5.40	1993	St. Peter	0		
Aug 16	107	1988	Madison SE	27	1976	Tower 3S	4.21	1926	Willmar	0		
Aug 17	105	1988	Campbell	29	1981	Tower 3S	6.00	2006	Argyle	0		
Aug 18	107	1976	Browns Valley	24	1975	Tower 3S	5.78	1935	Artichoke	0		
Aug 19	107	1976	Browns Valley	23	1907	Beardsley	15.10	2007	Hokah	0		
Aug 20	105	1976	Campbell	25	1934	Alborn	8.00	1913	Worthington	0		
Aug 21	103	1976	Milan	23	2004	Tower 3S	8.17	2007	Byron	0		
Aug 22	104	1971	Redwood Falls	26	1967	Cotton 3E	4.58	1999	Buffalo	0		
Aug 23	104	1922	Beardsley	25	1977	Tower 3S	5.82	1978	Remer	0		
Aug 24	107	1936	Pipestone	22	1977	Tower 3S	5.96	1940	Windom	0		
Aug 25	102	1886	St. Vincent	25	1915	Littlefork	5.08	1941	Pokegama Dam	0		
Aug 26	103	1973	Luverne	23	1915	Roseau 1E	6.72	1994	Stewart 6S	0		
Aug 27	103	1973	Tracy	22	1986	Tower 3S	5.63	1978	Winsted	0		
Aug 28	104	1937	Canby	21	1986	Tower 3S	6.00	1960	Litchfield	0		
Aug 29	103	1921	Beardsley	22	1976	Tower 3S	5.32	1980	Thorhult	0		
Aug 30	103	1976	Minneota	26	1935	Sawbill Camp	7.28	1977	MSP Airport	0		
Aug 31	100	1898	Beardsley	23	1970	Cotton 3E	5.85	1908	Crookston	Trace	1949	Duluth*
Sep 1	101	1913	Beardsley	23	1974	Tower 3S	7.70	1973	Nett Lake	0		
Sep 2	103	1929	Beardsley	22	1935	Sawbill Camp	5.91	1957	Halstad	0		
Sep 3	103	1925	New Ulm 2S	20	1997	Tower 3S	5.44	1996	Mankato	0		
Sep 4	103	1922	Beardsley	22	1918	Grand Rapids	5.53	2005	Chanhassen WSFO	0		
Sep 5	103	1922	Tracy	23	1885	Park Rapids	3.72	1946	Austin 3S	0		
Sep 6	105	1922	New Ulm 2S	23	1885, 2011	Park Rapids, Two Harbors	8.44	1990	Cloquet	0		
Sep 7	104	1931	Wadena 3S	20	1986	Tower 3S	8.70	2007	Tower 2S	0		
Sep 8	105	1931	New Ulm 2S	20	2000	Red Lake Indian Agency	5.54	1991	Young America	0		
Sep 9	105	1931	Beardsley	19	2006	Embarrass	4.75	1977	Gunflint Lake	0		
Sep 10	108	1931	Milan	17	1917	Roseau 1E	6.75	1947	Ely	0		
Sep 11	111	1931	Beardsley	22	1955	Ada	5.50	1900	Pleasant Mound	0		
Sep 12	102	1931	Beardsley	17	2000	Kelliher	6.50	1869	Komisk (Winsted)	0		
Sep 13	100	1939	Redwood Falls	17	1975	Roseau 1E	4.92	2005	Elk River	0		
Sep 14	103	1939	Redwood Falls	18	1964	Cook 18W	4.60	1994	Red Wing	0.3	1964	International Falls
Sep 15	100	1939	St. Peter	17	1964	Bigfork 5E	4.98	2004	Winona Dam	0.2	1916	Warroad
Sep 16	101	1891	Montevideo	17	1973	Karlstad	7.07	1992	Red Wing	0		
Sep 17	105	1895	Granite Falls	14	1929	Mahnomen	4.02	1955	Two Harbors	0		
Sep 18	100	1891	Montevideo	12	1929	Littlefork	7.25	1926	Albert Lea	2.4	1991	Duluth WSO
Sep 19	104	1885	Beardsley	16	1929	Alborn	6.08	1907	Red Wing	0.7	1946	Moorhead

*The only documentation for a trace of snow in August

Month Day	Maximum Temperature (°F)			Minimum Temperature (°F)			Precipitation (Inches)			Snowfall (Inches)		
	Value	Year	Location	Value	Year	Location	Value	Year	Location	Value	Year	Location
Sep 20	99	1891	Fairmont	14	1973	Karlstad	4.97	1983	Harmony	0.4	1945	International Falls
Sep 21	101	1937	Wheaton	13	1934	Alborn	3.95	1968	Lamberton	0.5	1995	Park Rapids
Sep 22	101	1936	Ada	10	1974	Thorhult	4.84	1968	Cambridge	2.0	1995	Baudette
Sep 23	99	1892	Granite Falls	14	2012	Goodridge 12 NNW	9.48	2010	Amboy	2.0	1942	Bigfork 5E
Sep 24	94	1935	Beardsley	15	2000	Red Lake Indian Agency	4.40	1869	White Earth	5.0	1912	Warren
Sep 25	95	1938	Angus 1N	11	1947	Alborn	8.64	2005	Winnebago	6.5	1912	Fosston
Sep 26	93	1974	Madison SE	11	1893	Crookston	3.50	2005	Rushford	7.5	1942	Long Prairie
Sep 27	97	1894	Beardsley	13	1893	Beardsley	3.50	1996	Wolf Ridge (Finland)	6.0	1942	Benson
Sep 28	97	1952	Argyle	15	1942	Alborn	3.65	1901	St. Peter	2.0	1899	Ada
Sep 29	96	1897	Moorhead	13	1899	Hallock	3.45	1925	New Ulm 2S	2.1	1899	Pokegama Dam
Sep 30	93	1897	Montevideo	10	1930	Big Falls	5.00	1995	Cook 18W	3.0	1985	Isabella
Oct 1	93	1976	Browns Valley	10	1886	Moorhead	4.25	1869	White Earth	8.5	1985	Isabella
Oct 2	95	1953	Wheaton	9	1974	Karlstad	4.33	1995	Sandy Lake	5.4	1999	Lakefield
Oct 3	95	1922	Ada	9	1999	Embarrass	4.50	1903	Pine River	0.3	1935	Virginia
Oct 4	94	1997	Albert Lea	10	1935	Ada	4.61	2005	Minneapolis	3.2	2012	Ada
Oct 5	98	1963	Beardsley	11	1988	Pine River	6.61	2005	Wild River SP	7.0	2012	Camp Norris DNR
Oct 6	94	1993	Madison SE	8	1935	Alborn	3.98	1998	Wolf Ridge (Finland)	3.0	1974	Caribou 2S
Oct 7	94	1993	Canby	11	1876	Fort Ripley	3.50	1931	Mankato	4.0	1894	Morris
Oct 8	91	2010	Milan 1NW	11	1917	Angus 1N	4.50	1860	Burlington	6.5	1925	Duluth Harbor
Oct 9	92	2010	Redwood Falls	10	2000	Embarrass	4.44	1973	Grand Rapids	7.5	1970	Slayton 9S
Oct 10	93	1928	Tracy	6	1932	Big Falls	6.13	1973	Vesta	10.0	1970	Oklee 5SW
Oct 11	92	1928	Canby	10	1935	Ada	3.28	1983	Litchfield	10.0	1909	Mount Iron
Oct 12	89	1975	North Mankato	0	1917	Fosston	2.62	1986	Harmony	7.0	1959	Bird Island
Oct 13	89	1958	Canby	2	1936	Sawbill Camp	4.71	1984	Fosston	7.0	2006	Warroad
Oct 14	91	1947	Redwood Falls	8	1937	Beardsley	4.45	1984	Mahnomen 1	4.1	1992	Argyle
Oct 15	93	1958	Madison SE	8	1937	Alborn	5.46	1966	Theilman	10.0	1966	Isabella
Oct 16	91	1958	Montevideo	4	1952	Bemidji	3.55	1998	Wadena 3S	10.0	1937	Bird Island
Oct 17	90	1910	Beardsley	2	1952	Bemidji	4.02	1971	Georgetown	7.0	1990	Cook 18W
Oct 18	87	1950	Chaska	5	1992	Crookston	3.25	1994	Deep Portage	16.0	1916	Baudette
Oct 19	90	2003	Milan	0	2005	Orr 3E	2.96	2007	Wolf Ridge (Finland)	9.0	1905	Lynd
Oct 20	91	1947	Canby	-1	1916	Argyle	3.95	1934	Chaska	10.0	1906	Detroit Lakes
Oct 21	91	1947	Little Falls	-2	1913	Roseau 1E	2.35	1982	Harmony	8.0	2002	Milaca
Oct 22	87	1899	Grand Meadow	0	1937	Meadowland	2.52	1979	Preston	8.4	1951	Virginia
Oct 23	91	1927	Chatfield	-10	1917	Grand Rapids	3.00	1995	Garrison	10.0	2001	Caribou 2S
Oct 24	88	1891	Fairmont	-5	1976	Isabela 1	2.65	1899	Faribault	12.0	1919	Itasca State Park
Oct 25	87	1927	New Ulm 2S	-10	1887	St. Vincent	3.22	1963	Lake City	15.0	1942	Sandy Lake
Oct 26	93	1927	Chatfield	-16	1936	Roseau 1E	3.49	2010	Two Harbors 7 NW	10.5	1913	Park Rapids
Oct 27	90	1927	Chatfield	-10	1919	Itasca State Park	4.25	1900	St. Charles	7.7	2010	Two Harbors 7NW

Month Day	MAXIMUM TEMPERATURE (°F)			MINIMUM TEMPERATURE (°F)			PRECIPITATION (INCHES)			SNOWFALL (INCHES)		
	Value	Year	Location	Value	Year	Location	Value	Year	Location	Value	Year	Location
Oct 28	83	1983	Browns Valley	-9	1919	Angus 1N	3.10	1900	Caledonia	9.5	1932	Big Falls
Oct 29	85	1937	Marshall	-3	1919	Campbell	2.88	2004	Montevideo	8.5	1932	Orr
Oct 30	90	1950	Canby	-8	1925	Duluth	3.15	1979	Glenwood	12.0	1951	Sandy Lake
Oct 31	86	1950	Worthington	-4	1913	Hallock	4.12	1979	Luverne	8.5	1991	New Hope
Nov 1	84	1950	Winona	-10	1919	Campbell	3.28	1991	Winona	24.1	1991	Duluth WSO
Nov 2	80	1965	Canby	-11	1951	Moose Lake	2.76	1961	Maple Plain	24.0	1991	Two Harbors
Nov 3	82	1909	Montevideo	-8	1951	Park Rapids	2.62	1919	International Falls	26.0	1991	Onamia Ranger Station
Nov 4	79	1975	Redwood Falls	-13	1919	Warren	1.84	1922	Fairmont	15.8	1982	Isabella
Nov 5	78	1975	Madison SE	-16	1951	Detroit Lakes	2.02	1948	Rushford	11.0	1993	Floodwood
Nov 6	79	1934	Montevideo	-16	1951	Moose Lake	2.15	1948	Pigeon River	12.0	1919	Cloquet
Nov 7	78	1931	Montevideo	-20	1936	Red Lake Falls	2.30	1991	Garrison	14.0	1943	Marshall
Nov 8	83	2006	New Ulm 2SE	-14	1991	Mankato	3.45	1945	Winona	16.0	1943	Vesta
Nov 9	83	1999, 2006	Springfield, Springfield 1NW	-15	1921	Milan	3.08	1983	Cloquet	26.0	1943	St. James
Nov 10	78	1999	Winona Dam	-15	1933	Big Falls	2.80	1975	Minnesota City	12.4	1919	Moorhead
Nov 11	75	2012	Winnebago	-22	1919	Itasca State Park	2.52	1940	Minneapolis	14.0	1940	Orr
Nov 12	74	1923	Faribault	-26	1995	Tower 3S	2.63	1940	Schroeder	16.0	1940	Farmington
Nov 13	79	1999	Fairmont	-24	1995	Tower 3S	3.80	1937	Sawbill Camp	15.0	1940	Taylors Falls
Nov 14	81	1999	St. James	-23	1911	Warroad	2.05	1951	Mankato	14.0	1909	Moorhead
Nov 15	76	1953	Madison SE	-36	1911	Angus 1N	2.68	1944	Stillwater	11.1	1956	Duluth WSO
Nov 16	75	1939	Marshall	-27	1933	Big Falls	4.10	1909	Two Harbors	18.0	1909	Fairmont
Nov 17	76	2001	Canby	-19	1914	Hallock	3.21	1996	Tower 3S	15.0	1996	Roseau 1E
Nov 18	75	1923	Faribault	-19	1940	Duluth	3.10	1996	Hinckley W	15.0	1998	Crookston
Nov 19	74	1897	Montevideo	-29	1896	Roseau 1E	2.85	1998	Grand Portage	19.0	1948	Dawson
Nov 20	74	1897	Faribault	-31	1896	Roseau 1E	3.23	1975	Canby	16.0	1975	Canby
Nov 21	72	1962, 2012	Tracy, Amboy	-25	1978	Tower 3S	3.04	1996	Lake City	16.0	1975	Montevideo
Nov 22	72	1990, 2012	Preston, St. James Filt Pit	-26	1896	Ada	2.18	1898	Willow River	13.0	1898	Pokegama Dam
Nov 23	68	2006	Browns Valley	-31	1898	Tower 3S	2.57	2012	Grand Marais	18.0	1983	Babbitt 2S
Nov 24	68	1984	Wheaton	-31	1898	Pokegama Dam	2.38	2001	Vesta	16.0	1983	Tower 3S
Nov 25	76	1933	Faribault	-36	1903	Pokegama Dam	3.00	1896	Le Sueur	16.7	1983	Island Lake
Nov 26	68	1914	Fairmont	-37	1903	Pokegama Dam	4.80	1896	Worthington	19.5	2001	Granite Falls
Nov 27	71	1998	Fairmont	-31	1887	Argyle	2.75	1988	Pine River	24.0	2001	New London
Nov 28	72	1998	Amboy	-36	1896	Bemidji	2.24	1960	Duluth WSO	16.0	1993	Campbell
Nov 29	68	1998	Albert Lea	-39	1896	Tower 3S	2.85	1934	Farmington	16.0	1991	Cambridge
Nov 30	68	1922	Montevideo	-45	1896	Pokegama Dam	2.80	2010	Moose Lake 1SSE	18.0	1985	Willmar
Dec 1	70	1998	Chaska	-51	1896	Pokegama Dam	2.12	1985	Two Harbors	16.0	1985	Winona
Dec 2	69	1998	Mankato	-47	1896	Pokegama Dam	2.51	1984	Caledonia	14.0	1985	Elbow Lake
Dec 3	72	1941	Canby	-38	1927	Itasca State Park	2.42	2013	Big Falls	17.7	2013	Two Harbors 7NW
Dec 4	71	1941	Long Prairie	-38	1873	Fort Ripley	3.00	2006	Thief River Falls	14.0	1955	Beardsley

Month Day	MAXIMUM TEMPERATURE (°F)			MINIMUM TEMPERATURE (°F)			PRECIPITATION (INCHES)			SNOWFALL (INCHES)		
	Value	Year	Location	Value	Year	Location	Value	Year	Location	Value	Year	Location
Dec 5	65	1998	Winona	-38	1873	Fort Ripley	2.23	1985	Milaca 1EN	12.0	1909	Little Falls
Dec 6	73	1939	Beardsley	-41	2013	Kelliher	2.00	1917, 2010	Worthington, Delano	23.2	1950	Duluth WSO
Dec 7	69	1913	Grand Marais	-42	1936	Pokegama Dam	1.31	1927	Lynd No. 2	12.0	1927	Chaska
Dec 8	67	1913	Grand Marais	-38	1932, 2013	Big Falls, Brimson 1E	2.02	1924	Babbitt 2S	14.0	1969	Isabella
Dec 9	74	1939	Wheaton	-39	1909	Warroad	1.31	1899	Minneapolis	17.0	2012	St. Francis
Dec 10	67	2006	Browns Valley	-41	1977	Thorhult	2.42	1911	New Richland	14.2	2009	Altura 5W
Dec 11	67	1913	Long Prairie	-41	1936	Pokegama Dam	1.70	1870	Beaver Bay	18.0	2010	Montgomery
Dec 12	64	1913	Tracy	-39	1995	International Falls	3.66	1899	Caledonia	18.0	2010	Altura 5W
Dec 13	62	1921	Lynd	-42	1901	Ada	1.92	1968	Hinckley	13.2	2010	Gaylord
Dec 14	60	1912	Preston	-48	1901	Detroit Lakes	2.38	1891	Red Wing	20.8	2005	Two Harbors 7NW
Dec 15	63	1891	Kinbrae	-47	1901	Pokegama Dam	2.25	1893	Grand Rapids	14.6	1996	Rockford
Dec 16	65	1939	St. Peter	-39	1903	Pokegama Dam	2.57	1984	Gunflint Lake	14.0	1940	Farmington
Dec 17	63	1939	Farmington	-44	1983	Mora	1.80	1984	Hinckley	15.0	1863	Beaver Bay
Dec 18	69	1908	Lynd	-52	1983	Mora	1.70	1977	Montevideo	8.7	1998	Two Harbors
Dec 19	60	2011	Madison Sewage Plt	-52	1983	Tower 3S	1.35	1902	Luverne	11.5	2008	Two Harbors 7NW
Dec 20	69	1923	Faribault	-49	1983	Tower 3S	1.50	1902	Worthington	13.0	1887	Le Sueur
Dec 21	64	1908	Lynd	-44	1916	Roseau 1E	1.45	1948	Bricelyn	12.2	2010	Two Harbors 7NW
Dec 22	62	1899	Two Harbors	-44	1963	Baudette	1.30	1920	Glencoe	16.0	1968	Artichoke
Dec 23	62	1923	Faribault	-48	1884	St. Vincent	2.10	1968	Cass Lake	13.0	1959	Isabella
Dec 24	57	1888	Northfield	-43	1884	St. Vincent	2.40	2010	Montgomery	15.5	1959, 2009	Isabella, Windom
Dec 25	62	1923	Faribault	-50	1933	Orr	2.66	1893	Wabasha	14.9	2009	Two Harbors 7NW
Dec 26	57	1936	Fairmont	-50	1933	Big Falls	2.50	1988	Marshall	15.0	1945	Bricelyn
Dec 27	54	1994	Canby	-50	1993	Tower 3S	2.50	1856	Fort Ridgley	22.0	2009	Island Lake 4E
Dec 28	59	1984	Winona	-51	1933	Big Falls	2.44	1982	Bricelyn	17.0	1982	St. Francis
Dec 29	61	1999	Montevideo	-47	1917	Itasca State Park	1.55	1982	Farmington	16.0	1982	Farmington
Dec 30	59	1999	Canby	-47	1910	Warroad	2.00	1936	Pigeon River	14.2	1887	Mankato
Dec 31	58	1921	St. Peter	-57	1898	Pokegama Dam	1.50	1887	Grand Meadow	18.4	1996	Two Harbors

BIBLIOGRAPHY

Ahrens, C. Donald. *Meteorology Today: An Introduction to Weather, Climate, and the Environment.* 6th ed. Pacific Grove, CA: Brooks/Cole, 2000.

Alley, R. B., J. Marotzke, W. D. Nordhaus, J. T. Overpeck, D. M. Peteet, R. A. Pielke, Jr., R. T. Pierrehumbert, P. B. Rhines, T. F. Stocker, L. D. Talley, and J. M. Wallace. "Abrupt Climate Change." *Science* 299 (2003).

Ashley, Walker S., and Thomas L. Mote. "Derecho Hazards in the United States." *Bulletin of the American Meteorological Society* 86.11 (2005).

Baker, D. G., J. C. Klink, and R. H. Skaggs. "A Singularity in Clear-Day Frequencies in the North-Central Region." *Monthly Weather Review* 111.4 (1983).

Baker, D. G., and E. L. Kuehnast. *Climate of Minnesota, Part VII: Areal Distribution and Probabilities of Precipitation in the Minneapolis–St. Paul Metropolitan Area.* St. Paul: University of Minnesota Agricultural Experiment Station, 1973.

Baker, D. G., E. L. Kuehnast, and J. A. Zandlo. *Climate of Minnesota, Part XV: Normal Temperatures (1951–1980) and Their Application.* St. Paul: University of Minnesota Agricultural Experiment Station, 1985.

Baker, D. G., W. W. Nelson, and E. L. Kuehnast. *Climate of Minnesota, Part XII: The Hydrologic Cycle and Soil Water.* St. Paul: University of Minnesota Agricultural Experiment Station, 1979.

Baker, D. G., and D. L. Ruschy. "Winter Albedo Characteristics of St. Paul, Minnesota." *Journal of Applied Meteorology* 28.3 (1989).

Baker, D. G., D. L. Ruschy, and R. H. Skaggs. "Agriculture and the Recent 'Benign Climate' in Minnesota." *Bulletin of the American Meteorological Society* 74.6 (1993).

——. *Climate of Minnesota, Part XVI: Incoming and Reflected Solar Radiation at St. Paul.* St. Paul: University of Minnesota Agricultural Experiment Station, 1987.

Baker, D. G., D. L. Ruschy, R. H. Skaggs, and D. B. Wall. "Air Temperature and Radiation Depressions Associated with a Snow Cover." *Journal of Applied Meteorology* 31.3 (1992).

Baker, D. G., B. F. Watson, and R. H. Skaggs. "The Minnesota Long-Term Temperature Record." *Journal of Climate Change* 7 (1985).

Baker, Donald G. *Climate of Minnesota, Part XIV: Wind Climatology and Wind Power.* St. Paul: University of Minnesota Agricultural Experiment Station, 1983.

Baker, J. M., and D. G. Baker. "Long-Term Ground Heat Flux and Heat Storage at a Mid-Latitude Site." *Climatic Change* 54 (2002).

Beddow, Jason, Philip Pardey, and Mark Seeley. "Changing Agricultural Climate: Implications for Innovation Policies." University of Minnesota Food Policy Research Center Fact Sheet, 2012.

Black, E., M. Blackburn, G. Harrison, and J. Methven. "Factors Contributing to the Summer 2003 European Heatwave." *Weather* 59.88 (2004).

Boulay, Pete. "Hey, How's the Weather." *Minnesota Conservation Volunteer* 66.389 (2003).

Brooks, G. R., S. St. George, C. F. M. Lewis, B. E. Medioli, E. Nielsen, S. Simpson, and L. H. Thorleifson. *Geoscientific Insights into Red River Flood Hazard in Manitoba.* Geological Survey of Canada Open File Report 4473 (2003).

Brooks, H. E. "Severe Thunderstorms and Climate Change." *Atmospheric Research* 123 (2013): 129–38.

Brooks, H. E., and N. Dotzek. "The Spatial Distribution of Severe Convective Storms and an Analysis of Their Secular Changes." In *Climate Extremes and Society,* edited by Henry F. Diaz and Richard J. Murnane. New York: Cambridge University Press, 2007.

Brown, Curt. *So Terrible a Storm:*

A Tale of Fury on Lake Superior. Minneapolis: Voyageur Press, 2011.

Burt, Christopher C. *Extreme Weather: A Guide and Record Book.* New York: W. W. Norton, 2004.

Carlson, R. E., J. W. Enz, and D. G. Baker. "Quality and Variability of Long Term Climate Data Relative to Agriculture." *Agricultural and Forest Meteorology* 69 (1994).

Changnon, S. A. "Comments on 'Secular Trends of Precipitation Amount, Frequency, and Intensity in the United States.'" *Bulletin of the American Meteorological Society* 79.11 (1998).

——. "Shifting Economic Impacts from Weather Extremes in the United States: A Result of Societal Changes, Not Global Warming." *Natural Hazards* 29 (2003).

Changnon, S. A., K. E. Kunkel, and Derek Winstanley. "Climate Factors that Caused the Unique Tall Grass Prairie in the Central United States." *Physical Geography* 23.4 (2002).

Changnon, Stanley A. "Damaging Thunderstorm Activity in the United States." *Bulletin of the American Meteorological Society* 82.4 (2001).

——. "Human Factors Explain the Increased Losses from Weather and Climate Extremes." *Bulletin of the American Meteorological Society* 81.3 (2000).

——. "Measures of Economic Impacts of Weather Extremes." *Bulletin of the American Meteorological Society* 84.9 (2003).

——. "Thunderstorm Rainfall in the Conterminous United States." *Bulletin of the American Meteorological Society* 82.9 (2001).

Changnon, Stanley A., ed. *The Great Flood of 1993: Causes, Impacts, and Responses.* Boulder, CO: Westview Press, 1996.

Changnon, Stanley A., Kenneth E. Kunkel, and Beth C. Reinke. "Impacts and Responses to the 1995 Heat Wave: A Call to Action." *Bulletin of the American Meteorological Society* 77.7 (1996).

Coues, Elliot, ed. *New Light on the Early History of the Greater Northwest: The Manuscript Journals of Alexander Henry.* 1897. Reprint: Minneapolis, MN: Ross & Haines, 1965.

Czuba, Christiana R., C. R. James, J. D. Fallon, and E. W. Kessler. "Floods of June 2012 in Northeastern Minnesota." U.S. Geological Survey Scientific Investigations Report 2012–5283, 2012.

Doesken, Nolan J., and Arthur Judson. *The Snow Booklet.* Fort Collins: Colorado State University, 1996.

Easterling, D. R., J. L. Evans, P. Ya. Groisman, T. R. Karl, K. E. Kunkel, and P. Ambenje. "Observed Variability and Trends in Extreme Climate Events: A Brief Review." *Bulletin of the American Meteorological Society* 81.3 (2000).

Edwards, Craig. *Nature's Messenger.* IUniverse, 2008.

Eiber, Tom. "Five Favorite Fall Tours." *Minnesota Conservation Volunteer* (Sep-Oct 1994).

Fisk, Charles J. "Reconstruction of Daily 1820–1872 Minneapolis–St. Paul, Minnesota, Temperature Observations." Master's thesis, University of Wisconsin–Madison, 1984.

Frohman, Carl E. "Spatial and Temporal Fluctuations in Minnesota's Growing Season Start, End, and Duration: 1899–2001." Master's thesis, University of Minnesota–Minneapolis, 2001.

Fujita, T. "Proposed Characterization of Tornadoes and Hurricanes by Area and Intensity." SMRP Research Paper 91. Chicago: University of Chicago, 1971.

Geer, Ira W., ed. *Glossary of Weather and Climate.* Boston, MA: American Meteorological Society, 1996.

Gettelman, Andrew. "The 'Information Divide' in the Climate Sciences." *Bulletin of the American Meteorological Society* 84:12 (2003).

Grazulis, Thomas P. *Significant Tornadoes: 1680–1991.* St. Johnsbury, VT: Environmental Films, 1993.

——. *Significant Tornadoes Update: 1992–1995.* St. Johnsbury, VT: Environmental Films, 1997.

Grice, Gary, and Pete Boulay. "History of Weather Observations: Fort Snelling, 1819–1892." St. Paul: Minnesota Department of Natural Resources, 2005.

Groisman, P. Ya, R. W. Knight, and T. R. Karl. "Heavy Precipitation and High Streamflow in the Contiguous United States: Trends in the Twentieth Century." *Bulletin of the American Meteorological Society* 82.2 (2001).

Haines, Donald A., and Earl L. Kuehnast. "When the Midwest Burned." *Weatherwise* 23.3 (1970).

Harding, Keith J. and Peter K. Snyder. "Modeling the Atmospheric Response to Irrigation in the Great Plains. Part I: General Impacts on Precipitation and the Energy Budget." *Journal of Hydrometeorology* 13.6 (2012): 1667–86.

——. "Modeling the Atmospheric

Response to Irrigation in the Great Plains. Part II: The Precipitation of Irrigated Water and Change in Precipitation Recycling." *Journal of Hydrometeorology* 13.6 (2012): 1687–1703.

Hershfield, David M. "Rainfall Frequency Atlas of the United States." U.S. Department of Agriculture, Soil Conservation Service, Technical Paper 40 (1961).

Hoff, Mary. "Catch the Wind." *Minnesota Conservation Volunteer* 66.391 (2003).

Huff, Floyd A., and James R. Angel. *Rainfall Frequency Atlas of the Midwest*. Champaign, IL: Midwest Climate Center and Illinois State Water Survey, 1992.

Hull, William H. *All Hell Broke Loose: Experiences of Young People During the Armistice Day 1940 Blizzard*. Edina, MN: Stanton Publications, 1985.

Huschke, Ralph E., ed. *Glossary of Meteorology*. Boston, MA: American Meteorological Society, 1959.

Karl, T. R., and R. W. Knight. "Secular Trends of Precipitation Amount, Frequency, and Intensity in the USA." *Bulletin of the American Meteorological Society* 79.2 (1998).

Karl, T. R., R. W. Knight, K. P. Gallo, T. C. Peterson, P. D. Jones, G. Kukla, N. Plummer, V. Razuvayev, J. Lindseay, and R. J. Charlson. "A New Perspective on Recent Global Warming: Asymmetric Trends of Daily Maximum and Minimum Temperature." *Bulletin of the American Meteorological Society* 74.6 (1993).

Karl, Thomas, Jerry Melillo, and Thomas Peterson, eds. "Global Climate Change Impacts in the United States." U.S. Global Change Research Program report. New York: Cambridge University Press, 2009.

Keen, Richard A. *Minnesota Weather*. Helena, MT: American and World Geographic Publishing, 1992.

Keillor, Steven J. *Grand Excursion: Antebellum America Discovers the Upper Mississippi*. Afton, MN: Afton Historical Society Press, 2004.

Klink, Katherine. "Trends and Interannual Variability of Wind Speed Distributions in Minnesota." *Journal of Climate* 15 (2002).

Krause, Todd. "Minnesota Tornado Statistics by Year." Personal communication to author, 2005.

Kuehnast, E. L., D. G. Baker, and J. A. Zandlo. *Climate of Minnesota, Part XIII: Duration and Depth of Snow Cover*. St. Paul: University of Minnesota Agricultural Experiment Station, 1982.

——. *Sixteen Year Study of Minnesota Flash Floods*. St. Paul: Minnesota Department of Natural Resources, 1988.

Kunkel, K. E. "Extreme Precipitation Trends of North America." *Natural Hazards* 29.2 (2003).

Kunkel, K. E., Karen Andsager, and D. R. Easterling. "Trends in Heavy Precipitation Events over the Continental United States." *Journal of Climate* 12.8 (1999).

Kunkel, K. E., D. R. Easterling, K. Redmond, and K. Hubbard. "Temporal Variations of Extreme Precipitation Events in the United States: 1895–2000." *Geophysical Research Letters* 30.17 (2003).

Kunkel, K. E., Michael A. Palecki, Leslie Ensor, David Easterling, Kenneth G. Hubbard, David Robinson, and Kelly Redmond. "Trends in Twentieth-Century U.S. Extreme Snowfall Seasons." *Journal of Climate* 22 (2009): 6204–16.

Kunkel, K. E., Roger Pielke, Jr., and S. A. Changnon. "Temporal Fluctuations in Weather and Climate Extremes that Cause Economic and Human Health Impacts: A Review." *Bulletin of the American Meteorological Society* 80.6 (1999).

Landsberg, Helmut E. *The Urban Climate*. New York: Academic Press, 1981.

Laskin, David. *The Children's Blizzard*. New York: Harper Collins, 2004.

Lass, William E. "Minnesota: An American Siberia?" *Minnesota History* 49 (Winter 1984).

Ludlum, David M. *The American Weather Book*. Boston, MA: Houghton Mifflin, 1982.

——. *Early American Winters*. Boston, MA: American Meteorological Society, 1966.

——. *Early American Winters, Vol. II, 1821–1870*. Boston, MA: American Meteorological Society, 1968.

——. *The Weather Factor*. Boston, MA: Houghton Mifflin, 1984.

Mahmood, R., K. G. Hubbard, and C. Carlson. "Modification of Growing-Season Surface Temperature Records in the Northern Plains Due to Land-Use Transformation." *International Journal of Climatology* 24 (2004).

Meehl, Gerald A., Thomas Karl, D. R. Easterling, S. Changnon, Roger Pielke, Jr., D. Changnon, J. Evans, Paval Ya Groisman, T. R. Knutson, L. O. Mearns, Camille Parmesan, R. Pulwarty, Terry Root, R. T. Sylves, P. Whetton, and Francis Wiers. "An Introduction to Trends in

Extreme Weather and Climate Events." *Bulletin of the American Meteorological Society* 81.3 (2000).

Melillo, Jerry M., Terese (T. C.) Richmond, and Gary W. Yohe, eds. *Climate Change Impacts in the United States: The Third National Climate Assessment.* U.S. Global Change Research Program, 2014.

Minnesota Department of Health. "Climate and Health." Available: http://www.health.state .mn.us/divs/climatechange/.

Minnesota Interagency Climate Adaptation Team. "Adapting to Climate Change in Minnesota." 2013.

Moran, Joseph M., and Michael D. Morgan. *Essentials of Weather.* Englewood Cliffs, NJ: Prentice Hall, 1995.

Narvestad, Carl T., and Amy Narvestad. *A History of Yellow Medicine County, Minnesota, 1872–1972.* Granite Falls, MN: Yellow Medicine County Historical Society, 1972.

National Oceanic and Atmospheric Administration, National Climatic Data Center. *Climatography of the United States No. 81: Monthly Normals of Temperature, Precipitation, and Heating and Cooling Degree Days, 1971–2000.* Washington, DC: U.S. Department of Commerce, 2001.

National Oceanic and Atmospheric Administration, Office of Hydrologic Development. *NOAA Atlas 14: Precipitation Frequency Atlas of the United States.* Vol. 8. National Oceanic and Atmospheric Administration, U.S. Department of Commerce, 2013.

Palecki, M. A., S. A. Changnon, and K. E. Kunkel. "The Nature and Impacts of the July 1999 Heat Wave in the Midwestern United States: Learning from the Lessons of 1995." *Bulletin of the American Meteorological Society* 82.7 (2001).

Petersen, William J. *Steamboating on the Upper Mississippi.* Iowa City: State Historical Society of Iowa, 1968.

Pielke, R. A. Jr., and D. Sarewitz. "Bringing Society Back into the Climate Debate." *Population and Environment* 26.3 (2005).

Ross, A. *The Red River Settlement, Its Rise, Progress, and Present State.* Edmonton: Hurtig Publishers, 1856. Reprint: 1972.

Rotberg, Robert I., and Theodore K. Rabb, eds. *Climate and History.* Studies in Interdisciplinary History. Princeton, NJ: Princeton University Press, 1981.

St. George, Scott, and Bill Rannie. "The Causes, Progression, and Magnitude of the 1826 Red River Flood in Manitoba." *Canadian Water Resources Journal* 28.1 (2003).

St. Martin, Thomas. *Minnesota Daily Climatological Record, 1819–1998.* Vols. 1–21. Woodbury, MN: The author, 2001.

Santer, B. D., T. M. L. Wigley, D. J. Gaffen, L. Bengtsson, C. Doutriaux, J. S. Boyle, M. Esch, J. J. Hnilo, P. D. Jones, G. A. Meehl, E. Roeckner, K. E. Taylor, and M. F. Wehner. "Interpreting Differential Temperature Trends at the Surface and in the Lower Troposphere." *Science* 287 (2000).

Schwartz, Robert, and Thomas W. Schmidlin. "Climatology of Blizzards in the Conterminous United States, 1959–2000." *Journal of Climate* 15.13 (2002).

Seeley, M. W. "The Future of Serving Agriculture with Weather/Climate Information and Forecasting: Some Indications and Observations." *Agriculture and Forest Meteorology* 69 (1994).

Seeley, Mark W. "Importance and Utilization of Agricultural Weather Information in Minnesota: A Survey." *Journal of Agronomic Education* 19.1 (1990).

———. "Some Applications of Temporal Climate Probabilities to Site-Specific Management of Agricultural Systems." *Proceedings of Site-Specific Management of Agricultural Systems.* Madison, WI: American Society of Agronomy, 1995.

Shelby, Ashley. *Red River Rising: The Anatomy of a Flood and the Survival of an American City.* St. Paul, MN: Borealis Books, 2004.

Skaggs, Richard H., Donald G. Baker, and David L. Ruschy. "Interannual Variability Characteristics of the Eastern Minnesota Temperature Record: Implications for Climate Change Studies." *Climate Research* 5 (1995).

Stott, Peter A., D. A. Stone, and M. R. Allen. "Human Contribution to the European Heatwave of 2003." *Nature* 432 (2004).

Stull, Roland B. *An Introduction to Boundary Layer Meteorology.* 1988. Reprint: Boston, MA: Kluwer Academic Publishers, 1999.

Tester, John R. *Minnesota's Natural Heritage: An Ecological Perspective.* Minneapolis: University of Minnesota Press, 1995.

Thiessen, Alfred H. *Weather Glossary.* Washington, DC: U.S. Department of Commerce, Weather Bureau, No. 1445, 1946.

Todhunter, Paul E. "Environmental Indices for the Twin Cities Metropolitan Area (Minnesota, USA) Urban Heat Island—1989." *Climate Research* 6 (1996).

Trenberth, Kevin E., Aiguo Dai, Roy M. Rasmussen, and David B. Parsons. "The Changing Character of Precipitation." *Bulletin of the American Meteorological Society* 84.9 (2003).

U.S. Department of Agriculture, National Agricultural Statistics Service. "Minnesota Agricultural Statistics 2005." St. Paul, MN: Minnesota Agricultural Statistics Service, 2005.

U.S. Department of Agriculture, Weather Bureau, Climate and Crop Service. "History of Early Floods in the Red River Valley." *Climate and Crop Report* 3.4 (1897).

U.S. Department of Commerce, Environmental Data Service. "Substation History: Minnesota." *Meteorological Records* 1.1 (1956).

U.S. Geological Survey, Department of the Interior. "Floods of June 2012 in Northeastern Minnesota." Scientific Investigations Report 2012-5283, 2013.

Watson, Bruce F. *Minnesota and Environs Weather Almanac 1976*. Navarre, MN: Freshwater Biological Research Foundation, 1975.

———. *WCCO Radio Weather Almanac 1975*. Navarre, MN: Freshwater Biological Research Foundation, 1974.

Weaver, C. P., and R. Avissar. "Atmospheric Disturbances Caused by Human Modification of the Landscape." *Bulletin of the American Meteorological Society* 82.2 (2001).

Whitnah, Donald R. *A History of the United States Weather Bureau*. Urbana: University of Illinois Press, 1961.

Williams, Jack. *The USA Today Weather Book: An Easy-to-Understand Guide to the USA's Weather*. 2nd edition. New York: Vintage Books, 1997.

Wright, George B. "Notes of a Remarkable Storm." *Minnesota Academy of Natural Sciences Bulletin* 1 (1876).

Yuan, Fei, and Martin Mitchell. "Long-Term Climate Change at Four Rural Stations in Minnesota, 1920–2010." *Journal of Geography and Geology* 6.3 (2014): 228–41.

⬥ WEATHER QUIZ ANSWERS

page 4 B

page 11 C; the results were
(1) 1930s dust bowl; (2) 1940
Armistice Day Blizzard; (3) 1991
Halloween Blizzard; (4) 1997 Red
and Minnesota river flooding;
(5) 1965 Fridley tornado outbreak
(tie); (5) 1965 Mississippi and
Minnesota river floods (tie)

page 23 C

page 30 B

page 33 A

page 38 B

page 42 C

page 53 B

page 56 A

page 64 A

page 87 C

page 95 C; December 16, 1874,
to March 8, 1875

page 99 B; Cook County at
Pigeon River Bridge along the
Canadian border recorded 170.5
inches of snowfall during the
winter of 1949–50, a state record

page 101 B

page 105 C

page 108 A

page 114 C

page 120 B; typically right-facing
triangles, adapted from military
maps showing an army front line

page 121 46.9 inches fell in
November 1991, starting with
the famous Halloween Blizzard

page 123 B

page 124 B

page 126 B; new formula includes
a wind speed estimate for 5
instead of 33 feet above ground

page 127 B

page 129 B; November: 46.9
inches in 1991; January: 46.4
inches in 1982; March: 40.0
inches in 1951

page 131 C; every two to three
years

page 134 C

page 135 C

page 137 B; 60 percent chance

page 138 A

page 147 C; 72°F, shattering the
old record of 56°F, set in 1937

page 148 B; a rarity that also
occurred in 1962, 1972, 1979, 1994,
1996, and 2014

page 152 June 4, 1935, when
1.5 inches fell at Mizpah in
Koochiching County

page 157 A

page 160 C

page 178 C

page 189 a dew point of 81°F

page 190 A; in 2010

page 191 B; 69

page 197 A; most recently:
Chandler in Murray County,
June 16, 1992

page 200 C

page 202 A

page 204 C

page 207 B

page 208 B; because low-level
advection of warm, moist air into
the storm is unimpeded by other
cloud formations, this last thunderstorm cell often has a higher
probability of becoming severe

page 211 C

page 212 B; or a bulge in a squall
line where winds are strong

page 215 A, B, or C, depending

on the county

page 218 A and sometimes C,
depending on the county

page 221 A

page 226 B

page 228 A

page 231 the smallest-ever daily
temperature range in July

page 247 B; 76°F (many other
places established new record
dew points that day, including
Faribault in Rice County, with
80°F)

page 248 A; at Tower in St. Louis
County on August 28, 1986

page 257 A; at Princeton in Mille
Lacs County on October 25, 1999

page 260 C; Appleton in Swift
County reported 90°F (tied by
Milan, Chippewa County, in
2003)

page 263 B; former record: 4.0
inches in 2003

page 264 C

page 266 C

page 268 A

page 271 C

page 272 B; at Bigfork, Itasca
County

page 273 B; up to 20-foot drifts
near Canby in Yellow Medicine
County

page 275 B; near Prinsburg
in Kandiyohi County, rare for
November

page 276 B

page 281 B

page 284 C

page 290 C

page 292 B

page 303 C

INDEX

Italic page numbers indicate maps and photographs; **boldface** page numbers indicate tables and figures.

Illustration Credits

Maps on pages 16, 24, 88, 89, 98, 188, 194, 245, and 307 created by David Deis with data provided by the MN–DNR–State Climatology Office and Allison Serakos

Photo on page 34 by Corey Babcock, DOE/NREL

Photos on pages 37 and 159 by Mark W. Seeley

Photo on page 222 by Joe Strub, NWS

Photo on page 223 by John Ahrens

Photo on page 232 courtesy Superior National Forest

Photos on pages 236 and 246 courtesy University of Minnesota Extension Service

Photo on page 291 courtesy the Department of Soil, Water, and Climate, University of Minnesota

All other images from Minnesota Historical Society collections